P9-ARR-066

*Feral Children
and Clever Animals*

Nora, companion to the painter Emilio Rendich, was among the earliest of animals taught under controlled conditions to converse with human beings.

Feral Children and Clever Animals

*Reflections on
Human Nature*

DOUGLAS KEITH CANDLAND

Oxford University Press · New York · Oxford · 1993

Oxford University Press

Oxford   New York   Toronto
Delhi   Bombay   Calcutta   Madras   Karachi
Kuala Lumpur   Singapore   Hong Kong   Tokyo
Nairobi   Dar es Salaam   Cape Town
Melbourne   Auckland   Madrid

and associated companies in
Berlin   Ibadan

Copyright © 1993 by Oxford University Press, Inc.

Published by Oxford University Press, Inc.,
200 Madison Avenue, New York, New York 10016

Library of Congress Cataloging-in-Publication Data
Candland, Douglas K.
Feral children and clever animals / Douglas Keith Candland.
p.   cm.   Includes bibliographical references and index.
ISBN 0-19-507468-8
1. Psychology—History.   2. Psychology—Research—History.
3. Feral children—Psychology—History.   4. Psychology,
Comparative—History.   I. Title.
BF95.C36   1993   150'.9—dc20   93-59

2 4 6 8 9 7 5 3 1

Printed in the United States of America
on acid-free paper

For Madame Guérin, Mrs. Singh, and
the unnamed native of Gabon

to honor them for their unappreciated
contributions to human understanding

Acknowledgments

MARY HOMRIGHAUSEN CANDLAND, wife extraordinaire, supplied the patience and encouragement necessary for me to prepare and write this book, along with the income from her own design and artwork, which supplemented fellowships and funds soon to be mentioned. As a capstone, she hit upon the title somewhere along the Santa Fe Trail. Our sons, Ian, Christopher, and Kevin, graciously asked about the work from time to time, but unwittingly played a major part in its origins, as their years of interactions with Kitty, Dante (Alighieri), Tory (Victoria Regina et Imperatrix), and Cleo (Cleo Lane) inspired my thinking about the interrelating of animals and people. I am grateful to these folk for giving me such a meaningful life.

Kay Ocker, secretary also without compare, now for twenty-five years, does everything well. I cannot count the number of embarrassments of language and taste she has kept me from demonstrating. My thanks to her are longstanding and much felt.

Bucknell's people, folk whom I have now known for over thirty years, always give all they have to make Bucknell work, indeed, to make it the first of its class. I am conscious of how much these people have done to make all of our work not merely easier but better. I am fortunate to have worked in such an environment for so long a period, and I regret only that my advice has not been taken on all opportunities provided. The text was written under the grace of a sabbatical leave from Bucknell University, for which part of the funds were supplied by the Charles Dana Foundation.

During the first stages of writing, the book was well served by Beau Beegle Vent, then a Bucknell student, whose library scholarship was outstanding. I profited from his views of the first drafts and from his scholarship. Aaron White, then a student at Muhlenburg College, gave the manuscript his attention during a summer of work. Jon Laguna, Alisa Bland, David Prybock, Amy Galloway, and Tom Mollerus, also Bucknell students, gave the manuscript the benefit of fresh readings. Their cogent remarks saved all of us much confusion.

Several people used their special talents to provide information or to assist in the preparation of the manuscript. They are Peter Judge, Susan Philhower, Douglas Eaton, Debra Cook, James Rice, Bill McGrew, Owen Anderson, Ruth Craven, and Thomas Carl. Betty Pessagno clarified the text through inspired copy-editing, and Allison Sole assisted in organizing photographs and credits. Julianne Means created the index. Several other people read parts of the work at the request of publishers who were considering the work. All of these reviews were of exceptional quality, and I thank them for sharing their professional scholarship even though the names are not known to me. At my request, and with the assistance of Oxford University Press, other scholars reviewed the manuscript in detail. The attention, advice, and scholarship of Donald Dewsbury, Duane Rumbaugh, and Sara Shettleworth are everywhere in evidence. I am grateful. The work was shepherded, edited, and produced, by Joan Bossert of Oxford University Press, and it is a pleasure to state publicly what many must know in private, that she is an editor of rare ability, foresight, and understanding.

The Ellen Clarke Bertrand Library of Bucknell University provides not merely books, documents, and a gracious environment in which to work, but that true contribution of any institution, an able staff of mind-readers. They not only find what is needed, but also give their own time and talents to whatever wants scholars and students think they might have. Through the courtesy of interlibrary loan, certain books were made available to me by other libraries. To these libraries, my additional gratitude: Bryn Mawr College, Clemson University, Dickinson College, Heidelberg College, Lafayette College, Lycoming College, Mount Holyoke College, Rice University, SUNY at Albany, Susquehanna University, Texas A&M University, and the University of Buffalo.

Ann Gibson, in charge of interlibrary loans at the Ellen Clarke Bertrand Library was a cheerful detective, as were Jean Bingman and James van Fleet. Tom Mattern, also of the library staff, showed himself equal to the questions I put to him. Throughout the writing of the manuscript, my respect grew for the contributions of four long-dead librarians, who, it became evident, were responsible for collecting and maintaining Bucknell's splendid collection of nineteenth-century materials. They were Professors Enoch Perrine (1885–1887), Freeman Loomis (1887–1894), William Martin (1894–1922), and Eliza Martin (1922–1938). I hope it is not too late to thank them.

I am grateful as well to the public library of Santa Fe, New Mexico, for space and the helpfulness of its staff while several drafts were undergoing revision.

Lewisburg, Pennsylvania   D. K. C.
Santa Fe, New Mexico
June 1993

Contents

Part III   The Mental Ladder

Chronology

The Chronology lists events mentioned and the temporal relationships among people, animals, experiments, and researches.

384 BCE	Aristotle born
322 BCE	Aristotle dies
1600 CE	East India Company founded
1660	George of Hanover born
1661	Lithuanian bear-boy observed
1672	Irish sheep-boy described
1683	Caroline of England born
1707	Carl von Linné [Linnaeus] born
1714	George I (of Hanover) becomes king of United Kingdom
1717	Cranenburg feral girl observed
1724	Immanuel Kant born Peter found in Hameln, Germany
1726	Peter becomes the possession of George I Swift meets Peter; finishes *Gulliver's Travels*
1727	George I of England dies
1731	Champagne feral girl observed
1737	Caroline, Princes of Wales, dies
1739	War of Jenkins' Ear
1746	Peter gets lost; found in Norfolk, England
1758	von Linné [Linnaeus] publishes tenth edition of *Systema naturae*
1775	Jean-Marc-Gaspard Itard born
1778	Linnaeus dies
1781	Immanuel Kant publishes *Critique of Pure Reason*

1785	Peter ["Wild Peter"] dies Fall of the Bastille
1799	Victor found in Caune Woods Victor meets Itard
1801	G. T. Fechner born Itard publishes first part of *The Wild Boy of Aveyron* Samuel Howe born in the United States
1804	Immanuel Kant dies
1806	Itard publishes second edition of *The Wild Boy of Aveyron*
1809	Charles Darwin born in England
1811	J. F. Blumenbach publishes *Vom Homo Sapiens Feru*
1819	Victoria, later Queen of England, born in Hesse
1826	Pinel dies
1828	Victor dies Kaspar discovered by civilization
1829	Laura Bridgman born in the United States
1833	Dash, the dog, born
1834	Ernst Heinrich Philipp August Haekel born John Lubbock born
1837	Victoria becomes Queen of the United Kingdom Laura Bridgman goes under the care of Dr. Samuel Howe
1838	Jean-Marc-Gaspard Itard dies Wilhelm Von Osten born
1840	Dash, the dog, dies
1842	William James born
1847	T. Wesley Mills born
1848	Carl Stumpf born Richard Lynch Garner born George Romanes born
1852	Conwy Lloyd Morgan born
1856	Sigmund Freud born
1857	Alfred Binet born Sepoy Rebellion
1859	Charles Darwin publishes *The Origin of Species* Edmund Husserl born
1861	Karl Krall born U.S. Civil War begins
1864	Leonard Trelawney Hobhouse born
1865	U.S. Civil War ends
1866	Von Osten moves to Berlin
1867	Lightner Witmer born

1869	Gandhi born
1870	Beginning of Franco-Prussian War
1871	Oscar Heinroth born End of Franco-Prussian War
1873	Edward Claparède born
1874	Edward Thorndike born Oskar Pfungst born
1875	Carl Jung born Melvin Haggerty born
1876	Robert Yerkes born Samuel Howe dies Victoria proclaimed Empress of India
1877	Gilbert Van Tassel Hamilton born
1878	John B. Watson born
1880	*Homo Sapiens Ferus* published by A. Rauber
1882	Charles Darwin dies
1886	Hugh Lofting born
1887	F.J.J. Buytendijk born G. T. Fechner dies
1890	Von Osten working with Hans I William James publishes *Principles of Psychology*
1891	Lloyd Morgan publishes *Animal Life and Intelligence*
1892	Garner arrives in West Africa Garner publishes *The Speech of Monkeys*
1893	Garner spends 112 days in a cage observing chimpanzees in French Gabon
1894	Georges Romanes dies
1895	Clever Hans (Hans II) born? J. B. Rhine born Romanes publishes *Animal Intelligence*
1896	Garner publishes *Gorillas and Chimpanzees*
1897	Edward Thorndike begins studying the mental processes of animals
1898	T. Wesley Mills publishes *The Nature and Development of Animal Intelligence*
1899	Mills publishes "The Nature of Animal Intelligence and the Methods of Investigating it" Robert Mowry Zingg born
1900	First public appearance of Clever Hans
1901	Victoria dies
1903	Herbert Graf (Little Hans) born

1903   Konrad Lorenz born

1904   B. F. Skinner born
        September Commission assesses that Clever Hans is not being
            signaled, but that more research is necessary

1907   Pfungst publishes *Clever Hans, the Horse of Mr. Von Osten*
        Yerkes investigates Roger the dog

1908   Lubbock publishes *On the Senses Instincts, and Intelligence of
            Animals with Special Reference to Insects*
        Witmer first sees Peter the chimpanzee in Boston
        Yerkes and "B.B.E." publish papers about Roger the dog in
            Century Magazine

1909   Freud publishes "Analysis of a Phobia in a Five-Year-Old Boy"
        Haggerty publishes "Imitation in Monkeys"
        Von Osten dies; Clever Hans becomes the property of Karl
            Krall
        Witmer publishes "A Monkey with a Mind"

1910   William James dies

1911   Alfred Binet dies
        Hamilton publishes "A Study of Trial and Error Reactions
            in Mammals"

1912   Claparède publishes "Les Chevaux Savants d'Elberfeld"
        Kamala born?
        Karl Krall publishes *Denkende Tiere (Thinking Animals)*
        First Tarzan book

1913   Claparède publishes "Encore Les Chevaux d'Elberfeld"
        John Lubbock dies

1914   War War I begins in Europe
        Hamilton publishes "A Study of Sexual Tendencies in Mon-
            keys and Baboons"

1915   T. Wesley Mills dies

1918   World War I ends

1919   Ernst Haeckel dies
        Amala born?
        John B. Watson publishes *Psychology, from the Standpoint of a
            Behaviorist*

1920   Hugh Lofting publishes *The Story of Doctor Dolittle*
        Richard Lynch Garner dies
        Gandhi begins political activity

1921   Amala dies

1922   Freud publishes a postscript to "Analysis of a Phobia in a
            Five-Year-Old Boy"

1924   Lady, the horse, born

1925   David Premack (Sarah's teacher) born

1927   Hamilton publishes *Comparative Psychology and Psychopathology*
        Lady Wonder, the horse, reads minds

1929   Hamilton publishes *A Research in Marriage*
        Leonard Trelawney Hobhouse dies
        Kamala dies
        Duane Rumbaugh, teacher of Lana and Kanzi, born

1930   (August 31) Donald Kellogg born
        (November 15) Gua born
        R. Allen Gardner, Washoe's co-sponsor and teacher, born

1932   George and Muriel Humphrey publish an English translation of *The Wild Boy of Aveyron*
        Oskar Pfungst dies

1933   Haggerty publishes *Children of the Depression*
        Beatrix T. Gardner, Washoe's co-sponsor and teacher, born

1934   Von Uexküll publishes *A Stroll Through the Worlds of Animals and Men, a Picture Book of Invisible Worlds*
        Author of this book, *Feral Children and Clever Animals,* born

1935   Lorenz publishes "Companionship in Bird Life, Fellow Members of the Species as Releasers of Social Behavior"

1936   Buytendijk publishes *The Mind of the Dog*
        Carl Stumpf dies
        Lloyd Morgan dies
        Herbert Terrace, Nim's teacher, born

1937   Melvin Everett Haggerty dies

1938   Edmund Husserl dies

1939   World War II begins in Europe
        Sigmund Freud dies
        Singh and Zingg publish *Wolf-Children and Feral Man*

1940   Edward Claparède dies

1941   Joseph Amrito Lal Singh dies
        United States enters World War II

1943   G. V. T. Hamilton dies

1945   World War II ends
        Oscar Heinroth dies

1946   Sue Savage-Rumbaugh, teacher of Austin, Sherman, and Kanzi, born

1947   Hugh Lofting dies
        India independence
        Viki born

1948   Gandhi murdered

1949   Edward Thorndike dies

1950	Konrad Lorenz publishes "Physiological Mechanisms in Animal Behaviour"
1951	Viki dies
1952	By means of a horse typewriter, Lady discloses the location of two dead children
1956	Robert Yerkes dies Lightner Witmer dies
1957	Robert Mowry Zingg dies
1958	John B. Watson dies
1961	Carl Jung dies
1965	Washoe born in West Africa
1968	Bruno born (half-brother to Nim)
1970	Lana born
1971	Koko (née Hanabi-Ko) born
1972	Moja born in New York
1973	Nim born in Oklahoma Sherman born in Georgia Herbert Graf (Little Hans) dies
1974	Austin born in Georgia F.J.J. Buytendijk dies
1975	Tatu born in Oklahoma
1976	Dar born in New Mexico
1978	Loulis born in Georgia
1980	J. B. Rhine dies
1989	Konrad Lorenz dies
1990	B. F. Skinner dies
1993	Report on Ai's counting
1993	Report on Kanzi and the child, Alia

Feral Children
and Clever Animals

Prelude

And the evening and the morning were the fifth day.

And God said, Let the earth bring forth the living creature after his kind, cattle, and creeping thing, and beast of the earth after his kind: and it was so.

And God made the beast of the earth after his kind, and cattle after their kind, and every thing that creepeth upon the earth after his kind; and God saw that it was good.

And God said, Let us make man in our image, after our likeness: and let them have dominion over the fish of the sea, and over the fowl of the air, and over the cattle, and over all the earth, and over every creeping thing that creepeth upon the earth.

Genesis

Introduction

HUMAN beings, such being their nature, desire to know. So wrote Aristotle, more or less, thereby describing one characteristic of being human and perhaps providing one possible distinction between human beings and their fellow creatures.

Inquiring human beings of our day have been taught that among the distinctions between humankind and animal life are, first, the ability to use language; second, the ability to make and use tools; third, a sense of consciousness about oneself; and, fourth, the ability to transmit culture. Each of these distinctions, as is true of all icons scientific and scriptural, has crumpled and fallen to make dust and detritus. As human beings have come to invest in the study of animal life, they have come to understand also that whatever may be thought to be unique and defining about human beings is also characteristic of other animals. The remaining candidates for distinction are metaphysical and spiritual, such as the idea that only human beings are conscious of death or that humankind, alone among animal life, desires to know. These icons may not be as solid as we think.

This book concerns the attempts by human beings to know about themselves by trying to understand the minds of other creatures, both those of other people and of other animals. How people have tried to do this tells us much about ourselves and the ways in which we human beings attempt to learn and to know. One way that reveals itself in European ideas of the eighteenth and nineteenth centuries was to examine "feral" children—that is, children presumably raised outside culture and civilization—who were presumably raised by animals. In this book, I first reexamine what is known about some of these so-called feral children: namely, Peter, Victor, and Kaspar Hauser, and the "wolf-girls" of India, Amala and Kamala. In Chapters 1 and 2 I examine not so much the dramatic aspects of the children, but why we human beings selected these forms of life for study in the expectation of reaching an understanding of what, precisely, inheritance gives and what society and socialization contribute. These children would be oddities in any age, but different times would want to know different things from them. For the first three children, the question was: What is the nature of the uncorrupted mind? For Amala and Kamala, the question appeared to be how and whether culture could be taught. Were we to find a feral child today, what question would we ask about its nature?

From the examples of feral children, we may see that in the last century or so the questions people ask about the workings of their minds have led to three ways of knowing about the human mind in a way we understand to be psychological.

One of these ways of knowing is psychoanalysis. This system, as is true of the system of understanding called evolution, interprets the past and asserts how matters came to be as they are now. In the case of Little Hans, a child whose fear of horses was reported to Sigmund Freud, Freud was led to develop and demonstrate certain concepts basic to the psychoanalytic understanding of the mind (Chapter 4). I illustrate the second method of understanding the mind with information about the careers of two horses, Clever Hans and Lady, whose mental abilities provided the opportunity and, perhaps, the motivation for this science of the mind that developed experimental principles of analysis (Chapters 5 and 6). The third way of knowing is phenomenology, a system with a long and rich history of application to human perception that is yet systematic enough to be applied to animal life. I emphasize those aspects of phenomenology that examine behavior by stressing the organization of perceptions. I believe that phenomenology is closely related to ethology, this being a specific way of studying animal behavior that emphasizes the categorizing of traits of animals. To illustrate these ways of thinking about the animal mind, I describe, chiefly, the mind of the dog (Chapter 7).

These three intellectual traditions associated with knowing the nature of the mind—psychoanalysis, behaviorism, and phenomenology—led historically to an emphasis on "testing" animals to determine their abilities and capacities. One such attempt (Chapter 8) resulted in the testing of the trained chimpanzee, Peter, to determine his mental capacities and his ability to write the alphabet. These tests were developed not in order to understand the mind of Peter so much as to promote a proposed social good; namely, the notion of testing children so they might be placed properly and efficiently into training and educational situations appropriate to their measured abilities. In other places, but at about the same time, a variety of animals, including chickens and monkeys, were being "tested" with the hope that the results would be useful in developing a "Mental Ladder." Such a ladder based on the principles of physical evolution might serve to show the mental relations among animal forms of life (Chapter 9).

Closer to our own times, the idea that we human beings might communicate with animal life directly has taken hold. The first attempts to teach humanlike speech to chimpanzees (Chapter 10) led to pessimistic results and predictions. Later such attempts, now using sign language rather than vocalizations alone, led to astonishing results as both chimpanzees and a gorilla appeared to project their thoughts to human beings, and the reverse (Chapters 11 and 12).

What are we to make of these seeming successes at interspecies communication? Do they reflect the great breakthrough in technique that now permits humankind to communicate with human and animal minds that have been previously silent, or at least noncomprehensible? Or are such results but one more example of a long tradition of humankind's misunderstanding of itself, of a wish to understand the mind of others, a wish so powerful that we forget ourselves and that it is we, not the animals, who create the myth of communication? The question is considered in the final chapter (Chapter 13).

For now, let us examine how previous societies became interested in silent minds through their interest in feral children.

What Feral Children Tell Us

PART I

Nature and Nurture: Children Without Human Parenting

1

 N A SUMMER day, at the time when hay was being harvested near the now German town of Hameln, Jürgen Meyer, who had been working the fields, met "a naked, brownish, black-haired creature, who was running up and down, and was about the size of a boy of twelve years old. It uttered no human sound, but was happily enticed [into the town] by its astonished discoverer showing it two apples in his hand, and entrapped within the Bridge-gate. There it was at first received by a mob of street boys, but was very soon afterwards placed for safe custody in the Hospital of the Holy Ghost, by order of the Burgomaster [Mayor] Severin."[1]

The year was 1724.

Peter, the name given by the street gangs to the boy, showed few signs of socialization or civility. When made comfortable in the hospital (more of a youth hostel, it would seem, than a place for the ill), he tried to leave in any way possible, sometimes by door and sometimes by window. Always alert and suspicious, he sat on his haunches or waited on all-fours, as would a four-footed animal. Seemingly unused to beds, he rolled back and forth on the straw pallet provided. He did not care for cooked foods but readily ate raw vegetables and grass. He captured birds, dismembered them, and ate the pieces. He showed approval of foodstuffs by beating his chest with his fists. When fitted with shoes, he learned to wear them but preferred not to do so. He liked having a cap put on his head and enjoyed tossing it in the water to see it "swim." He did not enjoy wearing clothes but learned to do so. His senses of hearing and smell were said to be sharp. He appeared to enjoy music.

Some thought Peter to be a true feral[2] man, a human being raised not by other human beings but by the natural state provided by the wild.[3] He was surely capable of learning, for he had learned from his experiences how to deal with the natural environment of the forest. Yet he was untouched by human contact, human demands, and human forms of socialization (Figure 1.1).

Peter's stay in the hospital was brief. We have available a contempo-

Figure 1.1   Wild Peter, said by one observer to be "more remarkable than the discovery of Uranus." Peter was examined by Dr. Arbuthnot and, seemingly, both Pope and Swift. It was hoped that Peter might provide evidence as to the composition of the human mind as it exists uncontaminated by society, for Peter came into human society from the forest, perhaps having been raised by wild animals.

rary description translated from the German: "Because of his wild manners a man was ordered to stay with him in the poor-house, who could watch his actions and projects as well as check his wildness. The man told me that he showed great fear of flogging, and when he threatened him with the rod, the boy behaved much more moderately, so that within three days he was much easier to handle, and that it sufficed to merely show him the rod in order to make him obey. Nevertheless, no one can deny that a wild nature is so deeply rooted in him that he always tries to run away."[4]

Peter's time in the town of Hameln was short. He lived with a cloth-maker, to whom he is said to have become more attached than to any other person. By October of 1725, fifteen months after his enticement into human society, he was sent to Hanover and then to London in February of 1726 as a "guest" of the royal house of Hanover, a family recently anointed as the ruling family of the United Kingdom. He became the "possession" of George I of England, the Duke of Hanover. We can only guess at their purpose in acquiring him, but the official explanation was a wish to offer Peter for scientific examination. Peter provided a seemingly unusual opportunity for his fellow human beings to examine the effects of socialization as separate from what humankind knows and does by nature. The determination of what we know by nature as contrasted with what we can learn by experience or by being taught is, after all, the root question of education, socialization, and some aspects of religious and philosophic thought. It was a question raised during the Enlightenment, and it ought to remain a significant psychological question in our own age.

Peter's presence in England was, as one observer put it, "more re-

markable than the discovery of Uranus."[5] Another observer, who saw
Peter in London, wrote to a friend: "I want to give you also news about
the wild boy, of whose education you wanted to take care. People have
given themselves all the trouble to teach him how to talk: so that one
would hear something from him about his past existence, if possible
something about his notions. But he has until today hardly learned enough
to ask in English for the most necessary things. His hearing is good, but
his pronunciation is more like babbling than like distinctive speech. He
does not know how to answer any question, and his memory is not as
good as an animal's instinct. In conclusion, his nature lacks humanness
and there is no hope that he will ever learn anything."[6]

Professor Zingg, describing the case in the late 1930s, adds that
"Peter was of middle size. . . . he had a respectable beard, and soon
accustomed himself to a mixed diet of flesh, etc., but retained all his life
his early love for onions. As he grew older he became more moderate
in his eating, since in the first year of his captivity he took enough for
two men. He relished a glass of brandy, he liked the fire, but showed all
of his life the most perfect indifference for money, and what proves,
above all, the more than brutish and invincible stupidity of Peter, just
as complete an indifference for the other sex."[7] And here Zingg adds
a note: "It [the indifference to the other sex] rather proved that Wild
Peter was feral, since no feral man shows sexuality. Tafel, another professor, [writes Zingg when analyzing the case with the advantage of modern insight] . . . also points out that this indicates that he was not a
natural idiot 'because it is proved by numerous examples that natural
idiots are very sensuous.' "[8] Views of feralness and retardation have
changed.

The decision to teach Peter to talk was so practical and therefore
obvious that we might overlook its subtlety. Of course, in order to understand Peter's experience, and thereby to advance our knowledge of
what is innate and what is acquired through experience, some means was
required to communicate with him. Our usual method is speech. By
teaching Peter to speak, it was thought to provide him with the means
and tools to communicate. As we will see, most dealings with presumably feral children have concentrated on teaching them to speak. At
the same time we must keep in mind that obviously the lack of speech
by no means implies that one is necessarily unthinking or uncommuni-
cative.

Peter was "given" to Caroline, the Princess of Wales, who loaned him
for investigatory purposes[9] to a certain Dr. Arbuthnot, friend of Alexander Pope and Jonathan Swift. Dr. Arbuthnot therefore became the first
principal investigator of a feral child. Of course, we want to know what
he saw, but for purposes of this book we want to know what the doctor
wanted to know. It is this question that allows us to reach across time to
understand the human minds of another culture living in another place
at another time; it is our way of investigating the human mind by investigating the psychology of the day.

We know something of the Doctor's conclusions, but nothing of his techniques. "Dr Arbuthnot soon found out that no instructive discoveries in psychology or anthropology were to be expected from this imbecile boy; and so, after two months, at the request of the philosophic physician, a sufficient pension was settled upon him [the boy], and he was placed first with a chamberwoman of the Queen, and then with a farmer in Hertfordshire, where at last he ended his vegetory existence as a kind of very old child, in Feb. 1785."[10]

"He [Peter] was never able to speak properly. *ki scho,* and *qui ca* [the last two words meaning the names of his two benefactors, King George and Queen Caroline] were the plainest of the few articulate sounds he was ever known to produce. He seemed to have a taste for music, and would hum over with satisfaction tunes of all kinds which he often heard. . . . No one, however, ever saw him laugh . . . he could be employed in all sorts of little domestic offices in the kitchen, or in the field. But they could not leave him alone to his own devices in these matters; for once when he was left alone by a cart of dung, which he had just been helping to load, he immediately on the same spot began diligently to unload it again."[11]

In 1746 Peter became lost, but he found his way to Norfolk, England. When found, Peter was sent to Norfolk prison for his safety, for there was political unrest and he did not speak. A fire broke out in the prison, and the gates were opened to permit the prisoners to escape the fire. When calm was restored, Peter was missing. He was found "sitting quietly at the back in his corner; he was enjoying the illumination and the agreeable warmth."[12] Peter was recognized and returned to Hertfordshire where he lived out his life. "Briefly, as an end to this tale, this pretended ideal of pure human nature, to which later sophists have elevated the wild Peter, was altogether nothing more than a dumb imbecile idiot."[13] Professor Zingg, using the accepted language of his times, means only that he had concluded that Peter was incapable of speech because of severe mental retardation.

We should not be disappointed that the questions regarding the state of Peter's natural, uncontaminated mind were not answered on the first try. Some would argue that Peter was so severely retarded that there was never hope to teach him language. Furthermore, some would say that had he learned a language, it would be discovered that in any event there was nothing in his mind. If nothing was found, of course, others would argue that the human mind was indeed a "blank slate" awaiting information. As the hypotheses were not true alternatives, the experiment could never decide which was true. That might well be the most important lesson for our times.

WHAT FERAL CHILDREN
MIGHT TELL US

In order to understand the professional, scientific, and literary excitement set off by Peter's presence, and the question it was thought communication with Peter might answer, we must turn our attention to the work of Linnaeus (Carl von Linné). Linné (1707–1778), originator of our present-day system for naming and classifying animals, derived the binomial method of naming by which animal forms are separated into genus and species. In his justifiably famous work *Systema naturae* Linnaeus separated all animal forms in this way, thereby providing the scientific world with a method to define, name, and compare species. The tenth edition of 1758 contained under *Homo* the classification

Under Primates, Man, and the Apes
Homo sapiens, homo diurnus, varians cultura, loco ferus.
[And, under the designation, *Loco ferus*,] [14]
Tetrapus, mutus, hirsutus [four-footed, dumb (in the sense of nonspeaking) and hairy].

As examples, Linnaeus listed nine cases including chronologically a wolf-boy from Hesse (1344),[15] a Lithuanian bear-boy (1661) said to have suckled at the teats of bears, an Irish sheep-boy (1672), then Peter, and two girls, one from Cranenburg (1717) and one from Champagne (1731). Peter, it seems, was not the first feral child to be discovered. However, he appeared at a point in the history of intellectual thought when there was a great deal of interest in the problem of separating the nature of the raw, unencumbered mind from the influences of socialization.

In addition to the merely curious, Peter aroused the attention of learned folk interested in natural science, especially among the adherents of the view that humankind was part of a Great Chain of plants and animals. For some, the interest was a concern about political philosophy and psychology, about how humankind comes to be social and political, and, thereby, about how to educate and instruct. These views overlap, of course, cojoining perspectives of how we understand our own humanness, our origins and natural limits, and, ultimately, how we can best govern ourselves.

What, and how much, of ourselves is innate, unlearned, clean of the effects of experience and of socialization, and how much and what can and do human beings learn from their experiences, their teachers, the environment? How we answer this question says much about our beliefs about our educational and political systems and our social expectations. Eighteenth-century Europe's interest in the possibility of democracy and the possible Rights of Mankind—whether these were given by reason, God, or the state—was the political side of the longstanding psychological question: What behavior is innate and what behavior is learned?

The answer might be found if only we could understand the content of Peter's silent mind. So thought European and British intellectual culture of the day; so, therefore, thought Dr. Arbuthnot, who was fortunate among the would-be investigators in being chosen to investigate Peter. Let us think a little about the ideas that guided him.

PHILOSOPHIES AND PSYCHOLOGIES
OF THE INTELLECT

At this period of time, in the Western world, psychology was a smallish branch of philosophy. The larger, more important, branch was metaphysics and speculation on how the mind works. Psychology was an even smaller twig from physiology (interest in the senses and how they relay and translate information to the mind itself). Modern Western psychology is descended from these two parents—philosophy and physiology— and the characteristics of both are to be seen clearly in the character of the offspring. The two parents have different ways of understanding, ways that have never become melded into a single subject matter or an agreed-to way of establishing validity. Accordingly, two kinds of psychology have arisen: that which favors investigation of the given, the genetic, the fixed; and that which is curious about what can be changed, what is flexible, and especially that which may be altered by teachers and by experience.

The issue as to what aspects of behavior can be attributed respectively to genes and to environment is the great divide in the intellectual history of humankind's thinking about itself. Our philosophy, our religions, our educational systems, our social beliefs—all of these divisions of belief and knowledge make important assumptions and suppositions about which behavior is innate and which is learned through experience. Think of the views of any philosopher, theologian, educator, or politician—indeed, any friend—and you will find tucked behind their notions a view, often tacit, about which aspects of the mind and behavior are innate and which are not.

Peter is the first of five children who occupy our attention in this first part of the book. The others are Victor, a boy whose background was akin to Peter's, except that he became a pupil of a young physician-teacher, Dr. Itard, who was interested in the boy's acculturation. While Itard undertook the training, he hired for Victor's care the loyal Madame Guérin. Dr. Itard was able to make systematic observations of Victor in a way that set the tone for many later studies of child and infant development, as well as studies of the development of chimpanzees.

Kaspar Hauser, a young man to be found a century later, differs from Peter in that there is no good reason to suppose that he was feral and much to suggest that he was not. It is evident, however, that at some time in his infancy he was captive. Most likely, he was placed in a dungeon as a child where he lived for several years. When Kaspar was discovered by civilization, he was reasonably well physically, but unsocialized. The teacher who worked with him, however, as well as the jurist who observed him, provide information on Kaspar's socialization. Kaspar did learn to speak, and from his reactions to perceptions and events, as well as his language, we know something about his socialization to the culture. He is unique for this reason, although one of the wolf-girls came to respond to words as well.

In this century, the two wolf-girls, Kamala and Amala, were found in northeastern India in a situation that suggested that they had been raised by wolves. Their removal to an orphanage, and the attention paid them by Mrs. Singh (and recorded by Reverend Singh), provide us with information on children evidently raised by animals of a species other than our own.

The fifth child, known as Little Hans, was neither feral nor raised by unusual parents, but a socialized and civilized child who became fearful of horses. He is included because his behavior contributed much to the beginnings of psychoanalysis. His situation came to the attention of Sigmund Freud, and it is from the documents in this case that Freud constructed some of the major hypotheses regarding the mind of the child. During the discussion of these cases, which occupy the first four chapters of this book, we pause in the middle to consider how the investigations of these children are related to the three kinds of thinking about the mind and behavior—namely, behaviorism, psychoanalysis, and phenomenology.

The stories to be told and analyzed concern human beliefs about what is learned and what is inherited. I tell the stories because they are informative samples of humankind's ways of thinking about our own minds and those of other animals. I also want to demonstrate that the intellectual history of how we view ourselves and our fellow human beings is but an elaboration of how we evaluate the fraction that is the equation, in our judgment, between innateness and experience. I tell the stories, as much as is practical, in the words of those who knew the children, for only the exact tone of the times can help us appreciate the subtle expression that says more than the author intends to be known. For each story, I have melded old information or found new material that revises our understanding of what is to be learned from the cases. But I have attempted to be as objective as possible, for encouraging your own understanding of these cases is of greater significance than my arguing my view to you.

This book is an account of teachers and students; both teachers of other human beings, and of animals and children. It is also a story about the psychology and philosophy of the mind, especially as our philosophy and psychology compel us to shape some questions about the nature of the mind in preference to others. Ultimately, the book is about your mind and my mind. The importance of feral children and clever animals is not that they are feral or thinking, but that we human beings ascribe characteristics to these situations. Such characteristics tell us much about ourselves, if less about the children and animals we study.

PETER AND THE QUESTION: PREWIRED OR BLANK BLACKBOARD?

As Linnaeus reported in *Systema naturae,* Wild Peter was not the first feral human being to be found and wondered at, but he was apparently the first person to have aroused interest because he might help answer the psychological question of whether learning derives from innate ideas or experience.

Toward the end of Peter's lifetime, in 1781, the philosopher Immanuel Kant, working and thinking not far from Peter's forest home in Germany, stated the idea that thoughts and ideas are of two kinds: the a priori (the given, those that exist without experience) and the a posteriori (acquired by experience but interpreted through the a priori). For example, among the a priori categories of thought are time and space. I literally *cannot* imagine anything that is not in both time and space. These concepts provide two coordinates, two categories, two filters that all must pass through. The human mind operates using such categories; it has no choice. The mind's ways of knowing are by experience which is filtered *through* these a priori categories. But we speak French or refer to certain wavelengths as "yellow" through a posteriori categories. We learn *of* them through experience, but we learn them on top of, or through, the a priori categories.

Other thinkers have suggested a different set of qualities for the mind. The concept of the blank slate, for example, also proposed at that time, argues that at birth there is nothing in or on the mind; all the categories of knowledge are learned by experience. Here we have the two alternative answers to the question. The one argues that the mind is composed of givens through which experience is mediated; the other maintains that the mind is blank at birth and that experiences become organized into it. Kant's supposition still guides our thinking about the interaction of innate sensations and innate ideas and their interaction with what we come to know through the perception of the environment.

With regard to innateness, what we know is what we interpret from our sensory systems, for these are our windows to the physical world. Note how well the idea of the blank slate or, in the terms of modern technology, the blackboard, fitted with the then-emerging eighteenth-century notion of individual rights, at least for males. If I am born with a brain that is no different from that of any one else, a brain that can be equipped, molded, and shaped by the environment, it follows that "all men are equal," or better put, "potentially equal." What makes men unequal is their education and their experience. What society can do is to provide the finest education possible for the abilities of each, thereby assisting each citizen to achieve his highest possible potential for understanding and intelligence. And, as men are born equal and each is educated to his maximum ability, so each should have an equal say in the affairs of the society. If we do not accept the notion of the blank slate or clean blackboard, we may instead accept the idea that the brain is pre-wired at birth, that ideas, understanding, and behavior are controlled genetically. It is one thing to point out that all behavior is genetic and quite another to see that genes also allow for variability in the individual. It is this variability that allows the changes on which natural selection works.

Some observers thought that Wild Peter would be able to supply some answers to the timeless question. If he could learn to talk, he could pre sumably communicate something about his experiences uncontaminated by civilization. Indeed, the very fact that he could learn to talk would be powerful evidence supporting the effort to demonstrate the importance of that form of education that emphasizes training. If he could not learn speech, those interested in showing the unmalleable effects of genetic structure would have evidence supporting their view.

Note the muddling of ideas in this simple analysis, for the muddling continues into our day. If Peter learns to speak, we learn something about the learning process and how to teach. If Peter does not learn to speak, we know nothing of the reason why. If Peter learns to speak and tells us of his feral experiences, we are still uninformed as to how much of his knowledge is given a priori and how much a posteriori. Even if he tells us every feral experience, we cannot know whether and if these two forms of categorization were available and used. Dr. Arbuthnot's work with Peter amounted to an observation, not an experiment. An experiment is designed to distinguish between alternatives. Although we think of alternative explanations of the nature of knowledge—it is innate, it is learned—Peter's investigators were naive regarding experimental design if they thought that teaching Peter to speak could help decide the issue.

Nonetheless, the scientific and intellectual interest in Peter was important. The observer who favorably compared the finding of Peter to the discovery of the planet Uranus was perceptive, for the discovery of the planet was less important than a solution to the question of the relative importance of innate knowledge and the knowledge gained through socialization. But the notion that Peter might learn to say something about his feral experience that would help us to decide the alternatives is illogical. Logically, the most one could do would be to show that Peter could learn tasks previously unfamiliar to him, such as speaking, that is, to demonstrate practical and successful ways of training. Eighteenth-century observers surely knew how to ask the right questions, for the questions remain with us in our times. However, they had not yet discovered the methods, the controlled techniques by which questions about the mind could be answered.

The question of how the brain is organized—whether prewired or clean blackboard—is not the same question as how does learning occur. To demonstrate learning (say, learning to speak) is to say nothing about its genetic or experiential origins. The logical confusion between genetic and experiential agents as causes of behavior seems to have been common in eighteenth-century thought, and they remain confused in modern psychology. The confusion has been assisted, to some degree, by inexact memories of Victor, the wild-boy of Aveyron.

VICTOR MEETS DR. ITARD

Dr. Jean-Marc-Gaspard Itard, writing of the events that came to his attention in 1799, seventy years after Peter was found, almost fifty years after Linnaeus's nomenclature of animals was published, and twenty years after Kant's work on the categories of the mind was issued, tells us the story:

"A child of eleven or twelve, who some years before had been seen completely naked in the Caune Woods seeking acorns and roots to eat, was met in the same place toward the end of September 1799 by three sportsmen who seized him as he was climbing into a tree to escape from their pursuit. Conducted to a neighboring hamlet and confided to the care of a widow, he broke loose at the end of a week and gained the mountains, where he wandered during the most rigorous winter weather, draped rather than covered with a tattered shirt. At night he retired to solitary places but during the day he approached the neighboring villages, where of his own accord he entered an inhabited house situated in the Canton of St. Sernin.

"There he was retaken, watched and cared for during two or three days and transferred to the hospital of Saint-Afrique, then to Rodez, where he was kept for several months. During his sojourn to these different places he remained equally wild and shy, impatient and restless, continually seeking to escape. He furnished material for most interesting observations, which were collected by credible witnesses whose accounts I shall not fail to report in this essay where they can be displayed to the best advantage. A minister of state with scientific interests believed that this event would throw some light upon the science of the mind. Orders were given that the child should be brought to Paris. He arrived there towards the end of September 1800 under the charge of a poor but respectable man who, obliged to part from the child shortly after, promised to come and take him again and act as a father to him should the Society ever abandon him.

"The most brilliant and irrational expectations preceded the arrival of the Savage of Aveyron at Paris."[16] George and Muriel Humphrey, a recent but modern commentator and translator, add: "The boy was brought to Paris and soon became a nine days' wonder. People of all classes thronged to see him, expecting to find, as Rousseau had told them, a pattern of man as he was 'When wild in woods the noble savage ran.'[17] What they did see was a degraded human being, human only in shape; a dirty, scarred, inarticulate creature who trotted and grunted like the beasts of the fields, ate with apparent pleasure the most filthy refuse, was apparently incapable of attention or even elementary perceptions such as heat or cold, and spent his time apathetically rocking himself backwards and forwards like the animals at the zoo. A 'man-animal,' whose only concern was to eat, sleep, and escape the unwelcome attentions of sightseers. Expert opinion was as usual somewhat derisive of popular attitude and expectations. The great Philippe Pinel (French educator and psychologist, now famous for teaching that confinement and chaining were of little use with the insane) examined the boy, declaring that his wildness was a fake and that he was an incurable idiot."[18]

Thus, as was true of Peter, a contemporary observer experienced with the insane and retarded judged Victor to be mentally incompetent.

Nonetheless, the wild-boy was examined in order to determine the nature of the unsocialized mind, just as Peter had been examined. In 1800, Pinel was fifty-five years of age; Itard twenty-six; and Victor, probably fifteen. Pinel was famous, and justifiably so, for his work as curator of the Bicêtre asylum. There he pioneered techniques and viewpoints still accepted today concerning the treatment and care of the insane and the "feebleminded." He believed they should be treated as were the physically ill; that cleanliness, healthy food, and fresh air were vital for treatment; and that such treatment itself could prove beneficial to both mind and body. He believed that such people were not only to be cared for while chained, but also cured by treatment. His view that Victor was feebleminded could not be dismissed as uninformed.

Itard and Victor saw one another for the first time in the Luxembourg Gardens in Paris.[19] The National Institute for Deaf-mutes was within sight, as it is today. Since it was summer, we can imagine the walks being lined with the neatly arranged blooming flowers that mark this park, and, if children did then as they do today, the pond was alive with model boats, some expensive, some makeshift. We do not know what Victor thought, or might have said about the sights in the park, or about his new companion, Dr. Itard. Itard, a twenty-six-year-old native of a province near the Pyrenees had just received his doctoral degree. The French Revolution was so recent that it was only Year Eight by the new French calendar.

What did Dr. Itard see? Harlan Lane, a modern scholar, tells us:

"The boy was twelve or thirteen years old, but only four-and-a-half feet tall. Light complexioned, his face was spotted with traces of small-

Figure 1.2   Dr. Itard and Victor. With his newly awarded doctorate in medicine, Itard had himself introduced to the child Victor with the hope of determining the construction of the pure mind, one unsullied by knowledge gained by experience. While the industrious Mme. Guérin cared for Victor's bodily needs, Itard studied the development of his mind.

pox and marked with several small scars, on his eyebrow, on his chin, on both cheeks. Like Itard, he had dark deep-set eyes, long eye-lashes, chestnut brown hair, and a long pointed nose; unlike Itard, the boy's hair was straight, his chin receding, his face round and childlike. His head jutted forward on a long graceful neck, which was disfigured by a thick scar slashed across his voice box. He was clothed only in a loose-fitting gray robe resembling a nightshirt, belted with a larger leather strap. The boy said nothing; he appeared to be deaf. He gazed distantly across the open spaces of the gardens, without focusing on Itard or, for that matter, on anything else. . . . .

"What he [Itard] saw," Itard wrote later, was "a disgustingly dirty child affected with spasmodic movements, and often convulsions, who swayed back and forth ceaselessly like certain animals in a zoo, who bit and scratched those who opposed him, who showed no affection for those who took care of him; and who was, in short, indifferent to everything and attentive to nothing."[20]

DR. ITARD'S INTENTIONS

The notion that Itard acted as a disinterested scientist in his treatment and relationship with Victor, or that Itard had no view on the nature-nurture question, can be dismissed with a reading of the foreword Itard wrote to a description of Victor:

"Cast upon this globe without physical strength or innate ideas, incapable in himself of obeying the fundamental laws of his nature which call him to the supreme place in the universe, it is only in the heart of society that man can attain the preëminent position which is his natural destiny. Without the aid of civilization he would be one of the feeblest and least intelligent of animals—a statement which has been many times repeated, it is true, but which has never yet been strictly proved."[21]

Note that, at this time, Itard is in the innate-ideas camp. He seems to assume that socialization is the adding of ideas to a barren mind. Unlike Dr. Arbuthnot, Itard does not assume that he can find the nature of innate ideas by teaching the savage to express himself. Itard has read on the subject: he considers evidence regarding the nature of man that has been derived from study of "wandering tribes." He believes that such information is not useful in the study of the nature of humankind, for "man is only what he is made." The "vagrant and barbarous" horse is so made, as is the civil gentleman. To understand wild man, the sort "who owes nothing to his peers," we must turn to those few cases of human beings raised in the woods. Itard does not suggest that Victor is interesting only because he was raised by animals, and he is wise to avoid this interpretation, for, of course, there is no evidence that either Peter or Victor was ever seen with animals. It will be our times that misunderstand the meaning of feralness and confuse it with presumed animal-parenting. Itard undertakes study of the wild-boy because he is interested

in the process of socialization, not because the boy may provide an opportunity to establish the correctness of ideas regarding the nature of the
mind.

Itard, freshly from medical school, tells us both the history of the feral
children and of the innocence of those who have studied them. His work
appears modern to us because he searches for the printed works of others
on the subject. He notes that, although feral children have been observed, the stories of them are unreliable, for such stories "reduce the
facts and are thereby unthorough." The unthoroughness leads to the
general conclusion that nothing can be done to educate these individuals.
To correct this bad reporting, and the conclusion that innate ideas are
fixed and render us unchangeable, Itard suggests that science has waited
for the chance (for him) to work with a feral person with the intention
". . . *to develop him physically and morally*." Or, at least, if this proved
impossible or fruitless, there would be found in this age of observation
someone who, "*carefully collecting the history of so surprising a creature,
would determine what he is and would deduce from what he lacks the hitherto
uncalculated sum of knowledge and ideas which man owes to his education*
[the italics are Itard's]."

"Dare I confess that I have proposed to myself both of these two great
undertakings?"[22]

But Victor was not yet Itard's ward: Victor was living at the Institute
where he was examined by several physicians, including Pinel. It was
Pinel whose work attracted attention; Pinel whose views were becoming
world-known; and Pinel who was the mature investigator of mental dis-
orders, insanity, and retardation. But it was Itard who undertook the
task of educating and training Victor, and Itard who wrote the two descriptions of these attempts from which the account is taken. For his
part, Pinel saw nothing to be learned from examining the wild-boy. He
was retarded, judged Pinel, and if the inexperienced Itard wanted to have
a go at it, fine, but the more experienced Pinel had warned Itard of the
hopelessness of the task.

Happily for us, Itard submitted two progress reports, written five years
apart, regarding the boy. In the first description, Itard tells us of the
circumstances and of his goals:

". . . I will venture here to give a careful analysis of the description
of the boy, in a meeting to which I had the honor of being admitted, by
a doctor whose genius for observation is as famous as his profound
knowledge of mental diseases. [The reference is to Pinel.]

"Proceeding first with an account of the sensory functions of the young
savage, Citizen Pinel showed that his senses were reduced to such a state
of inertia that the unfortunate creature was, according to his report, quite
inferior to some of our domestic animals. His eyes were unsteady,
expressionless, wandering vaguely from one object to another without
resting on anybody; they were so little experienced in other ways and so
little trained by the sense of touch, that they never distinguished an object in relief from one in a picture. His organ of hearing was equally

insensible to the loudest noises and to the most touching music. His voice was reduced to a state of complete muteness and only a uniform guttural sound escaped him. His sense of smell was so uncultivated that he was equally indifferent to the odor of perfumes and to the fetid exhalation of the dirt with which his bed was filled. . . . In a word, his whole life was a completely animal existence."[23]

Pinel's examination, as would be true of a neurological examination done today, consisted of examining the senses—the eyes, ears, touch, taste, and smell. The connection between the development of neurology and modern psychology is sometimes overlooked, but the two derive from common assumptions concerning the nature of the mind and its relation to the body. Specifically, these assumptions state that what is known by the mind comes by way of the senses; the senses are the windows to our universes. If the mind is a passive organ, one whose understandings are shaped by the experiences passed through the senses, then it follows that to understand the mind we begin by understanding the workings of the sensory systems. What do our eyes know when we are born; what do they learn from experience, and what can they tell the brain?

The study of psychology in the West, especially during the end of the nineteenth century, favored investigating the workings of the sensory systems. These investigators of psychology saw these workings in the same way that Pinel understood the purpose of measuring sensory capacity. The sensory systems were the pathways by which all or most human knowledge was transmitted; thus, an understanding of their workings was preliminary to understanding the stuff available to the mind.

Just as neurology both melded into other medical disciplines and evolved its own modern identity, so psychology, in contemporary times, has expanded its interest far beyond the workings of the sensory systems, to a variety of ways of measuring and evaluating behavior, such as mental testing, clinical counseling, and social behavior. Study of the sensory systems is now but a side issue to a psychology that sees its task as classifying and "helping" people. This is sad, for as Pinel and Itard well understood, an understanding of the sensory systems is the requisite first step for understanding the mind.

PINEL'S FIRST FINDINGS

Following his examination, Pinel had no doubt as to Victor's mental abilities. But Itard concluded that the older scholar, Pinel, was both brilliant and wrong: "Later, reporting several cases collected at Bicêtre (the asylum) of children incurably affected with idiocy, Citizen Pinel established very strict parallels between the condition of these unfortunate creatures and that of the child (Victor) now under consideration, and convincingly established a complete and perfect identity between these young idiots and the Savage of Aveyron. This identity led to the inevitable conclusion that, attacked by a malady hitherto regarded as incurable, he (Victor) was not capable of any kind of sociality or instruction." And now Itard moved to express his own view:

"I [Itard] never shared this unfavorable opinion and in spite of the truth of the picture and the justice of the parallels I dared to conceive certain hopes."[24] Itard now described the signs from Victor that give him (Itard) hope: Victor seemed to smell things that to us have no smell; he took a dead canary, investigated it, stripped it of the feathers, and tossed the body away. Itard noticed two matters that led him to believe that Victor could be civilized. One was that Victor had apparently been without human companionship and human socialization since the age of four. These early years are critical; and unlike other times in human development. As human beings are a highly social species, then, we would not expect Victor to show the socialized behaviors common to other human beings.

Second, Itard noticed signs of socialization during the few months Victor had been among socialized and civilized people. The two observations suggested to Itard that (1) there was no reason to suppose that Victor, or any one else so raised, could be better socialized, by human standards, and (2) there were signs that Victor had adjusted to human socialization since the time of his capture, this being time spent in human culture. Itard was encouraged.

"It can be seen," explains Itard, "why I argued favorably for the success of my treatment. Indeed, considering the short time he was among people, the Wild Boy of Aveyron was much less an adolescent imbecile than a child of ten or twelve months, and a child who would have the disadvantage of anti-social habits, a stubborn inattention, organs lacking in flexibility and a sensibility accidentally dulled. From this last point of view his situation became a purely medical case, and one the treatment of which belonged to mental science, that sublime art created in England and newly spread to France by the success and writings of Professor Pinel.

". . . I classified under five principal aims the mental and moral education of the Wild Boy of Aveyron.

"1st aim. To interest him in social life by rendering it more pleasant to him than the one he was just then leading, and above all more like the life which he had just left.

"2nd aim. To awaken his nervous sensibility by the most energetic stimulation, and occasionally by intense emotion.

"3rd aim. To extend the range of his ideas by giving him new needs and by increasing his social contacts.

"4th aim. To lead him to the use of speech by inducing the exercize of imitation. . . .

"5th aim. To make him exercise the simplest mental operations upon the objects of his physical needs over a period of time afterwards inducing the application of these mental processes to the objects of instruction."[25]

If these aims appear to be familiar, it is probably because they are repeated often, especially in college catalogs, as the purposes of education. In addition, they contain assumptions concerning how the human mind operates, whence it gathers the information it stores, and how it applies old knowledge to novel situations.

As we shall see, later in his life Itard put his goals into operation in regard to another project—the teaching of the deaf to communicate. It is for this application, perhaps more than for his work with Victor, that Itard is remembered and praised. For now, however, we want to see how Itard undertook to reach his goals with Victor, and how Victor responded and reacted to the strategies.

TEACHER AND PUPIL: ITARD AND VICTOR TOGETHER

Itard's first aim, as outlined, was to show Victor that a social existence was much more pleasant than a life lived amidst the rigors and loneliness of the forest. This aim also was fashioned from Itard's belief that while in captivity, Victor had regressed in some important ways. Although Victor became more tolerant of human companions, he also showed unpleasant emotions, occasioned perhaps by his being taken from the grace of natural surroundings. "His petulant activity had insensibly degenerated into a dull apathy which had produced yet more solitary habits. Thus, except for the occasions when hunger took him to the kitchen, he was always found to be squatting in a corner of the garden or hidden in the attic behind some builder's rubbish. It was in this unfortunate condition that certain inquisitive persons from Paris saw him, and after an examination of some minutes judged him to be sent to an asylum, as if society had the right to tear a child away from a free and innocent life, and send him to die of boredom in an institution, there to expiate the misfortune of having disappointed public curiosity."[26]

Evidently, a committee of some sort, charged with deciding what to do with Victor, visited and after a brief look, made its decision. Itard disagreed and criticized the committee's work, if somewhat opaquely.

Itard tells us that Victor's activities were of four kinds: sleeping, eating, doing nothing, and running about. In order to engage the senses and acquaint Victor with the fruits of socialization, he attempted to encourage Victor to explore the environment, although some of these explorations brought about unexpected rages rather than interest and appreciation. A heavy snowfall produced intense joy. Half-dressed, Victor tried every door impatiently before gaining entrance to the garden, where he then gambolled in the snow. Natural phenomena interested Victor: when the moon was up, he would spend hours staring at it, his own body motionless. This reinforced the view that he was raised by moon-watching animals.

At times, Itard reports, Victor would drop into a seeming melancholia, a sadness sometimes marked by awkward movements composed of sudden jerks, accompanied by a rhythmic rocking of the body and, at times, a physical immobility. As we now know, and as Itard surely suspected, these are signs of severe mental retardation, although their cause remains ill understood. Itard's goal in respect to Aim 1, an aim that required full-time work and attention, was to reduce the number of meals, to decrease the amount of sleep, and to increase the amount of interaction with the people in Victor's universe. How well Itard succeeded must await our considering the "progress report," written five years later.

Itard's second aim was to awaken the nervous system. Itard understood that Victor responded chiefly by means of the chemical senses, taste and smell, and the skin senses of touch, skin temperature, pressure, and the like. Itard judged (as Pinel had beforehand) that Victor's sense of hearing and vision were not as acute as those of socialized beings. Even a gun fired near the ear brought no reaction; objects in the visual environment failed to produce understanding of their function. Itard attributed this finding to the fact that hearing and seeing are refined senses that require specialized organs, the ear and the eye, to mature and perform, while touch and the chemical senses are more automatic, based more on elaborations of the skin itself. If so, Victor's perception of temperature was lacking or diminished. For example, he ate potatos fresh from the fire without wincing, and he sat in the courtyard without clothes in very cold weather. Some of Itard's time was spent protecting Victor from harming himself in the environment provided by society.

Because Itard had no way of measuring sensory perceptions (in a century, a method would be available, chiefly through the work of E. Weber and G. T. Fechner), his evaluation of improvement in the sensory systems is not that of a psychophysical measurement of thresholds. Rather, Itard merely noticed that Victor came to respond more suitably to objects and events within the climate of socialized society, a responsiveness that suggests improvement in the relationship between the brain and the sensory systems. But whether this improvement demonstrates learning on Victor's part or some reshaping of how the sensory systems operate, we do not know.

The third aim, extending the range of ideas by giving Victor "new needs and by increasing the social contacts," was not unrelated to the first aim. Attempts to interest Victor in toys, at least in toys of the sort that socialized children used, met with disinterest. Itard's descriptions of these attempts approaches resignation. Victor did, however, find a use for a set of nine-pins that Itard had left with him in the hope that Victor would come to enjoy setting them up and knocking them down. Instead, Victor tossed them into the fire and warmed himself from their burning remains. "Amusements" were a different matter. When taken outside the home for dinner in town, an event that appears to figure often in the catalog of interesting events described by Itard, a favorite amusement was the hiding of chestnuts under cups, a sort of eighteenth-century three-card-Monte. Many parents have often resorted to such an amusement with children to keep them occupied during longer meals. Victor enjoyed this game, or the chestnuts (long a favorite food); in this enjoyment Itard saw some hope of demonstrating Victor's ability to learn.

Nonetheless, Itard concludes, "With the exception of amusements which like this lent themselves to his physical wants, it has not been possible for me to inspire in him a taste for those of his age. I am almost certain that if I could have done so I should have had great success. To appreciate this, one should remember the powerful influence exerted on the first development of thought by the games of childhood as well as by the little pleasures of the palate."[27] This remark appears to say more about Itard than about Victor. Itard seems to be saying that Victor did not act as does a child: he did not play with toys, and he did not favor the morsels that attract people of culture. Itard was disappointed, rather as if these activities were signs of the civilized mind. Itard, for better or worse, was for Victor the definer and doyen of culture.

Perhaps Itard is making a more profound statement here. He may be suggesting that since the wild-boy did not learn to play at an age when children normally do, he could not learn now. Itard may also be saying that if such activities do not appear during a particular period in human development, such interest and ability are forever lost or stunted.

Here Itard may be adopting a "backup position" in his professional report to the authorities who, after all, decide whether money will be given to continue the project. If Victor does not learn, it is not because the assumption and theory are wrong, but because learning certain kinds of behavior requires they be done at a certain time in development. This notion of critical period of maturation is found in contemporary scientific thought, especially in the study of embryology. Itard uses the phrase "critical period." Rather than understand Itard's comment as a mere justification of the lack of evidence for his idea, we might see the germ of an important element of modern thought; namely, that the notions of the blank slate or the prewired system miss the point that the brain is neither, and both, depending on the particular level of maturation.

Attempts at socialization continued: Victor was taken to a farm, where he showed interest, Itard writes, in the horses and farm animals; he became separated from his caretaker, Madame Guérin, and cried upon finding her again. He spent much time fondling and hanging onto Itard, and Itard felt compelled to note that "People may say what they like, but I will confess that I lend myself without ceremony to all this childish play."[28] Itard's humanness began to be reflected in the formal documents describing Victor. Itard came to care for him, and as a result of his fondness the tone of Itard's observations shifts from the formal to the humane.

Itard's fourth aim was to teach speech and imitation. To the modern mind, imitation is a low form of intelligence that is no higher on our imagined Mental Ladder than, say, teaching the dog to bark for food. But from the eighteenth century forward, well into our times, the ability to imitate was regarded as an aptitude of animals of only the highest intelligence. At the turn of the twentieth century, the issue of whether and how animals came to imitate was of paramount importance in establishing which animals were placed on the ladder of intelligent thinking. The first studies of monkeys, for example, were concerned with assessing the ability to imitate.

To our human mind, as observed earlier, speech is the single most obvious act that separates us from other creatures. Because speech and the recording of it are the ways by which human beings not only transmit, but also store, information from now-dead generations, it is understandable that this capacity stands at the top of our list of intelligent activities. If we are to understand what Victor knew, and what he had learned in the forest without human culture, then we must rely on speech: we have no other ways of sharing his images and experiences.

Teaching the dumb to communicate was to be Itard's great interest and greatest achievement in life. As he was a young man at the time he began working with Victor, this experience may well have inspired his work with the deaf and the dumb to assist them to communicate in ways understandable to those who hear and speak. It is on the point of speech that Itard writes with greatest feeling and with a seemingly heartfelt attention to philosophic implications and his growing fondness for Victor.

Itard asks: *"Does the Savage speak? If he is not deaf why does he not speak?"* (The italics are Itard's.)

"It is easily conceived that in the midst of the forest and far from the society of all thinking beings, the sense of hearing of our savage did not experience any other impressions than those which a small number of noises made upon him, and particularly those which were connected to his physical needs. Under these circumstances, his ear was not an organ for the appreciation of sounds, their articulations and their combinations; it was nothing but a simple means of self-preservation which warned of the approach of a dangerous animal or the fall of the wild fruit. . . . When a chestnut or walnut was cracked without his knowledge and as gently as possible; if the key of the door which held him captive was merely touched, he never failed to turn quickly and run towards the place whence the sound came. If the organ of hearing did not show the same susceptibility for the sounds of the voice, even for the explosion of firearms, it was because he was necessarily little sensitive and attentive to all other impressions than those to which he had been long and exclusively accustomed." [29]

In other words, Itard says, the way a sensory system performs is not an adequate indication of its capacity. The ear of the socialized human and that of the savage may have like thresholds and be capable of equivalent discriminations, but the demonstration of discriminations alone does not indicate ability. Victor was indeed capable of hearing very well, but he demonstrated this ability only in relation to those sounds that had relevance to him. The firing of a gun near the head led to no response, because for Victor, the gunshot had no relevance; he had not learned to respond to the sound because he did not know its outcomes. In the forest and without human intervention, Victor had learned to respond to a set of sounds relevant to his needs, a set different from that of the socialized being.

Victor did not appear to perceive that the sounds available were produced by human beings, much less that they had meaning. The observation is not trivial: presumably one has to learn the origin of sounds, and if one merely hears sounds as sounds, Itard argues, some time will be required before the novice associates the sounds with what we know as human speech. Think of the human baby who must be trained to associate the source and location of sounds. If such training was never evident, as presumably it was not for Victor in the forest, why would we expect him suddenly to grasp the significance of these complex sounds? This explanation, in any event, sent Itard forward with energy.

In two instances, Victor showed promise. Once, when Victor heard two people talking outside his door, he went to the door and examined the lock. On another, when two people were talking in the kitchen, Victor, with his back to them, made no sign of recognition. A third party entered and said, "Oh, that is quite different!" and Itard noticed that Victor turned his head to the hearing of the "Oh." Indeed, Victor appeared to turn his head often whenever this person said "Oh," and this observation led to the choice of the name Victor. "This preference for "O" obliged me to give him a name which terminated with this vowel. I chose Victor. This name remains his and when it is called he rarely fails to turn his head or to run up."[30]

From our contemporary knowledge of the physiology of speech as well as reports that those who first saw Victor thought his tongue to be connected abnormally to the palate, we would suspect that Victor had a genetically deformed tongue. Itard suspected an abnormality of the larynx, although, of course, the positioning of the tongue would have been readily apparent in the presence of such an abnormality. Victor was able to return little more than a hiss to the association of objects and words that Itard tried to teach him. In vain did Itard repeatedly hold a glass of water to a thirsty Victor while he, Itard, formed the word *"eau."* And so Itard, noting the distress this little experiment caused, shifted to the word milk, *"lait."* Itard was rewarded, for he heard Victor "pronounce distinctly, though rather uncouthly it is true, the word *lait,* and he repeated it almost immediately. It was the first time that an articulate sound left his mouth and I did not hear it without the most intense satisfaction."[31]

A major concern, of course, is whether Victor used the term *lait* at times and places of his own choice. The answer is that, although Itard heard Victor say "lait" under different circumstances, he can offer no evidence that Victor demonstrated command over the word in the sense of using it to ask for something he wished. Victor's other "successful" words were *"Oh Dieu!"* which he stated in moments of happiness. Itard believed that he had acquired the word from the caretaker, Mme. Guérin, who used it often, as one can well imagine when we think of the constant and tiresome task of caring for Victor's needs.

Did Victor understand? Itard provides several examples of how Victor appeared to communicate his wants and needs to those around him, although the use of understandable speech was not evident. Victor used a specific call when he wanted a ride in a wheelbarrow, and he appeared to understand the request when he was wanted to do small chores: "Give me the water! Give me the comb!" And, as we will see, Victor was capable of setting the table on request.

As a result of Itard's work on Victor's speech, Itard heard Victor speak three words, *"lait"* and *"Oh Dieu."* As they were not always used appropriately, it is generous to think that Victor attached meaning to these words. Victor, in turn, communicated certain wants with grunts and howls, but it is not always certain whether it was the sounds themselves, or the sounds together with actions and nearby objects, that gave the hearer the impression that meaningful speech was being transmitted. Undaunted, Itard believes that Victor showed the same understanding that a child has when first saying "papa": first comes the word, next the association, then the precision of discrimination, and finally the meaning.

The fifth and final aim proposed by Itard was to induce Victor to employ "the simplest mental operations upon the objects of his physical needs over a period of time."[32] This aim brought great frustration to both Victor and Itard. Our correspondent is Itard, and he describes the games and puzzles that he used in the hope of teaching Victor "simple mental operations." Some days Victor made a response that gave Itard expectation, and, as any teacher of a difficult student can verify, even a very small promising move can give great hope. For example, when given the cards L, A, I, and T, and asked to arrange them while being shown a glass of milk, Victor arranged them this way: T, I, A, L. Itard was pleased: he thought that Victor had gotten them right, but only backward. One night, when visiting a friend of Itard's he took out the cards and, indeed spelled L, A, I, T, an act that brought the reward of a glass of milk from the friend and jubilation from Itard. (The practice of calculating the odds of a particular sequence did not appear until the nineteenth century. In this book we will note like occasions, especially when human beings listen to great apes, at which a single sequence has been taken as evidence for understanding.)

If happiness is related to the degree of frustration produced by the task, we can understand and empathize with Itard's moments of delight, hope, and expectation. Frankly, however, the time given to these tasks seemed to frustrate both persons. Victor became violent from time to time, and Itard, so human in his hopes and emotions, tells us this tale, which reflects Itard's essential humanity:

"Some time previously when Madame Guérin was with him [Victor] at the Observatory, she had taken him on the platform, which is, as is well known, very high. Scarcely had he come to within a short distance of the parapet when, seized with fright, trembling in every limb and his face covered with sweat, he returned to his governess [Mme. Guérin], whom he dragged by the arm toward the door, becoming somewhat calmer only when he got to the foot of the stairs. What could be the cause of such fright? That is not what I wanted to know. It was enough for me to know the effect to make it serve my purpose. The occasion soon offered itself in the instance of a most violent fit which was, I believe, caused by our resuming our exercises [mental exercises involving 'reading' cards]. Seizing the moment when the functions of the senses were not yet suspended, I violently threw back the window of his room which was situated on the fourth story and which opened perpendicularly on to a big stone court. I drew near to him with every appearance of anger and seizing him forcibly by the haunches held him out of the window, his head directly turned towards the bottom of the chasm. After some seconds I drew him in again. He was pale, covered with cold sweat, his eyes were rather tearful, and he still trembled a little, which I believed to be the effect of fear. I led him to his cards. I made him gather them up and replace them all. This was done, very slowly to be sure, and badly rather than well, but at least without impatience. Afterwards he went and threw himself on his bed and wept copiously.

"This was the first time, at least to my knowledge, that he shed tears." [33]

The five aims well describe the goals of working with disadvantaged human beings even today. We remain guided by the belief, if not the practice, that everyone can be educated, or, at least, trained. We believe that such education is accomplished by strengthening the senses, provided that contact with the noncaptive, civilized, and socialized world exists and that tasks can be taught that the individual can transfer to more advanced and general conceptualizations. The purpose of learning to place cards with letters on them is not, after all, to learn to place cards: it is to understand the relationship between these marks and words and the meaning of words. One does not expect to teach every word in the language, for the human mind comes to deduce and induce new things from what it already knows.

If Victor did not make the kind of progress that some scientists and citizens expected from their philosophic beliefs, this shows only that the situation was complex. As Itard explains, we should not expect the sensory systems of someone raised without human example to be like those of the cultivated person; time and development and maturation are needed. Nor should we expect such a person to be like us in all forms of behavior, for, after all, what has already been learned in the forest will influence what can be learned in society. And so the document, "First Developments of the Young Savage of Aveyron" concludes on a hopeful, though perhaps ominous, note. The story was retold and advanced in the pamphlet Itard published five years later, after a half-decade's experience with Victor. Here is the conclusion of the 1801 document:

"Consequently it devolves upon me to wait for more numerous and therefore more conclusive facts. A very similar reason has prevented me, when speaking of young Victor's varied development, from dwelling on the time of his puberty, which has shown itself almost explosively for some weeks, and the first phenomena of which cast much doubt upon the origin of certain tender emotions which we now regard as very 'natural.' Though here I have found it advisable to reserve judgement and conclusions; I am persuaded that it is impossible to allow too long a period for the ripening and subsequent confirmation of all considerations for they tend to destroy those prejudices which are possibly venerable and those illusions of social life which are the sweeter because they are the most consoling."[34]

FIVE YEARS LATER

In November 1806, when Itard transmitted his second report on Victor to the minister of the Interior, Itard had spent five years caring and educating Victor. He opens the report with a reminder of Victor's condition in 1797 (when he was captured) and in 1801 (when he was given into Itard's care). Itard reminds the minister (and thereby the public) to be mindful of Victor's condition: "To be judged fairly, this young man must only be compared with himself. Put beside another adolescent of the same age he is only an ill-favored creature, an outcast of nature as he was of society. . . . Victor is not more unlike the *Wild Boy of Aveyron* arriving at Paris than he is unlike other individuals of his same age and species."[35]

Itard elects to separate his report into three segments, each of which describes the changes in Victor: the senses, the intellect, and the emotional faculties. These three faculties are the same three that would dominate thinking about psychology until well into our time. The idea that the mind and behavior have three components and three functions (the senses, reason and intellect, and emotion and motivation) will be reorganized, both historically and in this book in Chapter 4, when we consider Freud and Little Hans. However, the idea of psychology as being composed of three functions that can be studied separately, and thereby in relation to one another, is to be seen in Itard's concept of Victor's mind.

Victor's Senses

In the intervening years, Itard had concentrated on Victor's sense of hearing, no doubt because of its direct relationship to speech. Itard performed a variety of thoughtful experiments in which he asked Victor to distinguish some sounds (a drum and a bell) and to indicate when two were alike. This approach would be taken up half a century later by G. T. Fechner, who invented and promoted it as the "psychophysical methods,"—these offering a way to determine the thresholds of the mind and, eventually, to compare our perception of the psychological world with our understanding of the "real" world.

Itard also asked Victor to learn to distinguish vowel sounds. Itard assigned each finger the sound of a vowel and asked Victor to indicate which sound he was hearing by showing the appropriate finger. But Victor was easily distracted from the requirements of the task, so Itard placed a bandage over his eyes. The bandage provoked nothing but laughter and giggling from Victor. Itard rapped Victor's hands lightly with a small stick when he erred in the task, and Victor, Itard states, ceased to think of these events as a game. Tears from insult came down Victor's cheeks. "I cannot describe how unhappy he looked [with the bandage now off] with his eyes thus closed and with tears escaping from them every now and then. Oh! how ready I was on this occasion, as on many others, to give up my self-imposed task and regard as wasted the time that I had already given to it! How many times did I regret ever having known this child, and freely condemn the sterile and inhuman curiosity of the men who first tore him from his innocent and happy life!"[36] Itard's written outburst is surely more informative than Victor's. What teacher has not felt the frustration of being unable to work with a student? What teacher has not come to mingle the act of teaching with the emotions of caring?

Order restored, the two continued their mutual instruction, Itard seemingly a little ashamed and eager to show the utility of education by kindness. But "I hoped in vain. All was useless. Thus vanished the brilliant expectations which I had founded."[37]

Similar results were found when Itard investigated Victor's sense of vision. It was not that Victor was blind; he navigated well, yet he lived in a visual universe that had little in common with that of Itard or others living in society. Itard moved on to touch by having Victor distinguish between objects of very different shapes. Some success was evident. Itard picked objects of more like shapes (a chestnut and an acorn), but Victor ceased to be able to make the distinctions expected. Yet he showed an ability to distinguish by touch and, in a way now to be described, left his mark forever on the history of human communication. "Thence," writes Itard, "I reached the point where he distinguished in the same way even the most similar metal letters, such as B and R, I and J, C and G."[38] Victor was not blind: but he had been taught to use his sense of touch, specifically that possessed by his fingers tracing shapes, to discriminate letters of the alphabet. This technique we know today as Braille reading. However, Victor never translated these letters into words, or words into meaning. The use of touch as a means of communication by language reappears in later times with Laura Bridgman (Chapter 7) and ape-human communication (Chapters 10 to 12).

Experiments concerning the chemical senses, taste and smell, were done chiefly at meals. Most meals worthy of report seem to have been taken in town rather than at home. Itard believed that Victor relied on smell, for he would pick up objects, such as stones, and sniff them before discarding them. Victor was seen to smell the hands and arms of his caretaker before following her. He came to enjoy wine but preferred water.

Itard concluded that, while hearing did not improve, the other senses "emerged from their long stupor." This comment reveals Itard's belief that, for a product to be noted, experience must act on the given, a priori material. Seeing that there was nothing more to be gained by studying Victor's sensory systems, he turned to the mental faculty.

Victor's Intellectual Functions

Itard was now convinced that Victor did not understand language. The word *"lait"* was no longer used to request milk. Itard hid the cup of milk when Victor was not looking, and although Victor searched for it, he could not use the word: "I concluded from the formation of this sign [the word *lait*], instead of being for the pupil the expression of his desire, was merely a sort of preliminary exercise with which he mechanically preceded the satisfaction of his appetite. It was necessary to retrace our steps and begin again."[39] Cleverly, Itard now made use of Victor's ability to distinguish letters by touch. Only now Itard hoped that words could be formed that would acquire meaning and that the meaning could be expressed by being spelled out by the touch system. Victor showed promise, for he distinguished objects from one another with suggestive accuracy. Itard wanted Victor to fetch the object shown by the written word. If Victor could do so, one could conclude that he could match the presence of an object with the spelling of its name. The assumption seems warranted, and Victor showed himself equal to the task, at least part of the time.

Yet, Itard was skeptical of Victor's ability. Without knowing precisely what Victor was using as a cue, he doubted that Victor was responding solely to the sight of the object. Itard hinted that he thought that Victor was responding to something that Itard himself was doing unthinkingly, giving Victor the clue as to which object to select. We will see that the issue of how and whether animals recognize cues, given unnoticed and unintentionally by the human, has not been resolved by modern thinking. We must give Itard credit for working so hard over so long a period to achieve a result important to all, yet still willing to be skeptical at the most difficult time to be so, when the data appeared to favor success. This condition of the teacher's skepticism of the pupil's learning will again arise in this book, but the issue will be one of human beings communicating with animals rather than two persons communicating with one another.

Itard's other attempts to reconstruct memory were unsuccessful and, now, for the first time in his writings, he saw the outline of the problem before him. After a time when Victor showed himself no longer able to remember the names for simple objects that he had known only the day before, Itard wrote to the minister, and to the future: " 'Unhappy creature,' I cried as if he could hear me, and with real anguish of heart, . . . 'since my labors are wasted and your efforts fruitless, take again the road to your forests and the taste for your primitive life. Or if your new needs make you depend on a society in which you have no place, go, expiate your misfortune, die of misery and boredom at Bicêtre [the asylum].' "[40]

Happily, as all parents and not a few professional teachers have learned, such outbursts from the teacher often have a beneficial effect on both pupil and teacher. Victor promptly showed memory of some things that he had seemingly forgotten, but the cost of learning, the price of attention to the task, was tears.

Reverting now to attempting to teach by imitation rather than by training, Itard did well with his student. Victor responded by imitation to pictures in a book and perhaps to the concepts of largeness and smallness. The procedures involved in imitation lend themselves well to drawing, and Itard used this compatibility by writing on a blackboard, hoping that Victor would imitate or copy the design. (Later, we shall consider the story of Peter the chimpanzee who is requested to do the same task.) The request was to make parallel lines, and Victor showed himself able to reproduce them. Itard tells us that with practice Victor was able to reproduce, on a blackboard, the outlines of certain words whose meaning he had attempted to teach Victor. However, Itard saw this imitation as a manual success, and not as a demonstration of writing or understanding. In this, as in many other matters of logic, Itard showed himself to be well ahead of the thinking of his times. Taken altogether, the five years of experiments designed to determine whether Victor could learn to write, to match objects and words, to distinguish letters, and to take lessons in reverse order, led Itard to this conclusion regarding Victor's intellectual functions: "Finally, however, seeing that the continuation of my efforts and the passing of time brought about no change, I resigned myself to the necessity of giving up any attempt to produce speech, and abandoned my pupil to incurable dumbness."[41]

Victor's Emotions

The concluding section of Itard's second, and final, report, was on emotions; it is short, accounting for only seven of the fifty-seven pages of the manuscript. Half of the seven pages on emotions are devoted to restating Victor's situation when he was found and when he was removed to Itard's care.

Victor, we are told, showed no sense of justice, of righteousness, or of wrongness of social habits, such as theft. He showed fondness for Madame Guérin and responded to her sadness at her husband's death. He showed signs of recognition and feeling to Itard, but the emotions were short-lived and seemed calculated to produce a result, writes Itard, chiefly food. Victor, as Itard tells us, was "essentially selfish."

Part of this selfishness was probably due to maturation, specifically to the onset of puberty. Itard discovered one aspect of this when he described that while bathing Victor and tickling his lumbar region he had produced an effect surprising to both of them, embarrassing to Itard and of more than passing interest to the boy. Itard tells us specifically what was occurring:

"I have not concealed the obstacles which arrested the development of certain of them [intellectual facilities], and I have made it my duty to describe exactly the gaps in his intelligence. Following the same plan in my account of this young man's emotions, I will disclose the animal side of his nature with the same fidelity as I have described the civilized side. I will suppress nothing. . . .

"But what appears still more astonishing . . . is his indifference to women in the midst of the violent physical changes attendant upon a very pronounced puberty. Looking forward to this period as a source of new sensations for my pupil and of interesting observations for myself, watching carefully all phenomena that are forerunners of this mental crisis, I waited each day until some breath of that universal sentiment which moves all creatures and causes them to multiply should come and animate Victor and enlarge his mental life. . . . I have seen him in a company of women attempting to relieve his uneasiness by sitting beside one of them and gently taking hold of her hand, her arms, and her knees until, feeling his restless desires increased instead of calmed by those odd caresses, and seeing no relief from his painful emotions in sight, he suddenly changed his attitude and petulantly pushed away the woman he had sought with a kind of eagerness. . . ." We remember that Peter's insensitivity to sex was remarked upon as well, and taken as evidence by both contemporaries and moderns of his intellectual retardation.

"Although the unhappy young man has been no less tormented by this natural ebullition, nevertheless he no longer seeks to relieve his restless desires by fruitless caresses. . . . When this storm of the senses breaks forth anew in spite of the help of baths, of a soothing diet and violent exercise, there follows a complete change in the naturally sweet character of the young man. Passing suddenly from sadness to anxiety, and from anxiety to fury, he takes a dislike to all his keenest enjoyments; he sighs, sheds tears, utters shrill cries, tears his clothes and sometimes goes as far as to scratch and bite. . . . I have been obliged, in order to remedy this state and because I could not or dared not do better, to attempt the use of bleeding, not, however, without many misgivings because I am persuaded that true education should cool and not extinguish this vital evolution. . . . the effect has only been transitory . . . the result of this continual desire, as violent as it is indeterminate, has been an habitual state of restlessness and suffering which has continually impeded the progress of this laborious education.

"Such has been the critical period which promised so much and which would, without doubt, have fulfilled all the hopes which we had entertained for it, if, instead of concentrating all its activity upon the senses it had also animated the moral system with the same fire and carried the torch of love into this benumbed heart. . . . Also I did not doubt that if I dared to reveal to this young man the secret of his restlessness and the aim of his desires, an incalculable benefit would have been accrued. But, on the other hand, suppose I had been permitted to try such an experiment, would I not have been afraid to make known to our savage a need which he would have sought to satisfy as publicly as his other wants and which would have led him to acts of revolting indecency. I was obliged to restrain myself and once more to see with resignation these hopes, like so many others, vanish before an unforeseen obstacle."[42]

We readers sense something of the teacher's disappointment: his struggle to gain some useful knowledge from the work, his growing affection for Victor coupled with the fact that Victor was also growing to be a chore, and his concern for Victor's future. The last point must have loomed important. Indeed, it may have been the cause of some of the more favorable interpretations Itard placed on the results when writing to the minister who also, of course, controlled the flow of money that kept the project alive and well funded.

But, in return, the office of the minister spoke in words that will astonish no one who submits requests for scientific or artistic support:

"This class of the Institute acknowledges that it was impossible for the instructor to put in his lessons, exercises, and experiments more intelligence, sagacity, patience, and courage; and that if he has not achieved greater success, it must be attributed not to any lack of zeal or talent but to the imperfections of the organs of the subject upon which he worked. . . . The pamphlet [the second used for this description] of M. Itard also contains the exposition of a series of extremely singular and interesting phenomena and of astute and judicious observations; and contains a combination of instructive procedures capable of furnishing science with new data, the knowledge of which can only be extremely useful to all persons engaged in the teaching of youth. In view of the foregoing, the class [meaning the minister] believes it desirable for your excellency to order the publication of M. Itard's memoir; and for the education of Victor, begun and pursued profitably to this day, not to be abandoned but for the government to continue its beneficence in behalf of this unfortunate young man."[43]

WHAT ITARD LEARNS

Our account focuses on Victor, treating Itard mostly as the observer. He is indeed an observer, but assuredly not a disinterested one. As Itard worked with Victor with the five aims in mind, his feelings for Victor underwent change from the day of the first interview in the Luxembourg gardens. He became angry with Victor, Itard lost patience, and at times he became cruel; he also grew in affection and concern for the object of his observation, and, by the time of the fifth-year report, empathy and understanding had become powerful emotions for Itard.

The lesson for us and there is no more important single lesson to be gleaned from this book is that no experiment, no observation, is a matter of one individual operating on another. The experiment changes *all* who are part of it: subject, experimenter, teacher, pupil, human, animal. When

Genesis gave humankind dominion, God did not thereby make humankind dispassionate in fulfilling its dominance.

Those who regard the experimental method as the present ultimate example of humankind's reason will consider Itard's responsivity to Victor a weakness. Some may see it as a blot on his work, whether it was characteristic of his age or of Itard himself. To think this is to overlook the view that human experiment always involves the measurement of change: the change in the presumed subject that is most easily arranged. What we forget to measure is the characteristic that most marks the human experiment: the change in the perceptions, emotions, and thoughts of the experimenter. It is this sort of change that is most difficult to locate or detect, chiefly because our view of the human experiment is that the experimenter's mind is not part of the experimental process. Reports assume the position that the experimenter is the unemotional (may we write unthinking?) machine who carries out the technique. Yet, if we can find evidence of psychological changes in the investigator as a result of experimentation, we may also find that the outcome of the experiment itself is so influenced. Let us tuck this notion to the side for a few chapters to see what evidence we might find for it.

If Itard did not answer the philosophic question of the day, he set the question in relief so that others could see its points and projections more clearly. Not all human beings are born equal, at least in the mental and physical sense, although they may be equal in some political sense. Itard appears to have come to the conclusion that Victor's ability to learn and remember was so limited that, however great Itard's disappointment, it was time for others to continue the education.

Itard would have other such pupils, although he would never again experience the companionable intensity that characterized his relationship with Victor. Itard would be responsible for both the assumptions and techniques that led to the education of deaf mutes to communicate. His contributions to our culture are considerable: the building of institutes in which the feebleminded and sensory impaired could learn to cope with a socialized world, where they might learn to communicate, and the idea for the Braille language. These reforms, if not his sole invention, were attributable largely to his insights, scientific care, and imagination. And, his concern for his patient, Victor, reaches us through the centuries in ways that Itard surely did not predict.

It is warming to think that Victor was the impetus for Itard's later successes in helping the blind to see and the deaf to comprehend.

2

Kaspar Hauser and the Wolf-Children

I F ITARD fell short of achieving his stated goal to civilize Victor, Victor taught Itard what Itard needed to know to forever change the lives of the blind and deaf. It is comforting to think that our failures may become successes; less so that probable or seeming successes may become failures.

Itard's hope to show the influence of education on a human being previously untrained and unaccustomed to human ways was not met. Victor was not to become socialized as a human being, despite Itard's teaching and Madame Guérin's life-long care. Itard worked for five years at the project, and Madame Guérin cared for Victor as boy and man for thirty; however, neither enabled Victor to enter the socialized and civilized activities of humankind. What Itard and Victor accomplished would serve as an example and a promise of what might be done for those born with sensory and mental defects. Victor did not change society's attitude toward these unfortunates, to be sure, any more than Itard singlehandedly invented Braille, sign language, and ways for the sensory-deprived to communicate both with one another and with those with full capacity for speech, seeing, and hearing. Yet Itard and Victor inspired later generations to take seriously the problem of how to communicate with those unable to do so in usual ways. Even more than this, the recent work that teaches apes communication owes much to Itard, and he, in turn, to the example of Peter, the wild-boy.

Itard's "progress report" is rich in its account of the techniques he used to try to reach Victor's intellect. Within a century, these techniques would become the foundations of an experimental psychology. The techniques, known today as matching to (or from) a sample, selection of the oddity, discriminating one sensory experience from others, making finer and finer discriminations, testing for the presence of concepts, such as long or large—these are all the examples of the "psychophysical methods" that Weber and Fechner would use fifty years later with the idea of establishing a measure of psychological experience both between and among people, between the perception of reality and the sensations imbedded in the mind.

At the time Itard placed Victor with Madame Guérin, effectively ending formal testing and treatment, those who favored prewiring as the explanation of how the mind is arranged—that is, those who thought that nature both determined and limited ability—could point to Victor's basic uneducableness as compelling proof that their view was correct. Those who argued for the impact and benefits of nurture and environment could only argue that Victor was not as opportune a subject as he appeared to be when first found. For them, the test with Victor was unproving.

Europe did not need to wait long for another seemingly perfect, yet puzzling, subject. His name was Kaspar Hauser. Like Victor, when "discovered" he was believed to have had an unusual living arrangement while an infant and child. Unlike Peter and Victor, he learned to speak, thereby fulfilling the original hope of finding a feral person able to describe his experiences.

KASPAR HAUSER

May 26, 1828 was a religious holiday in Nuremberg, one of the happier holidays of the Christian calendar, a time when people tend to be out of doors. But let the lawyer Feuerbach, then president of the Bavarian Courts of Appeal, tell the story[1]:

"A citizen, who lived at the so-called Unschlitt square, near the small and little frequented Haller-gate, was loitering before his door, and was about to proceed upon his intended ramble through the New-gate, when, looking around him, he remarked at a little distance a young man in a peasant dress, who was standing in a very singular posture, and, like an intoxicated person, was endeavouring to move forward, without being able to stand upright or govern the movement of his legs. The citizen approached the stranger, who held out to him a letter directed "To his honour the Captain of the 4th Esgataron of the Shwolishay regiment, Nuremberg.' As the captain, apparently referred to, lived near the New-gate, the citizen took the strange youth along with him to the guard-room, whence the latter was conducted to the dwelling of Captain von W. who at that time commanded the 4th squadron of the 6 regiment of Chavaux legers, and who lived in the neighborhood. The stranger advanced towards the captain's servant who opened the door, with his hat on his head and the letter in his hand, with the following words: "*Ae sechtene mocht ih waehn, wie mei Votta waehn is*" [which sounded like "I want to be a horseman as my father is"].[2] Repeated questions provoked only the answer, "Woas nit." ["Don't know."]

"He was, as the captain's servant declared in his disposition, so much fatigued that he could scarcely be said to walk, but rather to stagger. Weeping, and with the expression of excessive pain, he pointed to his feet, which were sinking under him; and he appeared to be suffering from hunger and thirst. A small piece of meat was handed to him; but scarcely had the first morsel touched his lips, when he shuddered, the muscles of his face were seized with convulsive spasms, and, with visible horror, he spit it out. He showed the same marks of aversion when a glass of beer was brought to him and he had tasted only a few drops of it. A bit of bread and a glass of fresh water he swallowed greedily and with extreme satisfaction. In the meantime, all attempts to gain any information respecting his person or his arrival were altogether fruitless. He seemed to hear without understanding, to see without perceiving, and to move his feet without knowing how to use them for the purpose of walking. His language consisted mostly of tears, moans, and unintelligible sounds, or of the words, which he frequently repeated: 'Reuta wahn, wie mei Votta wahn is.' ["I want to be a rider as my father is."] In the captain's house he was soon taken for a kind of savage, and, in expectation of the captain's return, he was conducted to the stable, where he immediately stretched himself on the straw, and fell into a profound sleep."[3]

When the captain returned and the young man was awakened (this, with much difficulty it seems), both went to the police station. The visit provoked consternation, for the young man seemed incapable of answering the simplest of procedural questions, such as providing a name or describing where he had been. Questions about his passport resulted only in utterances like those quoted. But unlike any one else the officers had interrogated, "He appeared neither to know or to suspect where he was. He betrayed neither fear nor astonishment, nor confusion; he rather showed an almost brutish dullness, which either leaves external objects entirely unnoticed, or stares at them without thought. . . . His whole conduct and demeanour seemed to be that of a child scarcely two or three years old, with the body of a young man.

"The only difference of opinion that seemed to exist among the greater part of these policemen, was, whether he should be considered as an idiot or a madman, or as a kind of savage."[4]

Deciding that nothing more could be accomplished by the interview, the police escorted the young man, who continued to have serious and obvious difficulties in walking, to the nearby jail. There, placed with another prisoner, he fell asleep, seemingly exhausted from the events.

THE MYSTERY

The boy was given the name "Kaspar Hauser." The first name is a sad joke, for "Kasper" means "clown." Establishing Kaspar Hauser's background was to become a major occupation, both because of a philosophic interest in what he might come to learn and because people were curious about where he had been for the period of his growth. Who had

had responsibility for him, where had he been maintained during these years, what had his life had been like, and who might he be? The story of Kaspar's life is a story of intrigue, deception, conspiracy, cunning, and crime—but it is unfair at this point to leap to events that will not occur for a few years. So for now we will continue with our description and analysis of Kaspar as the people of the times understood him.

We have some clues regarding Kaspar's origins. His clothes were of mixed origins. The trousers, for example, were of good quality, as was a kerchief with the initials K.H. The cap was stamped on the inside with a picture of Munich. The shoes did not fit: his toes stuck out from them. He had with him some blue and white pieces of cloth. Although they were not rags, they were not of any obvious use either. He also had a key, some gold sand (something one would not normally carry unless a transaction was to be settled), a rosary, a manuscript of Catholic prayers, and some pilgrim tracts associated with south Germany, one of which was ironically titled, "The Art of Regaining Lost Time and Years Misspent."[5] And he also carried the letter to the captain.

The letter begins: "I send you a boy who wishes faithfully to serve his king. This boy was left in my house the 7th day of October, 1812; and I am myself a poor day-labourer, who have also ten children, and have enough to do to maintain my own family. The mother of the child only put him in my house for the sake of having him brought up. But I have never been able to discover who his mother is; nor have I ever given information to the provincial court that such a child was placed in my house. I thought I ought to receive him as my son. I have given him a Christian education; and since 1812 I have never suffered him to take a single step out of my house."[6] The letter goes on to say that there was no sense in asking him to find his way home, for he had never left the house and was taken away at night; that he was a bright boy who learned easily and quickly; and that he should become a soldier.

To the letter was appended a statement and request, this in Latin script, but evidently by the same hand:

"The child is already baptized. You must give him a surname yourself. You must educate the child. His father was one of the Chevaux legers. When he is seventeen years old send him to Nuremberg to the sixth Chevaux-leger regiment. . . . He was born 30th April, 1812. I am a poor girl and cannot support him. His father is dead."[7] (Figure 2.1 depicts Kaspar Hauser in 1832.)

The letters do, of course, raise questions. Parts of the letters seem to be contradictory. The addendum in script seems to suggest aspects different from the main letter, and the questions of whether both the letters were dictated, or only one, or who wrote what, remain. If we take the dates offered to be true, Kaspar was born in April 1812, deposited with a family at six months of age, and released and found in May 1828, at which time he would have been just barely sixteen years old. The dates fit his physical being: he had a light down over his lips, the wisdom teeth appeared three years later; he was of slight build, and his feet showed no

Figure 2.1   When Kaspar Hauser was "discovered," he was thought to be mad
or a savage, but during his adolescence he was able to learn many of the ways
of society. The picture shows him as a young man, one already accustomed to
the clothing fashions of the day. The plate shows him shortly before he was
assassinated.

signs of ever having been bound by shoes; there was no horny skin on
the soles, which were reported to be as soft as his hands; the suggestion
is that he had not walked on his feet; both arms had been inoculated.
When handed paper and pen, he wrote *"Kaspar Hauser."*

While his face was described as brutish and inexpressive by all who
saw him when he was discovered, within a few months he was reported
to show humanlike expressions. He did not seem to know how to use
his fingers; his gait was that of an infant learning to talk. He walked by
placing both the ball and heel of the foot down at the same time. The
examining physician reported an abnormality of the bone structure of
the knee.[8]

Kaspar loathed meat and alcoholic drink, preferring bread and water.
Presented coffee, he would sweat and vomit. He refused milk. At night,
he lay on his straw bed; during the day, he sat on the floor with his feet
before him. When a mirror was given him, he caught the reflection, then
looked behind him, as if to find the person seen in the mirror.[9] During
a walk around the city (his companions hoped he might sight something
familiar), he responded to people generically as "Bua" and to animals as
"Ross," a word meaning horse. Black horses frightened him. A black hen
frightened him. The chiming bells at first elicited no attention; then he
began to notice their sound. A wedding procession and a military march
made sounds that appeared to interest him.

WOODEN HORSES

"Among the many remarkable phenomena which appeared in Kaspar's conduct, it was soon observed that the idea of *horses,* and particularly of *wooden horses,* was one which in his eyes, must have acquired no small degree of importance. The word 'Ross' (horse) appeared to fill the greatest space in his dictionary, which contained scarcely half a dozen words. . . . The words 'Ross! Ross!' which, also here, he so often repeated, suggested to one of the police soldiers . . . the idea of bringing him, at the guard-room, a toy of a wooden horse. Kaspar, who had hitherto on almost all occasions showed the greatest insensibility and indifference, and who generally seemed much dejected, appeared now to be, as it were, suddenly transformed, and conducted himself as if he had found in this little horse an old and long desired friend. Without noisy demonstrations of joy, but with a countenance smiling in his tears, he immediately seated himself on the floor by the side of the horse, stroked it, patted it, kept his eyes immovably fixed upon it, and endeavored to hang upon it all the variegated, glittering, and tinkling trifles which the benevolence of those about him had presented to him. Only now that he could decorate his little horse with them, all these things appeared to have acquired their true value. When the hour arrived when he was to leave the police guard-room, he endeavored to lift up the horse, in order to take it along with him; and he wept bitterly when he found that his arms and legs were so weak that he could not lift his favourite over the threshold of the door." [10]

A number of toy horses were provided by kind persons, and soon Kaspar was not merely decorating and playing with them, but giving them food and water as well. One horse, made of clay rather than of wood, began to melt around the mouth because of the amount of water poured down it, and a horse with a large open mouth was the recipient of an unending supply of crumbs.

KASPAR AT HOME

We are told that Kaspar distinguished women from men by their dress, the clothes of the former being colorful; that he himself preferred such clothes; and that he refused to believe that he would grow into an adult person. (It was necessary to mark his successive heights on the wall to demonstrate the truth of his growth.) Any number of clergymen failed to ignite a spark of religious interest in him. No animal, it was said with disappointment, could have been more unable to comprehend the "questions, discourses, and sermons" provided for Kaspar's edification. The hope was maintained that he might show an innate knowledge of aspects of the scriptures.

Hauser's keeper, Herr Hiltel, writes that he considered Kaspar to be a child, though not an idiot. By the fifth day in his home, Kaspar was given a room on the lower level where he could be observed constantly. There was nothing deceitful about Kaspar, and everywhere and always something innocent, records Hiltel. When Hiltel and his wife undressed Hauser in order to bathe him, Hiltel reports Hauser's innocence and lack of embarrassment at being undressed and unclothed.

The Hiltels's son, Julius, then eleven years of age (and sometimes their daughter Margaret, age three) was permitted to play with Kaspar, and it was Julius, writes Hiltel, who taught Kaspar to speak. Soon enough, Kaspar was sticking drawings and prints to the walls of his room. And, happily, soon enough Kaspar was sitting at the family table where, although he did not eat, he learned to sit, to use his fingers and hands, and to imitate customs. Julius was as amused by Kaspar as anyone and as willing to tease him; nonetheless, he grew fond of Kaspar and took seriously the opportunity to teach him to speak. Kaspar appears to have been treated rather as is a present-day exchange student deficient in the local language. Kaspar was also of interest to the mayor, the burgher-meister, Herr Binder, who most days had Kaspar to his house for conversation, and to a certain Professor Daumer, a teacher, who, much like Itard, was to devote his time to the education of the child.

The visits with the mayor led to the development of a more or less coherent study of Kaspar's whereabouts since birth. During their discussions, Herr Binder believed that he had communicated well enough with Kaspar, through constant questioning and careful listening to replies, to publish an account of Kaspar's history. This story is sometimes taken as the true explanation, although the only and entire source is this set of conversations. What Kaspar said, what he thought, what the mayor heard, how he phrased the questions—these are all questions that lead to the familiar problems of assessing information when two people communicate. Most people who knew Kaspar at the time did not think that his speech, much less his speaking, was far enough advanced for him to have provided a coherent story.

What the mayor heard Kaspar say was that he could only recall being in one place, a cell in Nuremberg. He lived with a man "with whom he had always been." He had always lived in this cell/hole/cage clothed in a shirt and breeches. He sat on the floor. In this cell, he never heard a sound, nor did he know day and night. When he awoke, he would always find water and bread. Sometimes the water had a bad taste, and when this was so, he slept for a long period. When he awoke, he found his clothes changed and his nails cut. In the cage were two wooden horses with ribbons. The man was good to him except once, shortly before he left the cage, when the man hit him with a stick. One day the man brought a table and paper and showed Kaspar how to write his name. Kaspar worked hard at imitating the writing (he did not know the meaning) and neglected the horses.

Other times, the man grasped him at the hips and made him raise himself up, seemingly trying to show him how to stand and walk. On one occasion he was bound and carried on a man's back. Several nights were spent on the journey. The man spoke to him only the sentence *"Reutra wahn . . ."* What Kaspar tells us later about the trip appears to be mostly dreams and thoughts that are perhaps not unlike the reverie in which many travelers reside during long journeys.[11]

Assuming that Kaspar was imprisoned somewhere for some time by somebody, where, when and by whom?

FEUERBACH VISITS KASPAR

Our knowledge of Kaspar comes principally through the recorded words of the jurist Feuerbach. On the one hand, Feuerbach reports the information dispassionately; on the other, his account lacks the sense of familiarity that taught us so much about the relationship between Itard and Victor. The difference between Itard and Feuerbach was that Feuerbach's interest in Kaspar derived from his desire to make a juridical point—namely, that captivity was a crime; that it was punishment and illegal, for it destroyed the humanness of the person. Feuerbach was a man with a view, a man seeking data to support his thesis.

Feuerbach went to Nuremberg on July 11, 1828, having heard (as it was the news of the day) of what he calls "the foundling." The scene in the prison in which Kaspar was then held was a dramatic one. Feuerbach records: ". . . from morning to night, Kaspar attracted scarcely fewer visitors than the kangaroo, or the tame hyena in the celebrated menagerie of M. von Aken. . . . We [another gentlemen, two women, and two children—two families at an outing, in other words] fortunately arrived there at an hour when no other visitors happened to be present. Kaspar's abode was in a small but cleanly and light room. . . . We found him with his feet bare, clothed, besides his shirt, only with a pair of old trousers. . . . The walls of his chamber had been decorated by Kaspar as high as he could reach with sheets of colored pictures."[12] He stuck the pictures up with saliva during the day, and took them down at night. The room was filled with toys contributed by the people of Nuremberg.

The description of Kaspar's linguistic ability is of particular interest. Feuerbach reports: "In all that he said, the conjunctions, participles, and adverbs were still almost entirely wanting; his conjugation embraced little more than the infinitive; and he was most of all deficient in respect to his syntax, which was in a state of miserable confusion. 'Kaspar very well,' instead of 'I am very well'; 'Kaspar shall Julius tell,' instead of 'I shall tell Julius.' The pronoun I occurred very rarely; he generally spoke of himself in the third person, calling himself Kaspar. . . . If you wished him to understand immediately who you meant, you must not say *you* to him, but Kaspar."[13] As is true of children coming to learn a language, both the narrowing and broadening of concepts led to some odd similes. As all hills were mountains, he referred to a man of some corpulence as "the man with the mountain." A lady, whose shawl dragged on the ground, was called "the lady with the tail."

Feuerbach's account gives us some insight into the workings of Kaspar's mind. For one thing, Kaspar was unable to say anything intelligible about his former life. For another, he was fond of the color red; he enjoyed looking at red wine, apples, and scenes through red-tinted glasses. His favorite beings were horses, both real and modeled, but he did not care for black horses. During the day, Kaspar no longer played, as he had done with the rocking horse, but preferred to write or draw pictures on paper. He was not always satisfied with his work. He did not care for new tastes or smells: put otherwise, he rejected unfamiliar chemical stimuli. Those he cared for were fennel, anise, and carraway, presumably the herbs that were incorporated into the bread he was fed in his former lifestyle. He found the smell of flower gardens offensive. Especially offensive were the smells of dead flesh, and, while walking in a cemetery, he had a violent reaction to what he said to be the smell.

He showed an ability to remember the names of people, an ability that astonished some who were present when Kaspar encountered persons he had met but once. He well remembered the names of flowers. He was respectful of authority, especially that of Herr Hiltel and Professor Daumer. He loved order and cleanliness; each trifle had its place. He wiped snuff grains from Feuerbach's collar. Feuerbach compared Kaspar to a blind man who had just recovered his sight—that is, to someone who found a new sense with which to interpret what had been well sensed before by other means. Kaspar did not understand certain illusions of the eye that experienced persons accept, if not understand. Relative height, for example, confused him: why should an object appear to be bigger or smaller depending on whether one saw it from above or below? A table was seen as the trapezoid it truly is, at least on the retina, not as the rectangle that we know it to be if we build it. Feuerbach thought of Kaspar as a person deprived of the sophistication of sociality and sensory experience; one who had now to learn how to interpret the ideas and notions that are innate to most of us.

On July 18, after an illness of a week whose severity frightened everyone, the recovered Kaspar went to live with Daumer. Here he encountered a bed for the first time, and he counted it among life's pleasures. At first, Daumer believed that Kaspar did not distinguish his dreams from reality, for in recounting the dreams, Kaspar appeared to believe that their content had happened. Of course, this interpretation may well rest entirely on Kaspar's primitive ways of speaking. Teaching Kaspar to eat meat was a chore, but the people around him emphasized it because they considered it a "given" that civilized people must eat dead flesh. Both Feuerbach and Daumer describe their difficulties in accustoming Kaspar to the pleasures and necessity of eating meat.

In the first few weeks of eating meat Feuerbach records that Kaspar grew at least two inches in height. Kaspar continued his writing and drawing and learned to play chess. He did not readily distinguish the animate from the inanimate: he worried that a tree struck with a branch would be hurt, and he treated carvings of animals as he would the animals themselves; at the game of ninepins, he was troubled that the pins were hurt by the ball, while he made no allowances for an animal's sense of responsibility, treating it as he would a human. Of interest to our knowledge of the development of visual perception is a description given by Feuerbach: "[Kaspar] had not yet learned by experience, that objects of sight appear smaller at a distance than they really are. He wondered that the trees of an alley in which we were walking became smaller and lower, and the walk narrower at a distance; so that it appeared as if at length it would be impossible to pass them."[14]

Kaspar is said to have first seen the stars of the night in August 1829 when he was shown the heavens. He asked who had put the candles in the sky, and ever after, Feuerbach writes, Kaspar was enamored of the star-filled night. Yet, unlike other people, he showed no fear of the dark, no hesitation of walking in the night, and saw no need for a torch to light his way. His eyesight was exceptional: for example, he could sight berries on a distant bush far more quickly than others could. Living with a family, he came to sense something of the relationships among them, according to Feuerbach, and was once thought to be depressed because, as Kaspar said, he had no mother or sisters or brothers.

Professor Daumer was interested in instilling higher mental processes in Kaspar or, at least, in uncovering what might already be in the mind. But no amount of discussions about God or theology or metaphysics had any result: Kaspar continued to claim that he could not understand these things. Notions of the "spirit," of "invisible will"—none of these concepts had meaning for Kaspar, although he once asked if God could "make time recede."[15] Church provided no great happiness or calm for Kaspar. "The crucifixes which he saw there excited a horrible shuddering in him; because, for a long time, he involuntarily ascribed life to images. The singing of the congregation seemed to him a repulsive bawling. 'First,' said he, after returning from attending a church, 'people bawl; and when they have done, the parson begins to bawl.'"[16]

"A DANGEROUS THING"

The move to Daumer's home was evidently made to provide Kaspar with round-the-clock training and education. Feuerbach recorded his unhappiness at Kaspar's having been sent to formal schooling. "One of the greatest errors committed in the education of this young man and in the formulation of his mind, was evidently, that, instead of forming his mind upon a model of common humanity suited to his individual peculiarities, he was sent a year or two ago to the Gymnasium [German high school], where he was besides, made to commence in a higher class. This poor neglected youth, who, but shortly before, had for the first time cast a look into the world, and who was still deficient in so much knowledge which other children acquire at their mother's breast or in the laps of their nurses, was at once obliged to torment his head with the Latin grammar and Latin exercises; with Cornelius Nepos, and, finally, even with Caesar's *Commentaries*. [The odd syntax represents the rendering of German by a translation into nineteeth-century English.]

"Screwed into the common form of school education, his mind suffered as it were its second imprisonment. As formerly the walls of his dungeon, so now, the walls of the school-room excluded him from nature and from life; instead of useful things he was made to learn words and phrases, the sense of which, and their relation to things and conceptions, he was unable to comprehend; and thus, his childhood was, in the most unnatural manner, lengthened. While he was thus wasting his time and his sufficiently scanty mental powers upon the dry trash of a grammar school, his mind continued to starve, for want of the most necessary knowledge of things which might have nourished and exhilarated it, which might have given some indemnification for the loss of his youth, and might have served as a foundation for some useful employment of his time in the future."[17]

"By the careful attention of Hr. Daumer's worthy family, by the use of the proper exercise, and by the judicious employment of his time, Kaspar Hauser's health had been greatly improved. He was diligent in learning, increased in knowledge, and made considerable progress in ciphering and writing: . . . and he was able, at the desire of those who directed his actions, to collect his recollections of his life in a written memoir."[18]

In fact, Kaspar's memoir consisted of nothing more than a page of scribbled and unclear notions. One could deduce nothing from it regarding Kaspar's former life. Nonetheless, news of the forthcoming work was to have its effect, an unintended and unwarranted one to be sure.

Most days Kaspar went from his house to Daumer's residence for his writing lesson between 11 A.M. and noon. On Saturday, October 17, he was told to stay home rather than go to his lesson, because he seemed unwell. Instead of teaching, the professor took a walk, leaving at home his (Professor Daumer's) mother and his sister, both of whom were sweeping the house. "A narrow house-door leads, by a passage, inclosing the yard on two sides, to the staircase belonging to Daumer's quarters; and, beside a wood-room, a place for poultry, and similar conveniences, there is in a corner, close under a winding staircase, a very low, small, and narrow water-closet. The small space in which this is, was rendered still smaller by a screen placed before it. Whoever is in the entry, upon a level with the ground (for instance, near the wood-room), is very well able to observe who comes down stairs and enters the water closet.

"About twelve o'clock the same day, when Professor Daumer's sister, Catherine, was busy sweeping the house, she observed, upon the staircase which leads from the first story to the yard, several spots of blood, and bloody foot-steps, which she immediately wiped away, without, on that account, thinking that anything extraordinary had happened. She supposed, that Kaspar might have been seized on the staircase with a bleeding at the nose, and she went to his chamber to ask him about it. She did not find Kaspar there; but she observed, also, in his room, near the door, a few bloody footsteps. After she had again gone down stairs, in order to sweep also the above-mentioned passage in the yard, single traces of blood again met her eye, upon the stone pavement of the passage. She went on to the water-closet where there lay a dense heap of clotted blood: this she showed to the daughter of the landlord, who had just come to the spot, and who was of the opinion that it was the blood of a cat.

"Daumer's sister, who immediately sponged off the blood, was now still more confirmed in the opinion that Hauser had stained the staircase: he must have trod upon this clot of blood, and neglected to wipe his feet before going up stairs. It was already past twelve o'clock; the table was laid, and Kaspar, who at other times had always punctually come to dinner; stayed this time away. The mother of Professor Daumer, therefore, went down from her chamber to call Kaspar, but was as unsuccessful in finding him, as her daughter had been before her.

"Mrs. Daumer was just in the act of going once more up into his chamber, when she was struck with observing something moist upon the cellar door, which appeared to her like blood. Fearing that some misfortune had happened, she lifted up the cellar door; she observed upon all the steps of the cellar drops or large spots of blood; she went down to the lowest step; and she saw, in a corner of the cellar, which was filled with water, something white, glimmering at a distance. Mrs. Daumer then hurried back, and requested the landlord's servant-maid to go into the cellar with a candle, to see what the white thing was that lay there. She had scarcely held the candle to the object pointed out to her, when she explained 'There lies Kaspar dead.'—The servant-maid, and the son of the landlord, who in the meantime had come to their assistance, now lifted Kaspar, who gave no signs of life, and whose face was pale as death and covered with blood, from the ground, and carried him out of the cellar.

"When he was brought up the stairs, the first sign of life that he gave was a deep groan; and he then exclaimed, with a hollow voice, "Man! Man!"—He was immediately put to bed; where, with his eyes shut, he, from time to time, cried out, or muttered to himself, the following words and broken sentences—'Mother!—tell professor!—man beat—black man, like sweep [meaning kitchen sweep, a man bathed in dirt from chimney cleaning]—tell mother—not found in my chamber—hide in the cellar.' "

During the next forty-eight hours, Kaspar experienced violent fits, accompanied by his mind seeming to be blank. Yet he uttered the following 'Tell it to the burghermaster.—Not lock up.—Man away!—Man comes!—Away bell!—I to Furth ride down. . . . I all men love; do no one anything. Why the man kill? I have done you nothing. . . . You should have first killed me, before I understand what it is to live. . . .

"The forehead of Hauser, who was lying in bed, was found to be hurt by a sharp wound in the middle of it, concerning the size and quality of which, the court's medical officer has given the following report. . . .

"The wound is upon the forehead, about 10.5 lines from the root of the nose; running across it; so that two thirds of the wound are on the right and one third of it on the left side of the forehead. The whole length of the wound, which runs in a straight line is 19.5 lines." [19]

After twenty-two days, Kaspar was thought to be out of danger. Later, he recounted what had happened. First, Kaspar recalled the noise of a door, something like that of the woodroom, and a bell, perhaps the sound of the housebell. He heard steps and, looking, saw a man with a "black hand" sneaking along the passage. The man was veiled, so he saw nothing of the face; the veil was a black handkerchief over the face. The next thing that Kaspar remembered was awakening and feeling the warm blood on his face. He felt frightened, looked for someone, opened the trap, and went down the basement steps. Seeing a dry spot on the basement floor, he decided to stay there. He thought himself once again entirely foresaken. And finally, he awoke in the bedroom where he was being tended. Feuerbach believed that Kaspar knew the nature of the weapon and that the weapon was intended to slice his throat. The assailant, noting the quantity of blood (writes Feuerbach) probably concluded that the deed was done properly and so it was that he abandoned the task after just one stroke with the weapon.

A witness saw such a man leaving the house and washing his hands in the street water trough. A person of similar description four days later approached a woman of the town and questioned her about news of Kaspar, news of which was posted on the village gates. No arrest was made, despite Feuerbach's opinion given later that the police examined every lead.

Kaspar continued living with the professor's family, and he continued to learn his numbers and spelling. Feuerbach complains that no systematic analysis was made of his ability or of the course of his learning, for Kaspar was not seen to be an experimental subject, as Victor was. No one seems to have seen his discovery as a means to test major hypotheses about educational philosophy. Perhaps Feuerbach, the lawyer and judge, was more comfortable with the rules of evidence and discovery than with the natural phenomena before him. Surely Feuerbach's chief reason for recording the story, and, one supposes, for investigating Kaspar and his circumstances, was to demonstrate the torture implicit in captivity.

Kaspar was now a problem and perhaps a serious one. Who could guarantee his security after this terrifying experience?

His "ownership" and place of abode were changed, as so often happens to "problem" children. He was taken under the protection of an English family, specifically by the Earl of Stanhope, who moved him to Ansbach and to another teacher who seemed to be less caring and less successful in Kaspar's education. In midafternoon on December 14, 1833, Kaspar left a friend's house, where preparations for the coming Christmas holiday were underway. Near a park, the Hofgarten, he was enticed by a stranger (we know this from Kaspar's later report of it) who said that he had word of Kaspar's mother. The stranger gave Kaspar a lady's handbag in which Kaspar later found a letter. The letter was in mirror writing; that is, it was written backwards by using the reflection from a mirror. It said:

> "Hauser can tell you exactly
> How I look and who I am.
> If Hauser will not take this trouble
> Then I will myself say
> I come [unreadable]
> I come from [unreadable]
> Of the Bavarian border [unreadable]
> At the river [unreadable]
> And I will even tell you my name.
> M.L.O."[20]

As Kaspar was searching the handbag and finding the note, the stranger stuck a thin pointed dagger into his chest near the heart. The stranger ran, and the snow left no tracks as the park was soon filled by people as they heard of the stabbing. Hauser managed to walk home, but when he told his teacher the story, the teacher didn't believe him. They then walked back to the park to check the story. The teacher eventually checked the chest and saw the wound whereupon Hauser was put to bed.

He died three days later.

Numerous conflicting explanations have been offered for Kaspar's murder. The most prevalent theory is that Kaspar Hauser was originally locked away because he stood in the way of a possible succession to the state of Baden. When the desired result was obtained, Kaspar was released, so goes the story, with the notes he carried with him being written to mask the true reason. When it became known that Kaspar was writing his memoirs, memoirs that never, it turned out, said anything of use by way of explanation, such news required Hauser's permanent removal—hence, the attempt, and the later successful assassination. It is a splendid theory, for it involves royalty, succession, conspiracy, conflicting emotions, and just the sort of combination of information and speculation that attract the detective in us all. (Zingg devotes a footnote of six pages to an analysis of the proposition and includes a genealogy that explains the problem of succession.[21])

PSYCHOLOGICAL ISSUES

The mystery and legal issues surrounding Kaspar's life have clouded examination of the psychological principles involved in his case. Unlike Peter or Victor, Kaspar came to speak and to write. He was not raised ferally but with human support. He did not, it would seem, range freely or learn about his environment. He was, to put it simply, caged. Whatever Kaspar learned, therefore, he learned either in the captive state, where he learned little about human society, or by the ministrations of the society he joined at the age of sixteen.

He showed an interest in writing and drawing. He learned to speak. He learned to use his senses, but it seems clear that he never reached the level of perception, especially visual perception, that allowed his mind to organize the aspects of his perceptions in ways that most of us do. We may see roads as narrowing in the distance, but we know this to be a movable illusion, and we continue along our way. To Kaspar, such a deduction was not evident. His successes speak to the value of education, although, unfortunately, for our analysis, nothing was taught systematically, for Kaspar was not seen as a subject. We cannot determine what Kaspar's senses were like, what he learned, or how he compared and judged. The techniques derived to help and study Victor were not needed for Kaspar, because Kaspar showed an ability to absorb information without the need for enormous repetition.

In later years, researchers of animal behavior would perform "Kaspar Hauser" experiments, as they came to be called, in order to assess the role learning and the role of innateness. The Kaspar Hauser experiment is one in which isolation is used to assess the course of learning and maturation. For example, if one wanted to know how a bird came to sing in the same manner as its relatives, the bird would be raised in acoustic isolation and then studied to see if it ever was able to make the appropriate call. The isolation experiment has flaws, among them the fact that it does not explain the interaction between environment and maturation.

Had Itard or Rousseau, or someone interested in political, psychological, or pedagogical issues, influenced Kaspar's development and training, we might be able to say more about Kaspar's perceptions when entering society and what he learned from society itself. What we find is a young man who came to learn about society, rather facilely compared to Victor or Peter, yet one who showed that in some respects the categories of perception he carried with him remained prominent when he was placed into society. Even with much contact with society, he misunderstood his direct perceptions: the road as narrowing, the trees and people as becoming smaller. The suggestion is that if some perceptions are not learned at one age, they are difficult to be taught at another: that there may after all be the "critical periods" that Itard suggested.

Kaspar's success was different from Victor's, even though it was Victor who received all the thoughtful and patient training and all the benefits of a well-developed hypothesis about the nature of knowledge. Those who believe in the genetic innateness of ability and the relative lack of importance of what is experienced, taught, and learned may point to Kaspar's experience as in no way disproving their notions, but the search for a clear test remains. Any comparison of Victor with Kaspar demonstrates that so many factors interact to produce characteristics in humans that it is simple-minded to think we can isolate any as unreservedly causative.

The question as to what is learned and what is innate is unanswerable, at least by reference to Victor and Kaspar, chiefly because we do not know their origins. Ironically, we do not know their intelligence at the outset, and our logic inevitably feeds back on itself and eats its own conclusion. From Kaspar we do find hints as to how to proceed in how to determine the ratio of innateness to experience in the forming of knowledge. Insofar as perception can be described by words, Kaspar describes a simple visual and auditory world, one that includes accepting the lies that our retina and cochlea tell us. Kaspar is not deceived, yet it is just such deception by our sensory systems that we who are civilized take to be representations of reality. Could he have learned how to use these deceptions as we do? Or are they the organization that our brains carry, organizations that we acquire by inheritance and that are thereby one of the characteristics of being human, or a primate, or a mammal?

For now, these two examples, show that any similarities between them is deceptive; that they are unalike (and therefore instructive). Probably of most importance, the two cases illustrate that what separates the situations is the kinds of questions Peter and Victor were thought to answer, the innateness of intelligence in the one and the influence of captivity in the other. The larger moral is simple: the answers we get are shaped by the nature of the question we ask, and the nature of such questions is very much attributable to the intellectual notions of the specific time. At times, the nature of the question is understandable from our grasp of the historical period involved. At other times, we understand the nature of the question because of the investigator's background and intentions.

To illustrate, we turn to a more recent situation, one that has elements of children that were not only wild but also observed to be raised by animals. In this situation too, the investigator's motives are more clearly relevant to us and the effects of the observed on the observer are also in sharper relief.

These are the wolf-children of Midnapore.

THE WOLF-CHILDREN

The Reverend J.A.L. Singh received his call to Christian education in 1912, but for the five years before that date, he had engaged in an anthropological quest that was, in time, to develop into a human experiment with religious significance. In a seemingly unconscious mimicry of the British, who then controlled India, the Reverend Singh set forth to the jungles and bush inhabited by tribals. Some thirty men would make up the party, which passed from village to village at night, seeking the array of animal life that then inhabited the Indian jungles. At times, representatives of the various tribes were "captured" (there seems to be no other word for the process) and taken to Midnapore, where the Reverend Singh and his staff taught them Christian ways and beliefs. When the training had been implanted, the reeducated souls were sent home to their fellow tribespeople to enrich others through their new knowledge.

Few, if any, areas of the world have witnessed as rapid a transformation in the environment in the last century as has India. Singh's description of the jungle reflects both his own sensitivity and the beauty to be found: "In the forest the sun could not be seen even during the day; only a ray of sunshine could be seen, here and there, during midday, peeping through the trees, bathing its focus with a golden hue, glorious and superb. . . . It was a glorious silence inside the jungle and the serene calm pervaded the whole area, bespeaking the divine presence in the wood. On a cloudy day you could see the beautiful peacocks dancing on the trees and on the creepers. . . . At times close to the big caverns, about one hundred to two hundred feet deep, you could see the Bengal tiger, sometimes with cubs, moving about and looking up at us."[22]

When tribals were found, their reactions to "discovery" were mixed. "At the beginning," Singh explains, "the men of the jungle could not bear our sight in our modern dress. They use to run away from us, disappointing us immensely. So we had to give up our costume, changing it into their mode of dressing and making ourselves just like them in appearance. This went on for several years till they got accustomed to us, and understood us to be their friends."[23]

This is what Reverend Singh chose to tell Professor Zingg of the United States about himself when Zingg inquired in 1937 for details about the wolf-children, children about whom the English-speaking world knew through ample newspaper accounts. The Reverend Singh himself had chosen not to make public his discovery and care of the children for these two reasons:

"1. They were girls, and if their rescue story became public, it would be difficult for us to settle them in life by marriage, when they attained that age.

2. We were afraid that such publication would lead to innumerable visits and queries, which would be a great drain on our time.

"Keeping these two considerations in our mind, myself and my wife guarded the secret of the rescue so that no one in the Orphanage [as the base in Midnapore came to be called] knew anything about their whereabouts and the wonderful discovery."[24] But this resolve had to be broken when one child was ill and a physician was called. The physician repeated the news outside Midnapore, with the result that the story spread from India, then to Britain, and finally the United States. Singh, transcriber of the story, is himself a person of some mystery. Some thought him but an opportunist, making the story of the girls public to serve his own ends. Others noted that by the time he did so, the tale was told and no one alive would suffer from its telling.

Zingg corresponded with Reverend Singh and acquired the diaries made at the time the girls were discovered. Zingg annotated and published these materials along with older accounts of other feral cases. Zingg never met Singh; he only met Mrs. Singh and not until after World War II by which time her health and memory had disintegrated. Yet Zingg did interact with the wolf-girls, for it was he who documented the story to the English-speaking West years after the event. That he achieved some modest fame for doing so does not imply that the diaries were untrue. It is surely to Zingg's credit that he recognized that his book was a piece of reporting, not scholarship, and that he included Singh as a co-author, even though the Reverend did not live to see the book through publication.

What became available to Professor Zingg, and thereby to us, was the Reverend Singh's diaries. Zingg examined them carefully for soundness and pronounced them to be original. He then sent copies of the diaries to a group of authorities on India, child care, and the like, and he recorded their questions about the diary and their comments. Zingg added his comments to the queries posed. He then published these along with some well-known, unedited accounts of other feral children. I rely on this account to tell the story. We join it at the point at which the Reverend Singh's party "discovered" the wolf-children:

"We took shelter in a man's cowshed in the village. The man's name was Chunarem and he was Kora by race [one of the aboriginal tribes in India]. At night the man came to us and reported in great fear about a man-ghost in the jungle close by. The Manush-Bagha (man-ghost) was like a man in his limbs with a hideous head of a ghost. The spot he cited was about seven miles from the village. He and his wife begged me to rid the place of it, as they were mortally frightened of it."

Singh became a little curious about the nature of the ghost and about the fear that had seized the village following news of its presence at Chunarem's house. He attempted to locate it with the intention of killing it.

September 24, 1920

"I got curious and wanted to see the ghost. We went out a little before dusk . . . but failed to see any sight of it. I thought it was all false, and did not care much. [Other people told of their being frightened by the ghost, and Singh recommended erecting a platform in a tree from which they could shoot the ghost.]

October 8 and 9, 1920

"We arrived at Godamuri on October 8 and stayed there with Chunarem. Early in the morning . . . we went out to see the machan and examine the haunts of the so-called ghost.

"It was a white-ant mound as high as a two-story building, rising from the ground in the shape of a Hindu temple. Round about, there were seven holes, afterwards found to be seven tunnels leading to the main hollow at the bottom of the mound. [There was a human path nearby, and at times, when walking near the mound, people saw hideous beasts. Because Singh and his party traveled with two guns and drums, it was thought that they might use them in the service of getting rid of the ghosts.]" Singh continues:

"Before dusk, . . . we steathily boarded the machan and anxiously waited there for an hour or so. All of a sudden, a grown-up wolf came out from one of the holes, which was very smooth on account of their constant egress and ingress. The animal was followed by another one of the same size and kind. The second one was followed by a third, closely followed by two cubs one after the other. The holes did not permit two together.

"Close after the cubs came the ghost—a hideous looking being—hand, foot, and body like a human being; but the head was a big ball of something covering the shoulders and the upper portion of the bust, leaving only a sharp contour of the face visible, and it was human. Close at its heels there came another awful creature exactly like the first, but smaller in size. Their eyes were bright and piercing, unlike human eyes. I at once came to the conclusion that these were human beings.

"The first ghost appeared on the ground up to its bust, and placing its elbows on the edge of the hole looked this side and that side, and jumped out. It was followed by another tiny ghost of the same kind, behaving in the same manner. Both of them ran on all fours.

"My friends at once leveled their guns to shoot at the ghosts. They would have killed them if they had not been dissuaded by me."[25]

Singh went to the tribals and asked for men to help dig out the antmound. They refused, pointing out that Singh and his men were there for the day, but they, the tribal villagers, would be left at the mercy of the ghosts. Persuaded, Singh and his friends therefore left Godamuri, but with the idea of returning to continue the hunt. On October 11, Singh found some men unacquainted with the tribe and the ghost story, and got their commitment to dig out the mound. They were not told the reason for digging; otherwise they presumably would not have joined the party. The mound was seven miles away, and work did not start until Sunday, October 17, 1920. Singh directed the workers; his friends mounted the ledge with guns but were told not to fire.

Sunday, October 17, 1920

"After a few strokes of the spade and shovel, one of the wolves came out hurriedly and ran for his life into the jungle. The second one appeared quickly, frightened for his life, and followed the footsteps of the former. A third appeared. It shot out like lightning on the surface of the plain and made for the diggers. It flew in again. Out it came instantly to chase the diggers—howling, racing about restlessly, scratching the ground furiously, and gnashing its teeth. . . .

"I had a great mind to capture it, because I guessed from its whole bearing on the spot that it must have been the mother wolf, whose nature was so ferocious and affection so sublime. It struck me with wonder. I was simply amazed to think that an animal had such a noble feeling surpassing even that of mankind—the highest form of creation—to be-

stow all the love and affection of a fond and ideal mother on these peculiar beings, which surely had once been brought in by her (or by the
other two grown-up wolves who appeared before her) as food for the
cubs. To permit them to live and be nurtured by them (wolves) in this
fashion is divine. I failed to realize the import of the circumstances and
became dumb and inert. In the meantime, the men pierced her through
with arrows, and she fell dead. A terrible sight!

"After the mother wolf was killed, it was an easy job. When the door
was cut out, the whole temple fell all around, very fortunately leaving
the central cave open to the sky, without disturbing the hollow inside.
The cave was a hollow in the shape of the bottom of a kettle. It was
plain and smooth, as if cemented. The place was so neat that not even a
piece of bone was visible anywhere, much less any evidence of their droppings and other uncleanliness. The cave had a peculiar smell, peculiar to
the wolves—that was all.

"There had lived the wolf family. The two cubs and the other two
hideous beings were there in one corner, all four clutching together in a
monkey-ball. It was really a task to separate them from one another. The
ghosts were more ferocious than the cubs, making faces, showing teeth,
making for us when too much disturbed, and running back to reform
the monkey-ball. We were at a loss and didn't know what to do.

"I thought of a device. I collected four big sheets from the men, called
in that region *Gilap* [the villagers' winter wrap]. I threw one of the sheets
on this ball of children and cubs and separated one from the other. In
this manner we separated all of them, each one tied up in a sheet, leaving
only the head free. We gave away the cubs to the diggers and paid them
their wages. They went away happy and sold the cubs in the Hat [fair]
for a good price.

"I took charge of the two human children, and came back to Chuna's
house in Godamuri. . . . I kept them in one corner of his courtyard in
a barricade, made of long sal [poles], not permitting the inmates to come
out. The area of the barricade was eight feet by eight feet. There were
two small earthen pots for rice and water placed on the side of the barricade, so that the keepers could pour in their food and drink from outside."[26]

The Reverend Singh left his find in this condition and continued his
planned tour during the next five days. When he returned, he learned of
the "miserable conditions" that had occurred during his absence. "The
children had been left to themselves without any food or drink. For Ghosts
to be living in Chuna's courtyard was more than enough to create a panic
in him and the family. They had left the place in hot haste, just after we
had left, and gone away to a place no one knew. The panic was so great
that it depopulated the whole village. . . .

"I . . . found the poor children lying in their own mess, panting for
breath through hunger, thirst, and fright. I really mourned for them, and
actually wept for my negligence. I sprinkled cold water on their faces.
They opened their mouths; I poured water in and they drank. I took

them up in my arms one by one and carried them to the bullock cart. I tried to make them drink some hot tea.

"The feeding was a problem. They would not receive anything into their mouths. I tried by spyhon [Singh's original spelling] action. I tore up my handkerchief and rolled it up to a wick. I dipped it in the tea cup; and when it was well soaked, I put one end into their mouth and the other end remained in the cup. To my great surprise, I found them suckling the wick like a baby. I thanked God most fervently for the great kindness in forgiving me my negligence in leaving the children under such a care. I thanked God doubly and many a time after this."[27]

The children were now tended for a few days. Raw milk was their only fare. Then came the seventy-five miles by bullock to Midnapore. The journey lasted from October 28 to November 4, 1920. Singh told only his wife the story of the origin of the children, although surely the other residents at the Orphanage must have been curious as to the origins of these foundlings. The girls were checked into the orphanage. Three weeks after arriving, the "matted balls" were cut from their heads. They now "looked very different." Singh guessed their ages at eight and a year and one-half.

Reverend Singh named the older Kamala; the younger, Amala.

THE YEARS IN THE ORPHANAGE

Figure 2.2 shows Reverend and Mrs. Singh and the personnel of the orphanage. Kamala sits at their feet. From the photograph, we learn something about the home—for example, that it housed and served both boys and girls. In an age such as ours, when missionary work is devalued because it is seen as forcing one's views on another, or as foolishly giving sociality and civility to those who may or may not want it (Rousseau and Itard would have understood this view; Feuerbach was critical of the kind of education given Kaspar), it is too easy to undervalue the nature of the missionary's work. The nineteenth century was an opportune time for such activity, for the colonizing of Africa and Asia by European states often involved a stated religious motive. Reverend and Mrs. Singh, we would think, saw themselves as educators (as Itard saw himself) and not as reporters (as Feuerbach saw himself). The Singhes' motive was to bring knowledge, especially Christian views and practices, to the tribals, because such knowledge might extend their Christian lives and make them valuable, to themselves and others. There are less admirable motives for doing research, to be sure.

In the early twentieth century, India had many tribes, just as it does today. Professor R. Ruggles Gates, who wrote one of the several prefaces to the Singh and Zingg account, notes: "The Native State of Travancore alone has some thirteen native peoples, such as the short Kanikars, who live in the malarial foothills; the Pulayas, who are somewhat Negroid in appearance; the Uralis of the Nilgiri Hills, who live in trees to escape from the elephants; and many others."[28] While some, today as in the

Figure 2.2   The personnel at the orphanage at Midnapore. The Reverend and Mrs. Singh are in the center, with the wolf-girl, Kamala, at their feet. Amala is not in the picture. Mrs. Singh and the staff cared for the wolf-girls, while Reverend Singh kept notes on their intellectual, emotional, and especially, their theological development.

past, want the tribals of India and elsewhere kept in their aboriginal state, others believe that they should be brought into modern society. Because of his concern with the orphans of the local tribals [these were tribals of Bengal, near the Gandak and Ghaghara[29]], Reverend and Mrs. Singh together with the staff taught Christian civility to those brought into its fold. Let us return to the diary and learn more about the two girls. Here is an entry for November 24, 1920:

"They were covered with a particular kind of sores all over the body. These sores ate up the big and extensive corns on the knee and on the palm of the hand near the wrist which had developed from walking on all fours. . . . At times, they [the sores] made them look like lepers. Our medicines were carbolic soap, carbolic lotion, tincture of iodine, zinc and boric acids. My wife and myself used to wash the sores with carbolic lotion and carbolic soap and bandage them with boric cotton, and when the sores commenced forming granulations, we washed them with tincture of iodine lotion, wiped them dry, sprinkled them with boric and zinc acids, and bandaged them with cotton. We did everything ourselves and did not call a doctor through fear of exposure.

December 5, 1920
"They got cured of the sores. . . . I found them very fond of raw meat and raw milk. . . . [They were not given much meat, as it was thought that it might make them more ferocious.]

December 19, 1920

"They would run very fast, just like squirrels, and it was really a business to overtake them. They were very attached to one of the foundlings, Benjamin by name. He had been picked up in the jungle from among the dry leaves. . . . [He was a year old and just starting to walk. They played together and seemed to be learning to walk together.]"

December 31, 1920

"But, unfortunately one day, on the thirty-first of December, Benjamin was bitten and roughly scratched by the wolf-children. After this Benjamin never came into their company and always tried to avoid them altogether. He was terribly frightened of them. . . .

"From the very beginning their aloofness was noticeable. They would crouch together in a corner of the room and sit there for hours on end facing the corner, as if meditating on some great problem. . . .

"We never kept them alone, but always purposely kept a few Orphanage children in the room. . . . They [the wolf-children] remained quite uninterested and indifferent.

". . . They wanted to be all by themselves, and they shunned human society altogether. If we approached them, they made faces and sometimes showed their teeth, as if unwilling to permit our touch or company. This was noticed at all times, even at night."

January 29, 1921

". . . When they wanted to get away, a girl, by the name Roda, tried to prevent their escape. They gave her such a bite and scratched her so that she was compelled to leave them alone. They ran out and entered the lantana bushes outside the compound. They ran fast like squirrels and could not be overtaken. . . . It was really a task to search them out, because they remained noiseless there till they were discovered. Such was their nature to shun human association. . . .

"[The other children] tried their utmost to allure them to play with them, but this they resented very much, and would frighten them by opening their jaws, showing their teeth, and at times making for them with a peculiar harsh noise. So it became almost impossible for us to bring about any sort of social relationship between them and the children."

November 1920 to January 1921

"At this stage of their association with us, for nearly three months . . . there was a complete disassociation and dislike, not only for us, but for their abode among us, for movement and play—in short, or everything human." [30]

Singh now recounts what he calls a change in appearance, this being the change following their being cleaned, their boils cured, and their hair cut:

"They looked [like] human children again. . . . [But the jawbones were prominent.] The jaws also had undergone some sort of change in the chewing of bones and constant biting at the meat attached to the bone. When they moved their jaws in chewing, the upper and lower jawbones appeared to part and close visibly, unlike human jaws. . . .

"They could sit on the ground squatting down, . . . but could not stand up at all. [The joints had lost their flexibility.] . . .

"Their eyes were somewhat round and had the look as if heavy with sleep during the day. But they were wide open at night after twelve o'clock. They had a peculiar blue glare, like that of a cat or dog, in the dark. At night . . . you saw only two blue lights sending forth rays in the dark. [This observation leads Zingg to a four-page footnote, and an interesting one. He considers the issue of whether the human retina may reflect light. Testimony is reported from a U.S. biologist who is involved in a shooting that results from the reflection at night from the retina of another human being. A fascinating issue, one well documented by the alert Zingg.] . . .

"They could see better by night than by day. . . . They could detect the existence of a man, child, animal, or bird, or any other object in the darkest place when and where human sight fails completely. . . .

"They had a powerful instinct and could smell meat or anything from a great distance like animals. . . .

"Kamala's instinct led her . . . to locate the entrails of a fowl thrown outside the compound, about eighty yards from the Orphanage dormitory, where she was caught red-handed eating them. . . . When any food was given them, they used to smell it before eating it. . . .

"The least sound drew their attention. . . .

"Their hands and arms were long, almost reaching to the knees. . . .

"The nails of the hand and foot were worn on the inside to a concave shape. This was due to scratching the ground with the fingers. . . .

"They used to eat or drink like dogs from the plate, lowering their mouths down to the plate. . . . When they were hungry they would come smelling to the place where food was kept and sit there. . . . The least smell of food or meat anywhere, even a dead animal or bird, would bring them to the spot at once. . . .

"Kamala and Amala could not walk like humans. They went on all fours. . . .

"[They] used to sleep like pigs or dog pups, overlapping one another. . . . They never slept after midnight and used to love to prowl at night fearlessly. . . .

"We were compelled to permit them to be naked all the while, except a loin cloth stitched behind them in such a fashion that they could not open it out. . . . They resented this very much at first. . . . They . . . never shivered or showed any sign of feeling cold.

"Fire they knew and would not go near it.

"They were fond of darkness. [They could not see well during the day when the hot sun was shining. They showed fear (one of the few times they ever did so) to a lighted match.] . . .

"The children used to pass urine or have bowel movements anywhere, and at any time. . . . How then, I thought, could they keep their cave so neat and clean?

"It is presumed that the cubs and the children too, when they were babies and could not go out with the wolves, served the calls of nature inside the cave, and the mother wolf used to lick the droppings up and take out the bones, etc., after they had eaten."[31] [Figures 2.3 to 2.6 are Singh's pictures of the girls.]

KAMALA AND AMALA: EMOTIONS AND THE INTELLECT

By November 1920, within a month of their capture, Singh noticed increasing signs of "a growing feeling, approaching human affection. Affection in this world is able to work wonders, if we have the divine patience to co-operate with it. Affection tames the pets in the house. . . . It is affection that binds one to submit to another, that binds a brave heart to a coward, a higher being to a lower creature, a noble to a slave, a man to an animal, and even an enemy to a bitter foe. The essence of human life is nothing but affection. This is represented in filial affection. The wolf-children looked for the same affection which they found in the mother wolf (in accordance with their crude animal nature) from their very infancy. That sympathy, or affection, or kindness, they were searching for among us, but they could not trust us at the beginning, and hence the delay in their progress in cultivating the human faculties."[32]

Reverend Singh recounts examples of times when the girls turned to Mrs. Singh for protection and affection. He offered the hypothesis that the children merely seemed retarded because they had not yet learned to trust. He takes the example forward by making a comparison to pets, a comparison that may seem a natural one considering the reactions of the girls to human civilization. "Pets do understand us much more than we think they are able to. It is in our houses that we teach them insincerity and build them up to grow and mature into it. In the jungle they are simple, and sincerely ferocious, which is already known by their species and kind. To wish the jungle beasts to be tame and to be our companions is beyond our expectation, from the very fact that their nature and ours are quite different."[33]

It is a sign of the change in intellectual thought from the time of Itard that Singh speaks of the animal instinct of the children while comparing them to the wilder animals. Itard nowhere makes an issue of Victor's presumed forest background or the possibility that his being raised with animals influenced his behavior in civilization. To Singh, that which is not human is animal, and the metaphor of taming the human and of domesticating the girls guides his thought and instructional methods. To Itard, the purpose of training is to determine what humanness resides therein; to Singh, the purpose of training is to drive out the animal instincts.

Figure 2.3   The wolf-children are able to crawl. This and the following photographs were taken by Reverend Singh with a box camera. The set of pictures is the only documentation of the wolf-girls' activities.

Figure 2.4   Reverend Singh's picture showing the "mode of eating." Singh explains their eating by writing, "The methods of eating were conditioned reflexes learned from the wolves."

Figure 2.5 A wolf-girl scratches at the door, evidently to indicate a wish for the door to be opened.

Figure 2.6 Reverend Singh's box camera catches the act of running. Notice the gait.

Figure 2.7   Kamala and Amala sleep overlapping one another.

Singh comes to a conclusion regarding the question that engaged Itard: how much can be learned from experience, and how much is given? Singh's view (although he had not set out to answer this particular question) was that their "human instincts were all lying dormant almost in a sub-conscious state." It was their animal instincts that had been given the chance for development, and develop they did, seemingly at the expense of the human instincts. What is necessary, he suggests, to make it possible for the human instincts to develop is love and trust, these being, evidently, the human capacities that mark living beings as beings uniquely human. Singh's evidence for this idea is that it took a year for the girls to show affection toward Mrs. Singh, but once they had shown affection, development was seen to occur in other intellectual and emotional areas as well. Singh describes the process as "taming," and it is a word that seems appropriate. Mrs. Singh gave ample devotion to the girls, encouraging play and physical contact. She was rewarded with an occasional grunt that, the Singhs hoped, would become human speech.

In early September 1921, both girls became ill. Dr. Sarbadhicari attended and found that he could do little without knowing something of the girls' background. The Singhs agreed to his request and, while asking for his cooperation in secrecy, told him the story of their discovery and capture. But, as we have noted, the story along with the physician, soon made the rounds of Midnapore and far beyond. The illness was serious.

"Diarrhea set in, followed by dysentery. Then round worms appeared on the twelfth of September 1921. The round worms were six inches long, red in color, and thick as the little finger of the hand, and almost all of them alive. When they were ejected Amala brought out eighteen such worms, whereas Kamala brought out 116.

"All these days and nights, from the sixth to the twenty-first of September 1921, Mrs. Singh remained with them [the girls] at their bedside nursing them. . . .

"Between these dates they were unconscious, cold, and motionless, only just breathing. . . . They just opened their mouths when the spoon containing some drink or medicine touched their lips. . . .

"The doctor could not give us any hope at all. Our only hope was prayer. . . .

"Thus the time passed, and Amala's case became hopeless. . . . Amala gave up the ghost on the twenty-first of September 1921. . . .

"Amala was baptized a little while before she expired.

"She was buried in the churchyard of St. John's Church, Midnapore, on the twenty-first of September 1921. Her death certificate ran as follows:

"This is to certify that Amala (Wolf-Child), a girl of the Rev. Singh's Orphanage, died of Nephritus on September 21, 1921. She was under my treatment. *September 21, 1921. S. P. Sarbadhicari, Indian Medical Service.*"[34]

Kamala, who evidently had regained her health before Amala's death, tried to wake Amala on the day of her death. "She touched her face, opened the lids of her eyes with her fingers, and parted her lips. She would not leave the body of Amala and had to be removed from the coffin. Two tears dropped from her eyes."[35] She sat in a corner, alone, for the next six days. She was seen to be smelling at all the places that Amala frequented. She ignored Mrs. Singh for some weeks, but eventually came to her bed and was stroked.

Now began a plan of exercises developed by Mrs. Singh to help Kamala use her body in human ways, to sit, stand, and run. For example, a swing was erected. In August of 1922, Kamala was seen to use the swing by herself, without help or prompting. The activity might not seem like much from our perspective, but from the view of those who lived and worked with her daily, it was an emotional and remarkable event.

KAMALA'S SOCIALIZATION

Peter, Victor, and Kaspar showed clear preferences for foods. So did Kamala. She ate with her mouth alone, not using her hands, and put the plate on the floor. In short, she ate wolf-fashion. She would eat and drink so much that she became ill from overeating. She much enjoyed meat, although this was not a staple of the orphanage. (Peter and Victor did not, it may be recalled.) She was not seen to kill animals, but if she found a dead bird or animal, she would take it. She ate the carrion and, on occasion, drove the vultures and crows of the field away from the carrion so that she might take it for eating. When finding a dead chicken, she took it and hid in the tall lantana. On one occasion a "feather and particles of meat were found on her lips and on the sides of her cheek." When she found a bone, she tore the remaining flesh from it. She was also fond of sweets but did not care for the flavor of salt. She would, at times, visit the dogs when they were being fed. The dogs did not growl at her but let her eat from their supply. Kamala's carnivorous activities no doubt helped persuade Reverend Singh of her animal nature.

Kamala still (1925) urinated wherever she was; however, if either Reverend Singh or Mrs. Singh were around when she needed to urinate, she would go to the dormitory bathroom. If they were not in sight, she urinated at will. She no longer objected strenuously to baths, but did not enjoy water. Evidently, she came to understand that urinating in the wrong places led to castigation, but only if the Singhes were around to see it. In short, she was not yet toilet-trained, but she was bright enough to use the presence, or absence, of the Singhes to discriminate her choice.

After five years in the Orphanage, Kamala demonstrated some intellectual functions. She knew some of the names of the babies housed there; she understood the concept of color; she accepted food only from her plate and knew her glass from among the others. On one occasion, she refused to go outside without being dressed, having complained with the sound "Fok" (frock?) to Mrs. Singh. Her vocabulary was thought to consist of thirty words. These words, however, were not those of common English speech, but rather sounds, often used by other children in other contexts, that had come to have meanings to the adults. For example, when food was offered, she used the word "Hoo" to indicate "yes," although the children used this word to express cold. ("Ha," however, is "yes" in Bengali.) A list of the words, and their Bengali and English equivalents, shows a similarity to the former.[36]

A visitor, Bishop H. Pakenham-Walsh, then of Coimbatore, provides an outsider's description of Kamala at this time in her life:

"When I saw Kamala she could speak, quite clearly and distinctly, about thirty words; when told to say what a certain object was, she would name it, but she never used her words in a spontaneous way. She would never, for instance, ask for anything she wanted by naming it, but would wait quietly till Mrs. Singh asked her, one by one, whether it was so and so she wanted, and when the right thing was named she would nod. She had a very sweet smile when spoken to, but immediately afterward her face resumed an appearance of unintelligence; and if she were left alone, she would retire to the darkest corner, crouch down, and remain with her face to the wall absolutely listless and with a perfectly blank expression on her face. . . .

"I saw her again two years later, and except that she had learned a good many more words, I did not notice any mental change. What interested me most was to find, from careful inquiry from Mr. and Mrs. Singh, that while the wolves had not been able to teach anything especially human to their little human cubs, so that there was no sense of humor, nor (except in the one case when Kamala wept when Amala died) of sorrow, very little curiosity, and no interest except in raw meat, neither had they [the wolves] taught them anything bad. . . . Human vices seem to have been as little inherited as human virtues, and this fact seems to me to have a very pertinent bearing on the consideration of what we mean by "Original Sin." It certainly shows the tremendous importance of the human environment, and of the training of little children."[37]

The bishop's stay was short yet motivated. Having heard the story of the Singhs' efforts to Christianize the wolf-children, his interest was both theological and psychological. The question of whether beliefs are innate or learned is of interest to theology. From where, for example, do humans acquire the concept of God? Must the concept be taught, or is it possible that human beings are born with the concept available? The bishop gives evidence for being among the "blank-slate" theologians ["Human vices are as little inherited . . ."] and appears to take it as evidence against Original Sin. He surely judges the child to be capable of learning, and judges the human environment to be instrumental in so doing.

The bishop was merely one of a number of potential visitors, and, no doubt, his ecclesiastical standing was responsible for his success in securing an invitation to meet Kamala. Kamala's (and the Singhs's) fame had spread. An invitation arrived from the Psychological Society of New York, USA[38] in 1928 offering to take Kamala to the United States where she could be presented to the public, but the invitation could not be considered, as Kamala grew weak. In any event, Singh records, after the fact, that he could not have subjected her to such inspection, even though, he muses, a trip to New York via London would have been interesting indeed. Her health grew less secure throughout the year. The attention of two doctors was constantly given, and the Reverend Singh was pleased to note that Kamala had different names for each physician, thereby showing that she distinguished the doctors. But the worrisome symptoms remained and, somewhat unexpectedly, Kamala *gave up the ghost at four* A.M. *on the fourteenth morning of November 1929. . . .*

On November 14, 1929 Dr. Sarbadhicari writes:

"This is to certify that Kamala (commonly known as the Wolf Girl), a girl of the Rev. Singh's Orphanage, expired this morning at 4. a.m. from Uraemia.

She was under my treatment for the latter part of her illness.

S.P. Sarbadhicari".[39]

If it is true that she was eight years old at the time of her discovery and capture, and that she lived for nine years in the orphanage, she was around seventeen years of age at the time of her death.[40]

FIVE CHILDREN: WHAT IS LEARNED AND WHAT IS INNATE?

In rethinking the five children described here—Peter, Victor, Kaspar, and Kamala and Amala—we may isolate two strands of thought that have dominated our thinking at various times about the origins of behavior. Before us, as before Itard, are a seeming bundle of beliefs and hypotheses that may be partially unraveled.

One source of the confusion of strands is that "genetic" is not the obverse of "experience," any more than "natural" is the reverse of "social," or "inherited" of "acquired." These terms are not reciprocals or alternatives of one another. The elimination or indifference of one does not imply the presence or importance of the other. Any number of thinkers have mentioned, and common sense dictates, that there is no experience without genes, none without inheritance, and that genes that are not acted upon by the environment are rarely of psychological interest. When we think of Rousseau's romanticism, his view that humankind loses its natural spirit once in a society of fellow human people, we think of "with" and "without." But all mammals have fathers who contribute genetic material if not their presence, and all surely have mothers who for some time at least contribute food and liquid and thereby warmth and comfort. To be sure, it would be helpful to know when Peter and Victor were deprived of human contact and what sort they had, but this information would not tell us the answer to Itard's question—How much can be learned, and how much is innate?—for this reason: The question itself implies a reciprocal relationship that does not exist. It implies that a set percentage is one; the remainder, the other. It is a false equation, rather like the notion that the unconscious is some set fraction of the mind.

The proper question, the answerable question, is: What can be learned at what age? Clearly, none of the four situations describes children of high intelligence. When their mental abilities are measured against those of their peers, all the children are lacking ability. This may be so because the aspects of the early environment masked or inhibited the opportunity for mental growth. It is also possible that no experiences would have affected the outcome, for the child was unable to learn from them.

Today human, but not animal, "intelligence" is measured by setting verbal, mechanical, or spatial tasks. Thus, we speak of a "mental age"— that is, the number of correctly done tasks that matches the number by the average person, divided by mental age (to equate for age) and multiplied by one hundred (to remove the decimal). If we believe that the intelligence was masked or inhibited in these children, we are suggesting that true IQ is a number which, although derived empirically from comparing the abilities of one person to the like abilities of a large number of like-aged individuals, is changeable and responsive to environmental factors. If we think the number to be relatively stable, one with which environment does not interfere, for the great majority of situations (excluding, perhaps, the most unusual, such as those recounted here), there is a tendency to believe that intelligence measures something genetic, something that is therefore permanent, inherited and inheritable.

The eighteenth-century philosopher of the intellect saw the need to show the malleability of intelligence, and to adapt educational and political structures accordingly. In our times we have witnessed adoption of the notion that intelligence is a measure of just such variability, which, at the same time, is assumed to be a stable measure of some inborn trait. Our use of numbers to quantify intelligence may mask the essence of the question yet being asked: That is, what is to be attributed to nature, what to environment? How much freedom and variability can education and political structure justify?

One idea surfaces often: It is that certain mental aspects can be learned at one age and under certain circumstances, but cannot be learned, at least not without commanding difficulty, if offered at other ages, or out of a certain sequence of learnings, or under inappropriate conditions. Itard speaks of these as critical periods, and the same description appears in Singh's description of the wolf-children. It is possible that the lack of such early education to these children meant that they could never learn. The deaf child who gains hearing at age ten will not speak or hear as a hearing child.

If the parables of separation from human parenting that have just been told has a single moral it is that, while we have much data from these situations, they contribute hardly a scrap of information to the question people thought was being asked. If we want to know what or how much is genetic and what or how much is learned, we can be assured that we will never have anything but a silly answer, for, as we have said, it is a nonsense question, one incapable of provoking a meaningful answer. The question can, of course, produce strings of words that sound like answers, and, in our confusion, we may accept these as so, especially if they fit our predetermined views as to how the world works.

But there may be a more powerful moral, and it is one that is explored repeatedly in this book. Indeed, it is the chief reason for beginning our study with the intriguing, if unsatisfying, cases of Peter, Victor, Kaspar, and Kamala, and Amala. It is that the reason each of these cases brought attention in its time, and continues to do in ours, is that people were able to see in the events a likely answer to a question about humanness and animalness that has long confused humanity. Each of us asks it in a different way, but essentially the question is—What is distinctive about being human? So distinctive, in fact, that we cannot confuse ourselves with the animal. What makes the animal mind different from the human mind, we ask wearily, suggesting that the answer may be "nothing."

The most instructive aspect of the people raised without human beings is how we human beings have chosen to interpret the situations; the data we choose to collect and the ways in which we choose to interpret them offer a clue as to how we understand ourselves. Another lesson to be derived from this book is this: The questions we ask about the silent minds tell us more about our own psychologies than about those who remain silent.

We understand Kaspar, Victor, and Peter as representatives of ourselves, as having minds like ours, yet, for whatever reason, not as the beings we assume ourselves to be. The animal-like behavior of the wolf-children may or may not be due to their being raised by wolves, but we are intrigued, as was Singh, at the carnivore wolfness, not only of the girls' behavior and preferences, but, in the long run, of ourselves. The questions we ask of these situations say much about how we define our humanness.

Four Psychologies

These children, perhaps truly feral but most likely not, remind us that it is we human beings who ask the questions of one another. The nature of these questions reveals much about the ways in which our own minds work. If we think of the mind as composed of both innate and acquired aspects, we ask one kind of question. If we think of the mind as primarily electrical, we ask questions based on the way electricity flows; if we think of the mind as computerlike, our questions reflect the activities of microchips.

Psychology likewise finds its questions in the models it uses. In this section, I suggest consideration of the four psychologies that are prominent in our century. These are the ideas of measuring mental ability, psychoanalysis, behaviorism, and phenomenology. Each is illustrated by a reaction of events that gave each model prominence.

Thinking About the Mind **3**

WITHOUT speech with which to communicate, how are we to imagine the experiences, the minds, of these silent subjects? The silent minds of feral children, whether human or animal, remain intriguing but awesome to us. The stories told of the five children provide empirical information about the pupils, but also tell us the teachers' views of human nature and of how we try to penetrate the human mind. The various views thus set the kinds of questions we ask of both silent children and the necessarily silent animals.

Few, if any, of the actors described in the first two chapters are likely to have considered that the questions they asked and not the answers they received would become data for later generations, saying something to us about the human mind at a particular time, in particular places, and in reaction to the psychological thought of the time. That innocence is why I quoted for you passages, rather than reported them; it is in these passages that the teachers described their methods and their pupils. By so doing, we were shown the kinds of questions, both explicit and tacit, that were being asked and, therefore, the kinds of answers that were necessarily to be forthcoming. I wanted the reader to come to understand how her or his own perceptions are additional overlays on the words written years and centuries before, and how those words, written by teachers such as Itard and Singh, are yet another overlay on any quest to understand the nature of the silent mind.

Another purpose is revealed: This book is about the human and animal mind, or, more precisely, about how we human beings think of the animal and human mind. The tales are explications of how, during the last two or three centuries or so, people have asked animals and some nonspeaking children to tell them something about their minds, about their mental abilities, and about their ways of thinking. They are also explications of how we attempt to decode these silent minds by seeing through a double filter: the filter provided by the teller of the tale and our own perception of those descriptions.

I use the examples of Peter, Victor, Kaspar, Kamala, and Amala to suggest that the ideas that guide their teachers are yet to be found in modern psychologies. "Modern" they may be, but longstanding they are as well. The questions posed by the teachers were characteristic of the intellectual notions of the time: Itard was interested in distinguishing the given from the innate; Singh in demonstrating the power of religion; and the bishop in determining what concepts might have been given to the mind by God.

In the following chapters of this section, I demonstrate how the observer comes to be part of the observation, just as Itard, Feuerbach, and the Singhs found themselves somehow changed by their experience with the subject of their investigations. I show, too, that four psychologies, four ways of thinking about the mind, have served to organize our ideas about how the minds of animals and human beings function.

THE FIRST PSYCHOLOGY: THE MENTAL LADDER

Human beings want to know about animal mentation in part because they are intrigued with the notion of building a Mental Ladder. This illustrative, if imaginative, ladder both compares and separates species in a fashion like the one built to show the physical relationships among animal forms. It is a ladder with people on top, other primates next, and so on "down" to the amoebae. Plato suggested the ladder of physique, but Carl von Linné, the eighteenth-century scholar constructed it far more sturdily. Linné categorized several types of human beings, including the feral type, that occupied our attention, if briefly, in Chapter 1.

The notion that the mental abilities of animal life can be expressed as a chain or ladder is not new. Both Aristotle and Darwin,[1] among others, showed us that animals are both alike and different in physical organization. Aristotle applied the notion of infinite gradations to animal life. Plato saw that the relationships could be expressed as a Great Chain of Being,[2] a chain uniting amoebae, people and all else between, a chain that extended to the heavens. The idea of a Great Chain has guided Western religion and science for two millennia.

On the imaginary Mental Ladder, while the poles represent the degree of development, the rungs represent different genera or species, each one capable of degrees or kinds of mental activity different from the others. Amoebae and such are on the lowest rung (presumably waiting to climb upward), and, as the ladder is built by people, the highest rung is reserved to describe ourselves. When upright, the ladder describes how animal life has developed from the simple creature, one capable of only the most elementary mental processes, to the most elegant of mentators—namely, we think, ourselves.

Although it may be perfectly rational for human beings to desire classification, to place animals, including human beings, into species and genera and onto a ladder of ability and mentation, the categorization is a dangerous one. The danger begins when we forget that the categorization is only a human way of organizing, naming, recalling, and remembering, and we add to the concept the notion of purpose. Once we classify, we tend to search to find meaning in the classification. Not content to name the planets, we search for some meaning in their names, placements, and relations to one another; this becomes astrology. In like fashion, having named the animals, we try to organize them in some pattern that speaks to us of some meaning attached to our organization. When we think of apes as similar but inferior to people, of tree shrews as lower class primates, of marsupials in some sense striving to become rodents, we read purpose and meaning into what was meant to be categorization alone.

Consider the philosopher Arthur Lovejoy's remarks that opened a set of lectures which then became a well-known book: "The title of this book [*The Great Chain of Being*], I find, seems to some not unlearned persons odd, and its subject unfamiliar. Yet the phrase which I have taken for the title was long one of the most famous in the vocabulary of Occidental philosophy, science, and reflective poetry; and the conception which in modern times came to be expressed by this or similar phrases has been one of the half-dozen most potent and persistent suppositions in Western thought. It was, in fact, until not much more than a century ago [i.e., 1830] probably the most widely familiar conception of the general *scheme* of things, of the constitutive pattern of the universe; and as such it necessarily predetermined current ideas on many other matters."[3]

Plato's idea appears to have suggested a complete chain, from firmaments in the heaven to creatures on earth. Each was a separate link, but each link was functionless unless attached, chained, to another. Plato probably used the scheme in the *Timaeus,* chiefly to explain what we moderns would subsume under astronomy. We moderns like to separate our knowledge; history in this pigeon-hole, literature in another, physics here, mechanics there; and the study of the origin of the universe is set apart as a separate study. Plato wanted the heavens and their origins to be related both obviously and significantly to us, to our planet, and to all the creatures that inhabit it. Modern Western thought has pushed this notion aside, leaving it to be seen as the quaint concern of Eastern thought.

It was Aristotle who, with this concept of Plato's, gave the Western world the legacy of the Great Chain, a gain whose value Lovejoy puts this way: "In spite of Aristotle's recognition of the multiplicity of possible systems of natural classification, it was he who chiefly suggested to naturalists and philosophers of later times the idea of arranging all animals in a single graded *scala naturae* according to their degree of 'perfection.' For the criterion of rank in this scale he sometimes took the degree of development reached by the offspring at birth [guinea pigs are born without fully developed sensory systems; horses stand and walk within a few minutes of birth]; there resulted, he conceived, eleven grades, with man at the top and zoophytes at the bottom."[4]

The idea that all matter, organic and inorganic, living and dead, was linked remained a lighthouse complete with shoals for Western approaches to knowledge through the Middle Ages (a time which, it needs to be said, translated Aristotle and Aquinas into our contemporary legal and religious system). Thereby to us, knowingly or not, as Alexander Pope writes:

> Vast chain of being! which from God began,
> Natures ethereal, human, angel, man,
> Beast, bird, fish, insect, what no eye can see,
> No glass can reach; from Infinite to thee,
> From thee to nothing.—
> From Nature's chain whatever link you strike,
> Tenth of thousandths, breaks the chain alike.[5]

Two thousand years after Aristotle set the concept, Darwin explained how the Chain may have come about: The production of many, slight, genetic variations within a population of newborn, coupled with the fact that more are born than can be supported by the environment, means that not all will survive to reproduce their genes. The physical difference between those forms that survive and those that do not leads to slight, but cumulative, changes in the structure of succeeding generations. Darwin, however, was careful to eliminate purposiveness from the Chain. By showing the ways in which the connections among species developed, it is sometimes thought that the explanation of natural selection encompasses purposiveness. However irresistible the notion of purpose is to the human mind, it is not a necessary attribute of either classification or of Darwin's system of evolution.

From the structure of animals we see most easily the relevance of the Chain, for after all it is the structure that we can verify most readily to be under genetic control. Darwin understood that the same principles, natural selection coupled with overpopulation, would lead to changes in mental processes as well. Indeed, what was to be his last book applied the principles to the expression of emotion in both people and animals.[6]

Darwin came to apply the same principles of evolution that determined animal structure and speciation to the evolution of the mind, of mentation. So it was that post–Darwin generations of psychological experimentalists began constructing a Chain or Ladder that demonstrated the *mental* relationships among and between animal forms that Aristotle, Plato, and Darwin suggested. The natural science of our time, especially the ways we name animals, the nomenclature, opted for the Aristotelian approach. Each animal is to have at least two names—the first its general or generic name, the second its specific or species name.

This form of nomenclature appears to fly against the notion of evolution and the Chain of Being. If every animal is of a species, and if each species has at least two names to distinguish it from other species and genera, as if each species were fully separate from each other, then how do species become other species? Of course, there is an answer to this riddle: nomenclature is for convenience, to do away with the muddle of common names, to show relationships among genera and between them. Yet it is surely true that this system of nomenclature leaves the notion of Aristotelian-like fixedness of the species. It suggests clearcut divisions and separations among animal forms, and is therefore a great comfort to those who wish to dismiss the myth of evolution. To the contrary, those who accept the concept of evolution must address the confusion that appears when seemingly fixed names are, in fact, supposed to be ever-changing. How can we have the continuity of the Great Chain while yet acknowledging and respecting the nature of the individual link?

One solution to this problem, though not a sound one, is to perceive the Chain as a Ladder; that is, to judge the links as not alike, even though they are linked, but to assume some change in quality along the Chain. Such a Chain then ceases to be circular and becomes a Ladder with a top and a bottom. We do indeed now arrange animal forms along a Ladder. Man, or humankind, graciously occupies the top rung, looked up to by the apes, baboons, monkeys, and, somewhere toward ground level, the fish, birds, and Pope's "unseen to the eye." When we use Aristotle's scalae, when we ourselves imagine and erect the Ladder, we want a scale that distinguishes not only by structure but also by mental ability. Separation by function and structure is the primary datum used by the evolutionist in reconstructing the Chain from links to rungs.

The first psychology among the four I offer is based on such comparisons. It is the the study of mentation through analogy to the principles of comparative anatomy. It is therefore called comparative psychology, although "comparative mentation" would do as well. Its earliest purpose was to arrange a Mental Ladder alongside the Ladder based on structure and function.[7] This section now describes the suppositions on which comparative psychology is based. Contemporary attempts to communicate with the silent minds of animals, as, for example, seen in teaching chimpanzees to use sign language, is based on these foundations and methods of inquiry.

Let us orient ourselves by giving thought, for a moment, to our human habit of separating the physical world from the psychological world. In some compelling ways, our view of the physical world is akin to our understanding of comparative anatomy, and our understanding of the psychological world to comparative psychology.

Physical World, Psychological World

To the first, the physical world—physical objects encompassing a speck of dust, a planet—we assign a reality that, however it may change, changes in an orderly fashion. We can plot the path of a planet and calculate what will change if circumstances also change, thereby making the changing reality constant and predictable. To the second, the psychological world, we assign variability and inexplicableness, believing these to be natural qualities of what it means to be a living being.

All human beings are, by nature, psychologists, for all human beings have theories about the minds, both those of our own and other species'. I think about why I do things: I think about what your capacities are, about what motivates you, about what you are likely to do. The natural psychologist[8] in us searches for traits that help to predict our behavior and that of others. We believe in the individuality of each living thing, while the "-ologist" in us, whether astrologist, anthropologist, psychologist, or sociologist, searches for relationships that explain. As all of us are already natural psychologists, we have some disdain for the -ologies of the social and behavioral sciences. Their data often seem inconsistent with our knowledge. Our experience as human beings is, we appear to think, superior to their statistics. This denial is, of course, one characteristic of being human: We think our senses and deductions to be superior. We also suppose that our common fate as human beings is that others sense, feel, and think more or less as we do. Although I can believe that there is no knowable reality other than my own mind, in the socially complex world we primates inhabit, to hold such a view consistently is one sure path to madness.

If pressed by a philosopher intent on making us use our resources of thinking, we discover that we are of at least two views regarding where the physical and psychological worlds join yet separate. To animals, including ourselves, we attribute physicalness. All have bodies subject to physical force, and we think of "lower" animals as being more subject to such force than "higher animals." We suppose them less in control of their behavior, more under the control of the prewiredness that we associate with instinctual behavior. Amoebae are moved by currents, by chemicals, by events they do not control but merely respond to. We, however, think of ourselves as having the opportunity to control our environment, as motivated, more than moved by, emotions, reasons, and the culture. We hold views about the millions of kinds of beings that comprise the whole of animal life; about who has consciousness and who does not; about who depends more on instinct and who less; about who is intelligent and who edible.

We think of the physical world as having a different sort of reality from that of ourselves and other living creatures. Planets do not become embarrassed; time does not love; space does not become jealous. Living creatures do these things. And so we think of other living creatures as being like us, as having the potential for control, and as having sensibility, self-awareness, and perhaps a consciousness we attribute to our own ideas and feelings.

One result of our mental separation of the physical world from the world of beings is that we are unsure of our fellow beings, not only those we classify as fellow human, but also those we classify as animals or as feral human beings, of whatever species. After all the philosophies are written, the religions explained, and the -ologies denounced, each of us lives in a very private universe of thinking and feeling. We ascribe like feelings and thoughts to most other human beings. We have no choice but to ascribe characteristics to animals in terms that are understandable to a human, for that is what we are. These are the only terms we know by which we can impose order on the world, whether it be the physical or psychological reality. Let us keep this issue in mind as we continue our description of how the notion of comparative psychology or comparative mentation developed.

For our instruction, we now examine the behavior of Dash, a dog. In this example, we use the one enormous advantage human beings enjoy that animals do not; namely, the ability to reach over years, to know what the now-dead saw and thought and wrote.

A Distinguished Member of the Humane Society

We learned something about the eighteenth-century Anglo-Saxon's mind by examining how Peter was examined by Dr. Arbuthnot; we sensed something about the French ways of thinking about human nature from Dr. Itard and Victor; we grasped something of the nineteenth-century German concern for justice and freedom from Feuerbach's purpose in observing Kaspar; and we sensed something of the religiously guided mind at work in the efforts of Reverend Singh and his wife and the observations of the bishop. The nineteenth-century British views of human nature are more evidently connected to our own way of understanding things, both because of recency and common language, but nineteenth-century England does not appear to have had a special event such as a seemingly feral child. Nonetheless, it is rich in evidence regarding views of the animal mind.

Consider the painting (see Figure 3.1), *A Distinguished Member of the Humane Society* by Edwin Landseer (1803–1873), the nineteenth-century British painter and sculptor whose works often depicted animals in what

Figure 3.1   *A Distinguished Member of the Humane Society,* painted by Edwin Landseer, also the sculptor of the lions in Trafalgar Square. The painting "disgusted" the real dog, Dash, thereby leading us to ask the nature of Dash's (and his owner's) emotions and intellect.

human beings might think of as emotional or sentimental ways. (The Trafalgar Square lions are his designs.) The engraving *A Distinguished Member* . . . is of value to us because we are to learn, not of a human being's reaction to it, but of the dog, Dash's, reaction to it.

Here is the background for the tale about to be told: In 1891 Conwy Lloyd Morgan (1852–1936) published a book, part-text, part-interpretation, called *Animal Life and Intelligence* in which he attempted to sort out, for a just barely post-Darwin, English-reading audience, ways in which animals experience, think, judge, and feel. In this book, he cites a number of reports of animal actions that would appear to show animal mental capacities akin to those we imagine for ourselves.

Dash's Disgust

The picture shown in Figure 3.1 is prominent in the anecdote quoted here:

"I quote [writes Morgan] from a letter received by Mr. Chattock: '. . . about our old spaniel Dash and the picture. I remember it well, though it must be somewhere about half a century ago [that is, around 1840]. We had just unpacked and placed on the old square pianoforte, which then stood at the end of the dining-room, the well known print of Landseer's *A Distinguished Member of the Humane Society*. When Dash [9] [the pet dog] came into the room and caught sight of it, he rushed forward, and jumped on the chair which stood near, and then on the pianoforte in a moment, and then turned away with an expression, as it seemed to us, of supreme disgust.' "[10] Morgan, a keen observer of nature and a thoughtful interpreter of Darwinism, as it was then called, uses this description to augment his statement that "The fact that dogs may be deceived by pictures shows that they may be led through the sense of sight to form false constructs, that is to say, constructs which examination shows to be false."[11]

Do dogs have mental constructs, as Morgan claims? Does any animal? Do people, for that matter? And what is meant by "construct," "false" or otherwise? Only by using our own, human, "constructs" to reconstruct the dog's mind, do we have any way to judge and answer these questions. And, what is the source of constructs? Are they innate, somehow connected by neurons and synapses in certain predictable ways, or are they acquired through experience? Or do constructs arise from some combination of these possibilities? As an example of innate and acquired constructs, note that most human beings have some form of color vision. My eye and brain appear to distinguish reds and blues at birth. In this sense, the construct "hue" appears to have an innate basis. Yet what I come to call these sensations of hue—red, or brot, or rouge, or rojo—appears to be acquired from those around me, from the language I come to speak and write.

What do we mean by "deceive"? In what human sense, much less in the dog's sense, is the dog deceived? Animals who feign death seemingly in order to avoid capture or being taken as prey seemingly deceive. Some animals and insects apparently use coloration to deceive predators. Is this deception the same as that practiced by human beings?

When we try to describe Dash's behavior, as the witness Mr. Chattock did for Morgan, we find that our language attributes to Dash a variety of humanlike mental qualities, such as motivation, thought, perception, intention. What other qualities, outside human qualities, could we name? We cannot name any, because human qualities are all we know and all we can ever know. All language conveys psychological notions, for what is language but a feeble attempt of the psychological world, the mind, to make contact with its perception of the real? The physical and the psychological worlds may thereby be forever separated, each running according to its own lawful ways, but in ways that allow no translation from one to the other. Perhaps. That is but one interpretation.

Perhaps, at best, we can acknowledge the psychologies, the feelings, the motivations inherent in the language we use, but then we are cornered in a room of mirrored reflections in which we use a linguistic babble of psychologies to describe our fellow beings. Consider Mr. Chattock's statement again:

"When Dash came into the room and caught sight of it [We do not know Dash's sensations and perceptions, any more than I know yours or you mine; we assume from the behavior], *he rushed forward* [Is "rushed" a description? Opinion? Is it measureable and distinguishable from unrushed?], *and jumped on the chair which stood near* ["Near?" According to whose perceptions of nearness?] *and then on to the pianoforte in a moment* [How long is a moment? How long to someone fearful, or to someone who waits?], *and then turned away with an expression, as it seemed to us, of supreme disgust."*[12]

Whether animals are capable of supreme disgust or any emotion, and, if so, whether they express it, and, if they do, whether we human beings perceive and understand its meaning properly, is an example of the chief topics of this book. The description of Dash's behavior, however much it attempts to be straightforward, nonjudgmental, and objective, must use categories such as time and space, in order to describe behavior in categories and contexts that we human beings share and in which we can communicate. Within the categories provided by time and space, Dash behaves. He moves; he jumps. The human observer records the motion and perceives an expression; from the expression the human observer makes a judgment regarding the feelings felt by the dog.

Remember Kaspar's mistakes: his interpretation of the stars in the heavens, of the "bawling" of the choir and minister, the color red, the importance of horses, the fear of living beings that were also black, as were some horses and chickens. In what essential way do these "mistakes" differ from Dash's mistake?"

Re-creating the Image

We, the readers, separated by one hundred fifty years from the scene just described, are able to re-create it in our minds, although no one now concerned with the act is living.[13] The re-creation, of course, is only in our mind, unless we think of the surviving engraving and Mr. Chattock's written description as being a reality separate from our minds' ability to see, read, and interpret.

We are able to ideate, to re-create, through memory that is not first-hand to us, the image of the writer, writing fifty years after the event [*"I remember it well, though it must be somewhere about half a century ago,"* this written around 1890.] How do we human beings accomplish this trans-time and trans-space re-creation? Whether other animals can or may do so appears to be unlikely, for this ability to use symbols to re-create events in time and space marks the great intellectual divide between human beings and other animals. At this moment, Mr. A. P. Chattock in this time and space is re-created in our experience. Mr. Chattock, Dash, maybe Morgan, assume a reality in our minds. They become presences that we did not have a few minutes ago, for we had never heard of them until we read these paragraphs, and we will probably not remember the names, a few minutes from now. Our minds create; our minds destroy. Mental life, as we know it, is, mostly, re-creation.

Mr. Chattock re-creates for us the movements of the dog, Dash. We can create the image of a square piano (Have we ever seen one?) As moderns, we marvel at the sort of person who unpacks a picture of "another" dog and gives it a prominent place. We are unlikely to travel with a picture of our dog, much less to exhibit it. We are touched, for the act is that of someone far from us not merely in time and space, but someone who *feels* in ways different from us. Time and space are evident categories for our perceptions and understandings: Where and how shall we categorize feelings?

". . . *an expression, as it seemed to us, of supreme disgust.*" Note the clear understanding by Mr. Chattock that the expression and any meaning it might have are those of the human observer. The observer gives meaning *to* the other being; meaning comes *from* the observer. What the other being perceives and feels, we have no way at present of knowing. Does Dash *feel* disgust at the sight of the Newfoundland? If so, why? How? How can we know?

Dash and William James

We have spoken of Dash's perception along with Mr. Chattock's perception as if the idea of a perception were common knowledge. The word may be in frequent usage, but an understanding of "perception" is anything but commonplace. Its importance is evident, and description of and by it is characteristic of one of the remaining three strategies. For now, let us bring to our discussion of the real dog and the psychological perception the views of the first modern American psychologist, William James (1842–1910).

A distinction between the sensation and the perception can be described no better than James did in 1890 writing at the same time, and for some of the same audience, as Morgan did in *Animal Life and Intelligence* (1891). James published a work in 1890 called *Principles of Psychology*, thereby creating from several well-established aspects of philosophy and biology a new discipline called psychology. James understood that epistemology, the study of how we know, was basic to the psychology he was creating. Analysis of the sensory systems is basic to the study of epistemology, for it is through the action of these systems that we receive and interpret our environment, and through them that we sense and perceive. James described the difference between sensation and perception in this way: "sensation is the raw stuff from which perception is made."[14]

The "raw stuff." James excelled in analogy, metaphor, and striking simile and often used such language to define. The "raw stuff" is a gloriously successful shorthand that describes all the electromagnetic wavelengths we see, hear, and feel; the chemical compounds we taste and smell; the molecular motion we use to sense the motion of our head and the deformation of the skin. When perceived, however, they tell us something about our location in the universe and our relation to some of the other objects within it. The "raw stuff" of sensation alone does little for us, it seems, because we require a brain to translate it into perceptions—a brain that organizes, categorizes, and thereby gives meaning. We share this brain in form, if not necessarily in function, with other mammals.

Humankind has had thousands of years to understand how we sense and perceive. We build tools, such as automobiles and airplanes, with greater ease than we have used to understand our own senses. The headway made is unimpressive. We are curiously uninterested in how our sensory systems work. We accept the perceptions with surprisingly few questions, and worry little that we neither know the true nature of sensation nor care how sensation becomes perception.

Surely the interface between the sensation and the perception would be among the very first matters the rational mind would want to understand about itself, but the interface has never been among the priorities of human beings. The irony is that we take for granted our perceptions for the reason that they *are* our reality, even while we acknowledge readily that imperfection of our perceptions. Unlike Kaspar, we "know" that the road does not "truly" come to a point in the distance; that people in the distance are not "truly" smaller than people near to us. We believe that, yes, our senses are inexact, even erroneous, and that it is not unusual for people to disagree about reality. It is as if we believe that one characteristic of being a living being is an individual, idiosyncratic notion of reality. Thereby we show no interest in how we perceive our universe in ways that are alike, and different from one another.

Mr. Chattock has a *sensation* of Dash; his *perception* of Dash is a different matter. For Mr. Chattock, the dog, the piano, the picture (we must use these nouns, for we have no other way of communicating our sensations), all have a meaning tucked in the perception; his sensory systems, here mostly vision, provide him with the raw stuff, the unmeaningful datum, from which the forms and perceptual categories arise. Nouns are a much simplified way of communicating aspects of perception. For the purpose of verbal communication, they must do; but for the purpose of describing a perception, they are deceptive and dangerous. What Mr. Chattock sees as a dog, what I see as a dog, and what you see as a dog are forever unknowable. Once meaning is added to the sensation—meaning from your experience with this dog and mine with that—the noun becomes so abstract as to destroy our ability to communicate our perceptions. The issues seem to become trivial: after all, a dog is a dog is a dog, we live with our perceptions, and that is the end of it.

Except for psychology. If we take seriously our desire to understand ourselves, if we are not satisfied with lives that are but reactions to the environment, then we will examine the meaning of our perceptions, for we are natural psychologists, working to give meaning to what we perceive. That is why Mr. Chattock assumes that Dash, too, both senses and perceives, for this is the point of Dash's reaction to Landseer's picture. Mr. Chattock presumes that the sensation of the picture is constructed by Dash's mind into some image that is to Dash complete and knowable. For Dash the perception has meanings and feelings from experience—namely, those of disgust. The feelings, in some way not known to us, produce an emotion and the emotion produces an expression: Mr. Chattock labels the expression "supreme disgust." Chattock, for us, overlays his perception of Dash with an account of Dash's perceptions, these prompting Dash's mental state. Dash perceives Landseer's picture; Mr. Chattock perceives Dash; and we, through time and space, perceive Mr. Chattock's perceptions of Dash which, third-hand, are that Dash perceives a picture of a dog and attaches meaning to it, that of "supreme disgust."

Where do perceptions, meaning, and feelings, come from? The question is seldom asked, or, more to the point, no one seems much interested in the answer. The question is among that set of issues that are, at once, of most importance and unanswerable. Modern thinking appears to accept the view that perception, meaning, and feeling have two likely origins. The first origin is that the human brain is prewired with categories—categories that it uses perceptually to sort the environment. Color is an example. Presumably, the nervous system knows red from green because of the way in which the visual receptors and their nervous system connections are arranged. Such is the sensation of color. At the level of the sensation, I have no names with which to identify, so that I cannot talk to you about red and green. I learn to use these words to identify aspects of my sensations, and the learning of these categories or labels for these sensations is the second way in which the categorization of knowledge occurs.

The philosopher Immanuel Kant[15] (1724–1804) referred to these as a priori and a posteriori categories, categories that are given and those that are acquired. A priori knowledge, insofar as we can ever know and talk about it, is given to us, is instinctual, is presumably unchangeable; it is the product of the sensory systems and brain. These sensations are reused in new categories as I learn about the environment. I learn to group aspects of sensation by language, by the concrete nouns (dog, Chattock) and by the abstract ones (liberty, justice). Your categories and my categories are sufficiently alike to permit us to communicate and, one suspects, sufficiently different to lead us to differ. Sometimes I understand your categorizations; sometimes they make me angry, or, as in the case of Mr. Chattock, they produce "supreme disgust." It is this separation of the mind and its perceptions that, I believe, compelled interest in Peter, Victor, Kaspar, Amala and Kamala. If so, it is a small exaggeration to write that modern psychology begins here. It begins at the very moment when humankind said, give me a feral mind, for from this mind I can ascertain what is given and what is acquired from the environment.

Dash and Epistemology

The approach Morgan used to investigate how animals think is a type of empirical epistemology. Here the task of the investigator is to determine the relationship between sensation and perception, between stimulus and response, between the raw stuff and the perception. This is the method most favored by many in our own day, but as we will see, it is not the only method.

Morgan begins a chapter of his 1891 book in this way: "It is part of the essential nature of an animal to be receptive and responsive. The forces of nature rain their influence upon it; and it reacts to their influence in certain special ways."[16] The metaphor of rain is a splendid one. The image is that of sensations dowsing the receptors with stimuli. Morgan could have pursued his metaphor as follows: sensations exist only if they strike a receptor prepared to accept those particular forms of stimulation. Stimuli are paired with receptors: a certain part of the electromagnetic spectrum stimulates the eye (I cannot sense these wavelengths with my ears or skin); the specific motion of certain molecules stimulates my ear (I cannot sense these with my eye or tongue). The body is closed from the universe except for certain windows that accept sensations. My sensing knowledge of reality depends on these small, specialized portholes. My body is like an old-style armored tank. The windows to the external world are but slits; the outside is but incompletely seen. I sense only a little and must piece these associations together and project them in order to make them into perceptions. I have very limited ways of knowing something about the world external to my body, but utility of these specialized ways tells me much that is useful about the real world and the objects within it.

The task of the experimental epistemologist, the experimental psychologist, is to reach an understanding of the relationship between sensation, the organ that receives it, and the way in which the organ and the mind combine to make sensations into perceptions. This approach to epistemology is unabashedly Aristotelian. It is this great thinker, compiler, and organizer of information who delineated the idea of five senses, each defined as a specialized receptor sensitive only to particular sensations. The ear is made for sound waves; the eye for electromagnetic waves within a small, but certain, range; the nose and the tongue are for certain molecules, and only these; the skin for pressure and touch. We assume that Dash "saw" the picture; that it was the sight that energized his emotions. And yet it is possible that instead he smelled something about the object, perhaps even the smell of Chattock's hand. From Chattock's description, we can only guess at the kind of sensation in Dash which evoked perception. We human beings assume that all animals use vision to the same extent as we do. Chattock assumes that Dash, as a human would, sees the picture with his eyes and goes to it because of the visual image. In fact, Dash's sense of smell was probably far keener than his sense of vision. Moreover, given our present-day knowledge of animal perception, it is unlikely that he could use the muscles of his eye to focus on the aspects of the picture and convert the dots of the photographic image into a perception called "other dog."

The sequence—first the sensation, then the organ of perception, finally the brain and the perception—is such an evident way to analyze how our sensory systems provide information that it seems almost churlish to suggest that the approach is chiefly semantic. We therefore think it to be natural and logical, characteristic of the normal, and indeed, even the superior mind. Morgan well represents the British approach at the end of the last century, and it is this approach that remains most prevalent today in North American classrooms, popular publications, and so-called acceptable science. The sequence has the advantage of appearing to be grounded in objective, scientific physiology; yet it may be grounded in nothing more than linguistic contrivance and convenience. The sequence is true because it represents the way we talk about ourselves and reality, but it may be the wrong sequence—or the idea of a sequence may itself be wrong.

Other epistemologies have appeared, but those of us reared in the English-speaking world find it difficult to grasp them, for they are based on different ways of understanding sensation and perception. Three that have received most attention are psychoanalysis, behaviorism, and phenomenology.

THE SECOND PSYCHOLOGY: PSYCHOANALYSIS AND LITTLE HANS

At the turn of the twentieth century, Freud wrote about the case of Little Hans, a five-year-old child whose anxiety at the sight and thought of horses was debilitating. Through this case, Freud was creating a "myth" that both explained and predicted, these qualities being two requisites of a sound hypothesis in science. Hans's feelings about his mother, father, sister, and playmates, all became involved with his feelings about horses.

Little Hans himself invented a "myth" which, for him, explained his fear. It is a myth that was at once individualistic, because the characters of the myth were of his experience, his place, and his time, and could be no one else's; and at once universal, for they came to stand for events that occur to every child, at every time, in every culture, but with different content.

Why horses? Why did little Hans not decide to fear telephones, or automobiles, or heights, or bugs? What did the horse represent in Little Hans's feelings and thoughts? This puzzle presents us with an opportunity both to see the assumptions and techniques of psychoanalysis at work and to grasp how one human mind imagines animals to think and feel. In the next chapter, Chapter 4, I review the details of Little Hans's situation, both to explicate how the epistemology of psychoanalysis does its work and to show how animals come to figure in human beings' construction of explanations for themselves about their own behavior.

THE THIRD PSYCHOLOGY:
BEHAVIORISM AND CLEVER HANS

Mr. Chattock seemed to have little doubt that Dash had thoughts and feelings not unlike those of human beings. A later generation of human thinkers would find this notion unduly romantic and generous, and an example of bad science. Morgan would argue that ascribing mental states to animals was not essential to the human being's understanding of animal mental processes. [17] He and many others rejected accounts of animal intelligence, criticizing their anecdotal nature, their unrepeatability, their lack of scrutiny for alternative explanations.

The idea that explanations and descriptions of behavior should be based on testing hypotheses of unambiguously defined concept would flower into behaviorism. This method and epistemology defined the subject matter of psychology and all mental concepts that could not be defined. Behaviorism took no stand on the issue of whether there *are* mental events, much less on whether animals have them. It could not take a stand, it was argued, because no method could be devised to determine the validity of mental processes in anyone, much less in nonconversant, silent animals. Behaviorism turned toward those aspects of animal life that could be defined, measured, and tabulated. Behaviorism is sometimes said to negate the existence of mental processes, but the charge is wrong. The behaviorist argument is simply that since these processes cannot be knowable, with them we can only create myths about animal mentation, and thereby only read our own thoughts onto the silent minds.

Behaviorism counted on the deductive power of empirical hypothesis-testing to find answers to its questions. In this technique, one sets a hypothesis from which deductions can be postulated and, by controlled experiment, their accuracy determined. It is a powerful method. Some would say that much human knowledge and all technical understanding are based on this basic method of science. Such hypothesis-testing was not new to nineteenth-century behaviorism; it was actually understood four hundred years before. But the steady application of these principles of analysis established the scientific knowledge that we understand today, and it is responsible for our knowledge of other planets, our own immune system, and the automobile. If such hypothesis-testing cannot determine the truth of all propositions, it has surely provided a means for understanding our universe in ways ungrasped before. Behaviorism offers a clear, logical, and readily coherent method for examining our own and our fellow creatures' behavior. If it leaves us unsatisfied, it may be that we are only "linguistically" unsatisfied; it may be our language, not our logic, that is barren and incomplete.

When Oskar Pfungst (1874–1932) found himself assigned to study a clever horse, a horse, like the child, named Hans, Pfungst was a student at the Psychological Clinic in Berlin. We do not know what he thought of the assignment, but we do know the technique he used to determine how Clever Hans was able to tap out the numerical answers to complex problems and to locate objects.

Pfungst used the new psychological science of hypothesis-testing then available to him. In a decade's time, the principles he used would be the watermark of behaviorism. To him, the experimental principles were but common sense, but to later students who identified with behaviorism, they would become doctrine. He designed and conducted a series of experiments with Clever Hans that were intended to test postulates derived from hypotheses predicting just how Clever Hans was able to perform these fantastic tasks. The results from these experiments showed that Hans was using cues given to him unknowingly by human beings. Or did they? The silent mind of Clever Hans gave what came to be called behaviorism a solid boost, for it showed that the methods of the experimentalist could explain the seemingly mysterious and that it could achieve this without ever referring to the mind.

At the time that Pfungst was studying Clever Hans, and most likely the reverse, professors and students in the United States were inviting apes and monkeys to join in their search to understand the mental processes of animal life. Because apes and monkeys were not "discovered" or named, much less observed and investigated, until the middle of the nineteenth century, investigation of representatives of this order was a bonanza. Almost nothing was known about their mental abilities, for only a few years before had European explorers of Africa reported on these beasts to the European-speaking world. Some reports offered evidence of humanlike activity, such as unified group aggression. The availability of monkeys and apes at seaports, where sailors sold their pets, provided an exceptional opportunity to advance our knowledge of the animal mind. Hence, the U.S. study of the monkey mind began in the port cities Boston and New York.

Ironically, then, our closest genetic relatives were the last order of nonhuman animals that we studied. Peter, the tuxedoed chimpanzee featured in Chapter 8, was the first great ape to be studied in a psychological clinic. The director of the clinic, having seen Peter perform on stage, wanted to determine the extent of Peter's mental capacities. Part of his interest, it would seem, stemmed from his desire to find the "missing link," for this was the proposed animal who, neither fully human nor only animal, was thought to be interposed on the Great Chain of Being, the Mental Ladder, between animal and humankind. The fossil record is weak in showing the relationship between monkey and humankind, and a living missing link would offer the opportunity for important study.

On occasion, apes were examined using the hypothesis-deduction method favored by Pfungst; other chimpanzees and monkeys were examined to determine their ability to learn to read and to write. Both qualitative and quantitative differences were measured. Some investigators were interested in the processes the chimpanzees used, while others were interested in the extent of the animals' capacities. Such information occupies this book, especially in Chapters 8 and 9. These investigations, in turn, set the stage for what would come. Now, nearly one hundred years later, attempts are being made to communicate directly with animals by using sign language or by having the animals learn a task that encourages such communication (Chapters 10 to 12).

The behavioristic approach to understanding the Mental Ladder of animal mental life was not to be tuned to the primates alone. In 1897 Edward Thorndike (1874–1949), an American, would begin studies of the mental processes of dogs, cats, and chickens, as well as monkeys. M. E. Haggerty and G. V. Hamilton would continue this work by examining the ability to learn by imitation and by comparing problem solving in animals and people (Chapter 9). The quest had begun: the Grail was to be the constructing of a Mental Ladder that separated species and genera by their intellectual abilities. That the attempt at construction suited well the view that the same sort of ladder should be built to distinguish human beings from one another was fortuitous. In its earliest years, the notion that the abilities of people should be separated was a progressive move: such information, it was thought, provided the opportunity to give each human being the most suitable education and training according to individual ability and need.

MYTH AS A WAY TO UNDERSTANDING

The two Hanses—one horse, one human being—gave substance and form to two of the chief ways of thinking about psychology in this century, at least in the English-speaking world, in the form of psychoanalysis and behaviorism. The first way provides exceptional opportunity to deduce the operations and causes of fear; the second prompts a reductive analysis, the kind of reduction that most think of when they think of the practice of science, in order to determine the cause and effect of a certain behavior. These two forms of thinking have guided our twentieth-century understanding of ourselves.

One form, psychoanalysis, builds a myth about ourselves. It is not a myth in the sense of a lie or an invention, nor a story like that of Santa Claus or the achievements of the stork. Rather, it is a myth that describes observable happenings in a way that tells a story as a way of explanation. Adults use myths as a shorthand way of describing events. The reality lies not in the veracity of the story, but in its ability to describe and make evident that which is difficult to say or imagine.

The theory of evolution is a myth in the same sense. The processes of evolution explain many observations that, each taken alone, would make little sense. These observations range from the origins of fossils and why shells are found beneath the sea, to why people have wisdom teeth. Evolution provides a single, simple, common explanation for the shape of the seashell and the migration of birds. Evolution explains the principle by which natural matter came to be as it is.

The greatest ideas of which humankind is capable are myths, these being stories that we retell because they describe aspects of our environment that are critical to our understanding of ourselves. Some myths become so ingrained that we take them as the essence of reality—deities, gravity, evolution. The symbol itself becomes the reality; the myth becomes apparently true. It is an aspect of being human that we live comfortably within the contradictions of many kinds of reality. I believe in a deity; I believe in evolution; I once believed in Santa Claus, and I still act as if I do, at a certain time of the year. I can adjust my realities; I coordinate and change them to fashion my own special view of reality.

Two great myths emerged in the nineteenth century. The first was the idea of evolution—that natural selection works on excess populations of genes and, by so doing, the structure and function of living beings are altered. The second great idea came from the thinkings and writings of Sigmund Freud. This myth, like that of evolution, describes how matters came to be the way they are, but it is not especially successful at predicting the future. Just as the story of evolution describes a mechanism that explains how genera and species came to be the way they are and, as such, explains the present, so Freud's myth—psychoanalysis—describes why people are the way they are and how they came to be.

Not surprisingly, the language we use is replete with words and concepts taken from both evolutionary theory and psychoanalysis. Think of the nineteenth-century mind of Mr. Chattock. We know little of it, of course, except what we might infer from his description of the dog Dash's reaction to the framed print of another dog. Almost certainly, in 1840 Mr. Chattock did not think of Dash as a genetic variation produced over hundreds of thousands of years, a variation of some form most like today's wolf. He thought of Dash as one more example of the excellence of God's creations. He did not, one imagines, think of Dash as co-evolving with human beings, as an animal whose relation to humankind led to each evolving separately, yet together, symbiotically, with each helping the other to adapt to inhospitable or new environments. In like fashion, one doubts that Mr. Chattock ever thought seriously about his own oedipal conflict and that, had it been explained to him, the teacher could have expected jail, an asylum, or a whipping.

But Mr. Chattock's physical being, as distinct for the moment from his mind, if alive today would appear no different from any of us living folk. Physical evolution does not appear to occur as rapidly as intellectual evolution. When Edweard Muybridge (1830–1904) [18] in 1872 practiced the then new gadgetry of photography by photographing plate after plate of men and women performing simple actions (climbing stairs, running, hitting a ball, carrying a pail) naked, so as to record the muscle movements, he gave us a record of the musculature and movements of people living over a century ago, doing many of the tasks and movements that we too perform. A hair style may suggest another time, but the movements of the body, the way a stair is climbed, a baseball hit, are easily characteristic of human beings regardless of fashion.

Alas, we have no photographs of the mind, none of the intellectual process, no fossils of behavior. Whether we like it or not, evolution and psychoanalysis are the current screens through which we examine the universe in place of photographs of the mind or perceptions. The body has changed little over the generations; the mind, however, is clearly placed in a time that has imposing, if imperfect, ways of viewing ourselves and nature. My body is the way it is not only because of its evolution, but also because of how nature and I have treated it. My mind, too, is under the influences of the principles of evolution. I understand it less well, for an ancestral mind is unavailable to me. I can compare my mind to that of others, searching for variation, and I can learn something about the evolution of the intellect. Psychoanalysis can be said to search heartily for the common elements, the common myth, to be found in the variation among human beings. Behaviorism, although it denies the utility of mental images and acts, gives to psychology prediction and precision and takes away confusion and conflict.

THE FOURTH PSYCHOLOGY:
PHENOMENOLOGY

On the continent of Europe, as distinct from the island of Britain, a way of thinking about how people in particular, but animals too, think and feel was evolving. It was an approach that emphasized perception and its meaning. It asked what the perception meant to the perceiver, and it was not much interested in how the sensation became a perception. It focused on meaning and feeling rather than on the connection between sensation and perception. Because it gave prominence to understanding, the nature of the phenomenon perceived, it became known as phenomenology. Its roots were in the metaphysics of Kant, and it would bloom later in the nineteenth century with the works of Edmund Husserl (1859–1938). The approach, though intended chiefly to describe human psychology, would become attractive to those who studied animals. Ironically, while the principles of behaviorism were often found by observing animals, behaviorism had no interest in the possibility of an animal mind. To the contrary, phenomenology based its data on human perceptions, as these were communicated by speech. At first, it was not seen as applicable to any other than the human mind, but in time a way has developed to evaluate animal perception.

Phenomenology begins its analysis of behavior by asking that the umwelt, the perceptual world, of the organism be plotted. The task is so difficult that the difficulty itself may explain the comparative lack of interest in phenomenological research, at least in the English-speaking world. If it is possible to know, and somehow to be able to represent, the complete perceptual world at one time, the next step is to plot it as it changes in response to changes in sensations.

In Chapter 7 I describe one way in which the phenomenological approach has been applied to animal behavior: here the goal is to determine the dog's umwelt and the way in which it uses its sense of smell to track. This is but an example, however, for other investigators, examining other animals, asking different questions, have also attempted to identify the components of the umwelt and establish how, and to what, it changes.

Both psychoanalysis and phenomenology are intellectual notions and ideas that emanated from continental Europeans. Both are epistemologies that provide us with a method for understanding ourselves and, in the process, others. Indeed, they are grand ways of thinking about thinking. While different, they are easily placed in contrast to behaviorism, an approach that usually eliminates the very perceptions that phenomenology and psychoanalysis value and think to be of greatest importance.

The modern origins of all three epistemologies can be found in late nineteenth-century thought. In Britain, caution about describing or measuring the mental life of animals and human beings was most powerfully expressed by Morgan (1891), whose account of Mr. Chattock and Dash has already been described. Morgan's refusal to admit mental explanations of behavior became a hallmark of behaviorism, a way of thinking that became powerful in the early years of this century through the work and theorizing of English, Russian, and American investigators. In the United States, John B. Watson (1878–1958) laid out the tenets of behaviorism in 1913. The resulting influence has been longstanding and remarkable: it gave the study of animal life a sure basis, a consistent methodology, and a way of being recognized as scientific and thereby a way to secure patronage for investigators. To the mind of the behaviorist, psychoanalysis and phenomenology lack the approach characteristic of good science; they are myths of the most misleading sort, those incapable of empirical verification.

The origins of psychoanalysis are in the German-speaking world at the turn of the nineteenth century, although its tenets were welcomed by the

English-speaking world as well. Phenomenology, too, is an intellectual development of Europe, especially that of German- and French-speaking scholars.

FIVE CHILDREN, FOUR PSYCHOLOGIES

Because psychology has more than one way of explaining its data, some observers believe it incapable of scientific findings. This view is without the benefit of logic. There may be as many psychologies as there are people to psychologize about it. Some may wish to make the Mental Ladder a myth on which to pin what data are collected; others may want to tease out the unconscious; still others may prefer to measure observable behavior; and yet others may look toward the nature of perceptions. These approaches to understanding are not alternatives: it is not that three are wrong and one is right. Each stresses a kind of data; each represents something broader than psychology itself—namely, an intellectual tradition that applies beyond behavior itself to literature, religion, education, government, philosophy, and prediction.

That these ways of knowing are not alternatives, but complementary aspects of understanding ourselves may be understood from their overlapping relationships. Psychoanalysis surely emphasizes the same perceptions that phenomenology uses in its inquiry, as it does when explaining the nature of human dreams; phenomenology is the immediate ancestor of a special and influential way of describing animal behavior, ethology; behaviorism is a stark method, but it is closely related to certain philosophic ways of thinking about modern physics—namely, logical positivism and operationism; while the Mental Ladder is a metaphor based on analogy with comparative physiology and anatomy.

We shall be forever disappointed in psychology if we insist on one true, final way to conceptualize the nature of the mind. Our minds are simply not made of such consistent matériel, and it is the seeming inconsistency that gives us hypotheses, suspicions, and data.

To explicate psychoanalysis, we turn now to an account of yet another child, the boy who came to be known as Little Hans. This child's youthful fears augmented and altered Freud's ideas, as if to show that the clinical case, like the experimental one, changes the observer as well as the observed.

4   *The Psychology of Psychoanalysis: Freud and Little Hans*

P SYCHOANALYSIS gains its insight and formulates its laws regarding behavior by detailed analysis of the individual case rather than by statistical measure of groups. By referring to psychoanalysis as a myth, I enlarge its importance to human beings, for it is by myths that we conduct our lives. These myths are about the nature of good and evil; about what is appropriate and what is not; about what gives pain and what pleasure; and about the probable outcomes of our actions.

The best way to uncover the meaning of a myth is to strip it to its original meaning, even though it is the overlay of meaning that gives the myth its form and vitality. We can, therefore, best understand the mechanics of psychoanalysis by examining how Freud came to understand Little Hans; how Freud translated the utterings of this child into a powerful myth about all human behavior.

LITTLE HANS

Hans and his family were known to Freud in several ways, and this knowledge may have come to be represented in the way that Freud understood the situation. Freud knew the boy's father, for the father attended weekly seminars with Freud. He was a music critic and is known to have been a literate and sensitive man. Hans's mother was "once treated by [Freud]. She . . . fell ill with a neurosis as a result of a conflict during her girlhood . . . and this had in fact been the beginning of my connection with Hans's parents," writes Freud.[1] Now, many years later, we know Hans's identity and we, of course, know how his life played itself out. We can judge the effect of the neurosis that Freud diagnosed. Freud introduces the story:

"The case history is not, strictly speaking, derived from my own observation. It is true that I laid down the general lines of the treatment, and that on one single occasion, when I had a conversation with the boy, I took a direct share in it; but the treatment itself was carried out by the child's father, and it is to him that I owe my sincerest thanks for allowing me to publish his notes upon the case. But his services go further than this. No one else, in my opinion, could possibly have prevailed on the child to make any such avowals; the special knowledge by means of which he was able to interpret the remarks made by his five-year-old son was indispensable. . . . It was only because the authority of a father and of a physician were united in a single person, and because in him both affectionate care and scientific interest were combined, that it was possible in this one instance to apply the method to use to which it would not otherwise have lent itself."[2]

So, we learn, Freud carries on the analysis of Little Hans's phobia chiefly through correspondence with the father. In reporting what Little Hans says through the father's testimony and not the child's voice, Freud is in some ways like Morgan quoting Mr. Chattock who describes Dash's emotions. The written account regarding Hans is a very early one in the history of psychoanalysis, and reveals the ideas that would later form, in rudimentary shape, the central aspects of psychoanalysis.

Freud is a writer of clear and engaging prose. An admirable characteristic of his thinking and writing is his willingness to present alternative arguments for the reader and then to comment fairly on the likelihood of their validity. He does not saddle the reader with argument or augmented words; he presents and interprets. As is also true of the writer of a good mystery, he makes certain that the reader has the facts and then proceeds to show the one inevitable explanation, to his mind, that can contain and explain all the observations.

Hans was a child similar to any other child brought up in a well-meaning, interacting, and loving family. The family took outings, sometimes to visit relatives, including other children, at Gmuden. Of special significance to this story, the family would visit the zoo at Schönbrun, the then imperial palace grounds, outside Vienna. The zoo became important to the phobia, as did some of the animals in it, as well as the horses that worked the streets. Vienna in the first decade of the century was a horse-powered city, a fact we must remember if we are to understand the importance of the phobia. For Hans developed a phobia of horses; it was the sort of phobia that at first makes us mildly uncomfortable in the presence of some object, but then compels us to leave the situation or not to face it at all. The seriousness of Hans's phobia of horses would be like that of a modern child who grew fearful of cars. With cars so dominant a part of our social scene—cars passing on the street, cars on TV and in films, cars and their variants, trucks, and all motorized vehicles—such a fear would be socially debilitating today, just as the fear of horses was to Hans in Vienna at the beginning of the century.

Our story does not start with the fear of horses, for our goal is to examine the development of the phobia, and, as always, the onset of the development predates recognition of the fear. The story starts with Hans, now three years old, showing interest in his "Wiwimacher." (The English translation is given as "widdler.") He wants to know if his mother has one (she says, "yes," she does); he associates the penis with a cow's being milked, with an engine blowing steam; he fondles it, and at bath he tells his mother that it feels good to the touch. On one occasion in the bath, his mother says that he must not do that, or he will be sent to Dr. A. who will snip it off.[3] "Then what will you widdle with?" she asks.

"With my bottom," replies Hans.

At age three and three-quarters, he asks his father if he, the father, has a widdler too. "Yes, of course," replies his father. Another time, Hans watches the mother undressing but can find no widdler. "I was looking to see if you've got a widdler," he explains. "Of course, I do," she says. He explains, "I thought you were so big that you'd have a widdler like a horse."

All who have been around young children will recognize the confusions and generalities about the availability and function of bodily parts. However amusing the varying interpretations may be to the adult, they have in common the attempt by the young mind to find constancy. A concept develops of the function—the giving of urine, of steam, of milk. We see the normal and usual process by which any being forms categories of thought, whether the concept be penis, hue, or time and space. There is an attempt to establish likenesses, followed by an attempt to show differences. Hans's questions are answered directly, although modern parents might be more careful about the "snipping off" suggested by the mother, but only because we think and at some level accept these Freudian conceptualizations. For his age, Hans is acting as any other three-year-old boy in any Western culture would: he is learning how to separate the sexes physically and is discovering to which set he belongs.

A sister is born, and when Hans expresses confusion regarding its source, the busy and otherwise occupied household credits the stork. With the birth occurring at home, unlike today's disappearance of most mothers at birth time (in the eyes of the child), he hears the noises, senses the illness, sees his mother afterward, comments that the blood in the bowls did not come from *his* widdler (the reader may speculate on the line of thought leading to *that* comment), and declares the baby to be beautiful while being critical of its lack of teeth. He is displeased that the baby does not talk. He is overheard in his sleep to say, "But I don't want a little sister!"[4]

The mysteries of bodily functions and birth itself are confusing indeed to the forming mind. What is the source of these liquids, solids, and human bodies?

HANS'S DREAM

When Hans reaches four years of age, his family moves. Next door to their new home lives a little girl of seven or eight. Hans cannot help watching her. He finds a post from which he is apt to see her; and he keeps to it. He talks at home of the girls at Gmunden and of the stork. His chief interest is turned away from his own bodily function and toward his feelings for others. Around this time, Hans begins to make remarks that express apprehension of horses, giraffes, and the larger, hoofed animals at the zoo. He no longer wants to visit the giraffe or elephant houses, although previously these animals interested him the most. Instead, he takes a new liking to the smaller animals, to birds, pelicans, and the like. When traveling to the zoo, he seems fearful when the steam trolley he is on passes horses.

After several visits to the zoo at Schönbrun, Hans asks his father to draw a giraffe. When the father finishes (Figure 4.1), Hans asks him to "Draw its widdler too." "Draw it yourself," answers the father encouragingly, with the results shown. Hans draws the top vertical stroke, and then adds the bottom, saying "Its widdler's longer."

Hans's dream of the crumpled giraffe marks the onset, to our minds, of the phobia. Here is the dream:

"In the night there was a big, and a crumpled, giraffe in the room; and the big one called out because I took the crumpled one away from it. Then it stopped calling out; and then I sat down on top of the crumpled one."[5]

Within the story as told so far, we can find two themes that are basic to psychoanalytic theory. The first of these is the acknowledgment of attention to sexual matters by children. The second is the salient place of

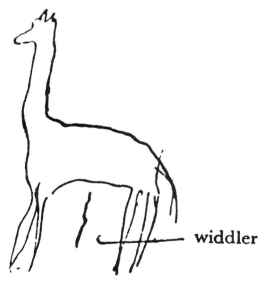

Figure 4.1   Little Hans's (Herbert Graf's) drawing of the giraffe at the zoo. Little Hans's father (Max Graf) asked his son the nature of the scribble under the animal. When Hans described it as the giraffe's "widdler," the father so identified the scribble on the drawing. The image of the giraffe figures in Hans's story of "the crumpled giraffe," a story that describes his version of what Hans's parents were doing in their bedroom.

dreams in the interpretation of one's desires. The idea that the young engage in what adults perceive as sexual thoughts was not a welcome one to either the British or Continental mind in the 1900s, while the notion that we human beings have thoughts that are not readily available to the conscious mind was, if not wildly popular, at least not unknown.

Most people living today accept some degree of unconscious awareness as a fact of mental life. The belief has become part of our language and our way of explaining human behavior. Our age assumes that we as natural psychologists are motivated by ideas and wishes of which we are not aware. True, not everyone accepts the notions of certain extensions of the idea of unconscious thought—the descriptions of oedipal periods, castration fears, fixations, and the like—but we more readily accept the presence of an unconscious, of defensive mechanisms, than the seemingly stark postulates of what occurs during the development of sexuality. We moderns appear to be only half-convinced of the tenets of psychoanalysis: we accept the notion of an unconscious and permit it to be applied to ourselves, while often rejecting the role of the unconscious in the development of our own sexuality. In short, our own unconscious thoughts rise to say, "Believe only the idea, not the conclusions!"

Do we grant similar mental mechanisms for animals? When the dog makes noises during sleep, we ponder whether the dream is of the chase or the meal, but we probably do not imagine the pursuit to be of a loved one. We think of animal sex as mechanistic—necessary, seasonal, hormonal, but probably untouched by "superegos." The Freudian analysis of the mind is an analysis of the human mind, not the animal mind, for as encompassing as the theory is, psychoanalytic explanations demand the use of language for use as data. Language can express descriptions of mental life, and it is on this kind of language that the psychoanalytic explanation of life depends. The concepts of psychoanalysis do not form a statement of how all mental life came to be, or of its contents: it is restricted functionally to that which can be expressed in humanlike language.

THE CRUMPLED GIRAFFE

Toward the end of his life, thirty years after he wrote about the events regarding Hans, Freud, speaking to an interviewer, allowed that one of the great mysteries of life to him was "Why sex?"[6]

Why sex, indeed—meaning, why is sex so important in human mental life? Sex is as important to the reproduction and success of a species as is breathing to the life of the individual, but no one spends much time thinking about breathing. It is automatic and not of concern unless it is prevented. Nor does breathing, disguised or not, occupy much of one's dream life or organize and control relations with other people. Why should sex be of such importance, then, when it is a biological necessity comparable to breathing? Freud had no answer, nor do we.

When Hans's father wrote to Freud, he interpreted the dream as follows:

"The big giraffe is myself, or rather the big penis (the long neck), and the crumpled giraffe is my wife, or rather her genital organ. . . .

"The whole thing is a reproduction of a scene which has been gone through almost every morning for the last few days. Hans always comes in to us in the early morning, and my wife cannot resist taking him into bed with her for a few minutes. Thereupon I always begin to warn her not to take him into bed with her ("the big one called out because I'd taken the crumpled one away from it"); and she answers now and then, that after all one minute is of no importance, and so on. Then Hans stays with her a little while. "('Then the big giraffe stopped calling out; and then I sat down on top of the crumpled one.')"[7]

Freud, in reply, accepts the father's interpretation, adding explanation here and there. The growing fear of horses in the street and of hoofed stock in the zoo continues. Hans speaks tentatively of his fears, and when noting this, Freud makes a point that will be seen as increasingly important in the years ahead as psychoanalytic explanations become refined and applied. "We know [writes Freud] that this portion of Hans's anxiety had two constituents: there was fear *of* his father and fear *for* his father. The former was derived from his hostility toward his father, and the latter from the conflict between his affection, which was exaggerated at this point by way of compensation, and his hostility."[8]

The father's interpretation of the dream shapes his conversations with Hans, at least those conversations reported in the father's letters to Freud. He tells Hans that he (Hans) will not be cured of his fear of horses if he (Hans) continues to visit the parents' bedroom in the mornings. Hans knows of the father's letters to Freud and has visited Freud in the company of his father. Hans begins to speak clearly of his fear of horses; of horses in the street, of horses running fast rather than slowly, of horses of one color more than another, of those hauling buses or visiting the warehouse across the street. His clarity regarding the source of his fear leads Freud to make the insightful remark that "I should be inclined to say that, in consequence of the analysis, not only the patient but his phobia too had plucked up courage and was venturing to show itself."[9]

In such expression is the fecund soil of fear. Now that the subject is broachable, one day the following discussion ensues between father and son:

I: [the father] "Which horses are you actually most afraid of?"
Hans: "All of them."
  I: "That's not true."
Hans: "I'm most afraid of horses with a thing on their mouths."
  I: "What do you mean? The piece of iron they have in their mouths?"
Hans: "No. They have something black on their mouths." (He covered his mouth with his hand.)

 I: "What? A moustache, perhaps?"

 Hans (laughing): "Oh no!"

  I: "Have they all got it?"

 Hans: "No, only a few of them."

  I: "What is it that they've got on their mouths?"

 Hans: "A black thing."

(I [the father] think in reality it must be the thick piece of harness that draw-horses wear over their noses.)"[10] (See Figure 4.2.)

HANS'S WISH

Later in the same conversation, Hans tells his father of the time when he, Hans, went shopping with his mother for a waistcoat. It is a story that the mother later confirms to the father. A bus-pulling horse "fell down." Hans was frightened. He lay on the ground and kicked his feet in the air to demonstrate the actions of the horse. Hans says that the horse was dead and then that it was not, saying that the first interpretation was only a joke. (The horse died; no, it didn't: I was only joking. Note the trying of alternatives and the denial.) The horse that fell, Hans says, was black, big, and fat.

> *I:* [the father] "When the horse fell down, did you think of your daddy?"
>
> *Hans:* "Perhaps. Yes. It's possible."[11]

 Freud believes that the event that precipitated the phobia of horses (and the generalization to other hoofed animals) was the horse falling, kicking, and possibly dying. The mental association is between the horse and Hans's unconscious wish that his father might die, that he might be

Figure 4.2 Little Hans's description of the horse that terrifies him into inaction. Hans's father and Sigmund Freud come to understand that the horse's head represents the father. From this identification springs the psychoanalytic notion of the oedipal complex.

gone or might be supplanted. It is not an acceptable conscious wish, although the wish that a parent, a controlling figure, would be gone is to be found in every child's thoughts. The father continues the questioning by suggesting to Hans that aspects of the father and the horse are associated, such as the father's moustache and the black bolt on the draw-horses. One morning Hans visits his father whose shirt is off.

> *Hans:*"Daddy, you *are* lovely! You're so white."
> *I:* "Yes. Like a white horse."
> *Hans:*"The only black thing's your moustache." (Continuing) "Or perhaps it's a black muzzle?"[12]

Another time, Hans's mother returns home after purchasing a pair of yellow pants. When she is showing them to her husband, Hans throws himself around in a little fit. When asked why, he says that he doesn't like the yellow pants. Hans says that the color reminds him of excrement and cheese and breakfast and his, Hans's, kicking (like the horse?) when he is placed on the toilet. We are reminded that bodily functions and sexual activity are not as distinct in the mind of the child as in the mind of the adult. Hans reminds the father of a time when he, Hans, was "playing horse" with his child friends on visits to Gmunden. Some connection is made between the "horse play" and the horses having "widdlers" or Hans's curiosity about the female playmates having "widdlers," but Hans declines this interpretation.

"We," continues the father, "went out in front of the house. He was in very good spirits and was prancing about all the time like a horse." (Notice the father's interpretation of how he was "prancing.") So I said: "Now, who is it that's the bus-horse? Me, you or Mommy?"

> *Hans* (promptly): "I am; I'm a young horse."[13]

With this spoken identification of person and animal of horse and child, Hans's interactions with his father and, most probably, with his mother, playmates, and animals takes on a different setting. Within a month, Hans reports to his father the following dream:

"The plumber came; and first he took away my behind with a pair of pincers, and then gave me another, and then the same with my widdler. He said: "Let me see your behind!" and I had to turn around, and he took it away; and then he said: "let me see your widdler!"

> *I [father]:* "He gave you a *bigger* widdler and a *bigger* behind?"
> *Hans:* "Yes."
> *I:* "Like Daddy's; because you'd like to be Daddy."
> *Hans:* "Yes, and I'd like to have a moustache like yours and hairs like yours." [He pointed to the hairs on my chest.]"[14]

Freud concludes the presentation of the case, this being the father's letters to Freud and Freud's sparse comments added to the father's interpretations, with this forward-looking comment:

"With Hans's last fantasy the anxiety which arose from his castration complex was also overcome, and his painful expectations were given a happier turn. Yes, the Doctor [the plumber] *did* come, he *did* take away his penis—but only to give him a bigger one in exchange for it. For the rest, our young investigator has merely come somewhat early upon the discovery that all knowledge is patchwork, and that each step forward leaves an unresolved residue behind."[15]

As is true of the myth of evolution, the myth of analysis may be said to explain the past without predicting the future. For this reason, both ways of thinking may be judged to be imperfect by the standards of contemporary science, which looks to prediction as the ultimate test of soundness, whether or not such a test is fair. Psychoanalysis predicts nothing about Hans's future from this fear and father-son exchange, although, as we shall learn, something is known about what is to become of Hans.

ANXIETY: HANS AND VICTOR COMPARED

The interpretation of Hans's fear introduced the twentieth century to an explanation of anxiety that it has embraced in its myth about itself. It is, after all, the age of anxiety, some would say. Anxiety became and perhaps is a compelling myth of our times. We treat it as a sickness that we are to be cured of, although, of course, anxiety may be adaptive in the sense that it warns and prepares us. As a bodily reaction, on the one hand it does so, and on the other it can sometimes render us inoperative and unable to respond.

From anxiety the psychoanalyst draws concepts illustrated by Hans—castration anxiety: oedipal anxiety, fear, and phobias. These concepts are organized into a story, a myth, about how the developing human being senses, perceives, and organizes the phenomena she or he experiences. The myth is that one's body is separate and unique, that one is of this sex or that, that a child is under the power of adults, but that even so, one can control the behavior of others by one's actions. This is the stuff of which the development of the individual is constructed; to fear, to identify, to separate oneself while desiring others; this too is the stuff of which humanness is made.

The myth used to explain Hans's anxiety and, by extension, the development of all children, is that the wish to have the father gone presents contrary emotions, and so must be sent into the unconscious, where it seems not to be bothersome. But once there, it strives to be heard again and resents its captivity. As it strives for escape, it learns to work in disguise in the hope of lulling or fooling the active, conscious mind. For Hans, it takes on the shape of the animals, the giraffes and stock of the zoo, the horses of the streets. An association is made between a physical characteristic of the father and those of the horses. The horses come to represent the father, and Hans transfers the burden of hate and love existing simultaneously onto them. When this cannot be done (one must, in that society, see horses), the system breaks down. The anxiety remains, surfaces, with no object to attach itself on, and is now seen as an unrealistic one. We call this state of affairs neurosis.

To satisfy the anxiety, we give it a home. We develop other relationships that serve to help us build a set of truths about the world we experience. To us, these are reality; to the outsider, they are the myths. To Freud and psychoanalysis, these myths are not only common but are also required in the process of human development. Had Itard been born two hundred years later, he would not have asked "What innate? What learned?" But, did Victor pass through stages of the myths now called oedipal or castrative? How might the oedipal period begin, proceed, and resolve itself in a feral being? Are they programmed by the brain, or products of civilization?

It is not, therefore, meaningful to ask if oedipal periods or castration fears exist in the sense that we can point at them or define them in some evident way. The value of the concepts lies not with their veracity but with their utility. To be sure, Hans's behavior is physical and thereby can be named and measured; he kicks, he crawls in bed, he speaks certain sentences. We can define, name, and quantify these aspects of behavior, but naming the behavior does not complete the task of those who would understand the mind. "Castration fear" is a choice of words that describes the presumed workings of the unconscious. Since the unconscious by definition, is that which we are not at present aware, the mental and feeling world of beings works at the level of understanding approachable only by grasping at myth, not by pointing at the object and measuring its frequency.

Freud asks the contemporary mind to understand that myth has a reality different from the reality composed of objects that we can point to and thereby distinguish. We are accustomed to the belief that the physical world is lawful and, that once understood, the laws predict. The learning of these physical laws is the basis of learning in Western society which we think to be the most objective and important. Freud and those who came after him are asking us to believe that the mental and feeling world of individuals is just as lawful and uniform. Let us examine the nature of that lawfulness and uniformity by examining Freud's comments on Hans's experiences.

FREUD'S ANALYSIS

First, let us return to Freud's letters in reply to Hans's father. After presenting selected parts of the father's letters regarding Hans's fears and along with some commentary, Freud turns to his analysis of what has occurred. The levels of communication in this story are as complex as they are in life, for the information we have is translated by the father from Hans's conversation, translated to Freud by letter, at which point

Freud then selects that which he considers to be germane to the interpretation. We, the readers, now almost a century removed, are far from the events. The play is rather like Dash to Chattock to Morgan, for we are reading Hans's father's account to Freud of what Hans said to his father. Freud is writing the account in 1909, a few years after the events. Much of the lawful nature of analysis has been laid down in previous publications, as in "Three Essays on the Theory of Sexuality."[16] Refinement of the ideas is yet to come during the remaining thirty years of Freud's life. Here Freud used the information gathered about Hans and his family to illustrate several fundamental aspects of the developing theory of psychoanalysis: that is, the way in which the developing child comes to understand self as a piece of reality, the role of the unconscious, the importance of anxiety, and the symbolic stories used to build myths that help us to understand ourselves.

At the onset of the discussion, Freud cautions us about "suggestion." He notes that "suggestion" has become a catchword for the psychologies of the day (1909). "Nobody knows and nobody cares what suggestion is, where it comes from, or when it arises—it is enough that everything awkward in the region of psychology can be labelled 'suggestion.' "[17] for a time, and yet another term at another time.

One pervasive aspect of psychology is that it discards very little human experience. Freud does not care for "suggestion" as an explanation because he has a bigger and better myth. His myth is a Great Myth—one, like evolution, that binds so much material to itself that it describes and explains many events and observations that had been seen previously as separate and distinct from one another. To the contemporary mind, evolution binds rocks and fossils to chickens and shrimp to trees and plants and flowers to weather and cosmology. Such is the power of the myth. Because it binds so many once disparate elements into something common, the myth gives us an understanding of our world and its history—of how it and we came to be the way we are and, sometimes, of what will become of us.

Now, writes Freud, "I do not share the view which is at present fashionable that assertions made by children are invariably arbitrary and untrustworthy. The arbitrary has no existence in mental life. The untrustworthiness of the assertions of children is due to the predominance of their imagination, just as the untrustworthiness of the assertions of grown-up people is due to the predominance of their prejudices."[18]

Freud was writing at a time when children were thought to be—well, frankly, "born liars." Thus, it was apparently thought that children needed help in seeing reality and that reality, being a characteristic of the adult mind, was to be gained through experience and education. The pinnacle of this hierarchy was likely the adult, male, white, European mind. This may have been what Freud meant when he contrasts the imagination of the child with the prejudice of the adult—and here he appears to mean "prejudice" in the original meaning of the word, "to prejudge."

Historically, two extreme positions have become prevalent in explaining the child's mind and perhaps the animal mind. One is that the child has a perfect view of reality: it comes into the world equipped to sense accurately, but society, parents, and the educational system impose another reality. This view governed contemporary observations of Peter and Victor. To the nineteenth-century European, this perspective became the romantic view—we are born "good," naturalness is the desired state, and we are born well equipped to sense reality perfectly, but we trade such purity for the questionable rewards of civilization. This notion did not die with the nineteenth century; it survives in contemporary views of proper diet, behavior toward the environment, and investigations involving animals.

The second view is that the infant comes without imprinted or prearranged concepts. At birth the human mind is a blank slate, a blackboard, on which nothing exists until something is written through the training provided by parents, teachers, and society generally. Of course, the Western World has leaned heavily on this notion, particularly as it postulates the means and purposes of education. We can speculate that this view motivated Dr. Singh to see Kamala and Amala civilized.

Freud is not so much interested in which of these two views is correct as in showing us that the mind of the child is not unrealistic. It imagines, and it is an adult prejudice to say that the child confuses reality and imagination. The reality of the child is as real to the child as my reality is to me. That I do not accept the idea that horses fly does not mean that it is an unreal or incorrect statement to the believer. When the father cannot accept Hans's views of reality, it is not that Hans's views do not grasp reality, but that he grasps it differently from the father.

HANS'S MIND AND BODY

Hans's reality is now occupied with separating his mind from his body and once understood as separate, distinguishing this body from those of others. He is imagining, trying out hypotheses, testing what it means to be himself. Of some interest to him is the "widdler," a part of the body that in urinating seems to respond to the call of some other agent and to have a mind of its own. Later in life he will understand the import of that independence. The universe of widdlers is itself circumscribed by Hans's observations—for example, the observation that large animals have large widdlers. This view is supported by his inspection of zoo animals at Schönbrun and, of course, by his examination of the many horses who animate the streets. Presumably, he transfers this generality to his father, either through deduction or observation. He expects his mother to have a very large one, but she does not. The falling apart of the principle is unnerving, but he guesses that it is connected to having a sister. The parents acting either separately or together—he is not sure which—have the ability to take away his widdler. He may assume that they have done this to the baby sister or that it was done to his mother. Of the two, it is the father who is the most terrifying in this regard, the one most likely to do the act in retribution for Hans's wishing him dead or gone.

Is this terrifying story pathologically imaginative? No, deduces Freud. The story is actually healthy, for every male child develops and learns to subsume the myth of castration into the normal personality. It would be pathological *not* to wish the father gone; *not* to notice the collective widdlers; *not* to fear the power of the father. The myth is a necessary one for the individual who creates his own successful identity, independent of others and yet dependent on others in some important ways.

How does one make oneself independent when one sees oneself as clearly dependent? The child's problem is a monstrous one: it must learn how to maintain a relationship, especially with the mother and father, those nurturant but demanding, encouraging yet correcting, persons. To the child, as to many adults, the contradictions are real: it must establish itself by withholding itself and yet give of itself in a way that procures a like return. How is one to love and be loved in return? That need is seen when Hans, like all children, would, in play, hit his father and then "kiss it better." The father and the concept of horse would appear to share a contradiction in feeling that becomes too powerful for Hans to live with successfully. It is possible to live with contradictions of reality, perception, and feeling, Freud writes, because seldom are alternatives conscious at the same time: they successfully repress one another, first this way and then the reverse. Until repression and, perhaps, neurosis come along, contradictions exist happily together. But when contradictions appear at the same time, one must give in to the other—or a new principle that holds them both must be forthcoming. The finding of that principle and its development may lead to neurosis and madness, but most often it leads to health, maturity, and a learning of where one is and was in the world.

The birth of Hans's sister accelerates forward for Hans the struggle both to attain more independence and to understand sexuality. For a time he is frightened of the bath, perhaps because of its association with the bowls of blood he saw after the childbirth. But the most dramatic anxiety begins in the street—dramatic because Hans appears to have no particular associations with horses. It became clear that the anxiety was produced by horses, but at the onset Hans himself can make no such connection: the anxiety is all-powerful, overwhelming in effect. Hans is a fortunate child, for, as Freud reminds us, his parents neither laughed at nor bullied him. His mother, it will be recalled, had been a short-term patient of Freud's and the father was sympathetic to Freud's way of thinking. Given our contemporary understanding of the meld between parents and child, we are tempted to inquire into their motivations, unconscious or otherwise. Our focus should remain on Hans, however, both because the information we have is transmitted at the first level by the father and because psychoanalysis focuses on the mental developments of childhood, where, after all, the repression and therefore the neurosis begin.

In the dream of the two giraffes, Hans expresses something of his hypotheses regarding his parents; these hypotheses are based, of course, on perceptions and experience, and are perhaps partial, but nonetheless, they are important to the imagination. Freud tells us that, stripped of its distortion, the story is of the mother and the father; they are the two giraffes, the large and the small. Characters in a dream may appear random or inconsequential, but they are not. Characters come from experience, and the question must be answered: Why choose a giraffe and not a moose? Hans had been to the zoo shortly before; the image was fresh and, although Freud does not remark upon it, we know that picture of giraffes were framed above Hans's bed. Freud suggests the long neck is suggestive of the widdler, and we can recall Hans's picture of the giraffe with the drawing of the widdler that we added. One giraffe sits on the other, crumpling it. At a rudimentary level he understands. The dream both re-creates and tries out the hypothesis: it asks what if this is the way the act is performed, and then the consequences are tried out in the relatively safe environment of the dream.

Hans's connection between the horses and his father are made evident by the emphasis on the bit and moustache, the moustache being one characteristic of the adult male. One reaction by Hans at this time is aggressive; he kicks, he imagines two fantasies in which he hurts people or things and about which the father fails to remonstrate. The aggression is truly directed toward the father; occasionally it is manifested directly through hitting during play, but far more often indirectly, by means of fantasy and perhaps by words not reported to us. The bind for Hans is how to become like his father while yet increasing his independence. It seems hopeless for Hans, yet it is a problem everyone strives to overcome; indeed, it may be at the root of all phobias and the producer of anxiety. Hans responds by substituting horses for the father, at first with the idea that horses might bite him but later that they would fall and die. The wish is that the father be removed, in the extreme, to die; it matters not in our fantasy how it is done. The wish is painful and unacceptable; it is protested and denied. The psychic result of wish and repression is that anxiety grows.

The order in which the father reports these events and conversations to Freud leads us to understand that Hans's interests turn toward an explanation of defecation. Hans associates the color of his mother's pants with "lump," and his father's discussion leads to an earlier memory regarding "playing horse" with the playmates at Gmunden. Hans, we are told, also "played horse," with his father. Most children do, we assume. "Playing horse" (although Freud does not develop the comment) contains elements of aggression, of kicking, of falling down, and it is at times dangerous and physically punishing.

Hans takes the analysis into his own hands when he provides the dream of the plumber who in the bath takes his behind and gives him the larger penis. The dream, wonderful in its simplicity, is of growth. It resolves the fear of castration; it shows accomplishment, for the wanted is acquired; and it permits Hans to identify with his father. Hans does not accomplish these things only symbolically. He does so in a more real fashion because it is *his* dream. He is no longer the passive person whose words and questions are put forth for analysis. He is the doer, the person who dreams the dream, the possessor of the mind that uses the dream to come to terms with the anxieties.

The anxieties will remain, in much modified form, because they are now residual, and the experience of them can be altered, though not canceled. Analysis alone does not cure; it provides the information whereby the person can rethink and again feel associations. The power of analysis is that it educates by encouraging people to pull from their own mind what is to be found there. It does not instruct actively, but it expects the individual to re-sort associations. Psychoanalysis is in that romantic tradition, being a tradition traceable to Plato's ideas, that learning is a capacity of the individual who, given information and the opportunity, will educate her or himself. Knowledge about oneself and the world will appear and reappear when prompted but will be shaped, of course, by how the myths are played out.

HANS'S LIFE

Hans's experience is of longstanding importance because it represents so well the Freudian myth of how our mental images and associations are created and manipulated. His experience is the prototype of all children's experiences.

Hans grew to adulthood undamaged, it would seem, by the attentions paid by the mother, father, and Freud himself. As noted earlier, Hans visited Freud only once during the period described by the father and did not visit him again until the age of nineteen, identifying himself as a person who was the object of the original interpretation. In 1922 Freud published a postscript that described their meeting and conversation. Freud pointed out primarily that Hans had not become a monster, "robbed of his innocence," as some had predicted. Indeed, Freud tells us, Hans could not remember any of the events described in the father's letters. Freud compares Hans's memory to the common experience of the dream, for the events, like the dream, meld into experience and are promptly forgotten, but are always reshaped.

Freud and Little Hans had at least one other meeting. As has been stated, Freud was a family friend, and the father, a music critic, was a frequent visitor at Freud's weekly sessions at his home which covered topics in the development of psychoanalysis. We also now know that Hans's name was Herbert Graf, born in Vienna on April 10, 1903. We know that Freud took Hans a present on his third birthday in April 1906. Freud tells us of carrying it, puffing all the way upstairs, to the little boy. The gift was a wooden rocking horse.

Herbert Graf, by the way, became a designer and stage manager of opera, holding such positions at the Metropolitan in New York from 1936 to 1960. He left Germany in 1934 to work with the Philadelphia Orchestra and became a U.S. citizen in 1943. He managed opera houses in Geneva and Zurich, married twice, and died of cancer in Geneva on April 5, 1973.[19]

PSYCHOANALYSIS: RELIABILITY AND VALIDITY

One way to evaluate various approaches to knowledge, such as psycho-analysis, behaviorism, and phenomenology, is to examine their reliability and validity. By the first is meant whether measurements can be made that are free of error. For example, I have a physical thing, a tabletop, or a psychological trait, such as intelligence, to be measured. Assuming that I have a tool with which to do so, a ruler or a test, I can apply the tool to the object or capacity. A reliable tool will produce the same result time after time with an acceptable error; an unreliable tool will not. Nonetheless, there is a twist worthy of our consideration.

If I measure the tabletop or a person's IQ one hundred times, I might arrive at a mean of 6.5 meters (for the table) or 105 (for the IQ). Are these figures reliable? There is no way to answer this question without measuring the error. Let us say that my error of measurement for both the table and the IQ is 0.05. Depending on your purposes, this degree of error may, or may not, be acceptable. If I want the table to fit into a space 6.5 meters, the error may be too much, but the figure may be fine indeed for the IQ. Notice in this example that the error is my error; that is, we are measuring the error from my doing the measurement many times. Assume that rather than measuring one table one hundred times, or the IQ of one person by giving the test one hundred times, I measure one hundred tables or the intelligence of one hundred people. Now the error becomes informative in a different way, for it tells me the range of measurements and the likelihood that any one table or person will be at a certain point within the range. What was error is now variability. What under one condition of measurement I saw as error of measurement, I now see as telling me something very important about the variability within the population of tables or people measured.

Is psychoanalysis reliable? If we asked one hundred psychoanalysts to interpret the story of Little Hans, would each tell us the same interpre-tation, with only minor deviation? This is a sensible way of measuring reliability, for if they could not agree, the set of data would be unreliable and, it follows, that no interpretation can be any better than the degree of error. Of course, psychoanalysts would agree on the story of Little Hans, for an agreed-to myth explains the story. Reliability of interpreta-tion of such classic tales should be high: reliability of unknown cases, of cases not yet worked into the general myth, might be lower. But myths can be as reliable as measurements of tables and intelligences.

Validity is the second way we have available to measure the accuracy and utility of theories and methods. A theory is valid if it does what it sets out to do. A test of mechanical ability that successful mechanics fail and the clumsy ones pass is not valid. An IQ test may give the same scores over and over again, but not predict IQ. SATs may be reliable for an individual but predict college success poorly.

Is psychoanalysis valid? It depends, of course, on what we expect psychoanalysis to do. If our goal is to cure the sick, it may work only on a small part of the population. If our goal is to understand ourselves and others with whom we deal, however, it may be more or less valid. In other words, it all depends on why we compose and develop it. It may be valid for some purposes and not for others.

Psychoanalysis is easily and often faulted; yet it has a staying power for us human beings. It is easy to shoot holes in a myth, because any myth is obviously untrue when it is compared aside perceived reality. Yet we use the concepts of psychoanalysis routinely; even if we think the concepts nonsense, we accept the notion of an unconscious and of events experienced in infancy imprinting themselves in such a way that future behavior is affected. The logic of psychoanalysis is, surprisingly, most like that of behaviorism in one important respect, for both prefer careful analysis of a single, or few, subjects, and do not trust means and standard deviations of groups to tell us much about the mind or behavior. In short, reliability comes from thorough investigation of a single subject rather than measurement of variance in large populations.

The case of another Hans, the horse Clever Hans, speaks to us of the power of empirical methods and of hypothesis-testing and, not accidentally, offers another model of the mind. In this case, described in the following two chapters, the roles are reversed: Here the model is that of the mind of a horse, and from that example we draw a model of the human mind. But we will look at the achievements of psychoanalysis again. For now it is enough to realize that psychoanalysis is among the most successful and compelling psychologies of all time.

We cannot escape feeling its power.

The Psychology
of Experimentalism
and Behaviorism: Clever Hans
and Lady Wonder

5

T HE TERM *behaviorism* is now used to describe almost any epistemology that sets hypotheses and tests them by measuring behavior itself. Often it is incorrectly used to describe *any* experimental work with animals. By measuring the behavior only, the status of mind is put outside the scheme. At first, it strikes us odd to have a science of psychology that is uninterested in the mind, but upon consideration we can understand the logic. Mind is yet one more example of the human invention of categories. Unless we can define it accurately and meaningfully, the concept can become a trap. The trap is one not unlike Freud's description of "suggestion," and we moderns may add "instinct" to the list of examples.

Behaviorism does not have a simple, specifiable, time of birth. But the analysis of the horse, Clever Hans, illustrates so well the tenets and powers of the experimental method when applied to behavior that it is the centerpiece of this chapter. The chapter that follows pursues the story of Clever Hans, not for what it says about behaviorism, but for what the end of Hans's life tells us about human and animal interaction.

Over the years, with its numerous retellings, the story of Clever Hans has become distorted. The story of Clever Hans is often cited by those who practice sign language with chimpanzees (for example, researchers who appear in Chapters 10 to 12), but only a few of such accounts are reliable and some are astonishingly wrong. The point is important because the ape sign-language trainers are sometimes accused of the same errors that were made in the study of Clever Hans.

The second part of the story as told in Chapter 6 concerning Clever Hans's life after his role as an experimental subject ended is mostly unknown, I think, chiefly because the documents did not survive World War I in quantity. Or perhaps the reason is that the parable to be extracted from the tale did not fit the psychologies of the times any more than "suggestion" suited Freud.

Let us examine what happened.

BERLIN, 1904

In the summer and fall of 1904, when Little Hans was celebrating his first birthday in Vienna, anyone in Berlin might have visited a courtyard in the north of the city, one surrounded by high apartment houses. Noon is the daily hour at which the hosts are available. There is no fee; all are welcome.

Here we find man and a horse (see Figure 5.1). The man is sixty-six-year-old Hr. von Osten. He is white-haired and white-bearded, and he sometimes wears a large, floppy, black hat. The horse is named Hans. You will notice that the relationship between Hr. von Osten and his horse, at least in public is a gentle one. No whip is used on the horse—he receives only gentle encouragement supplemented by bread and carrots.[1]

You may ask questions of the horse, but only in German, for this is the language to which Hans responds. If Hans understands the question and is ready to answer, he nods his head; if not, he shakes his head, inviting further discourse. He is somewhat more recalcitrant with strangers than with von Osten, his longtime associate. Hans does not speak, as is reportedly possible among some animals, notably dogs and pigs, but Hans taps with his right hoof and, as you will see, he answers questions by the number of taps. In addition, he uses his muzzle: "Yes" is a nod; "No" is

Figure 5.1.   Clever Hans II, von Osten, and an early version of the writing-board used by Hans, evidently to answer questions about harmony, mathematics, dates, and the location of objects. Hans I died before reaching much fluency, but Hans II became the toast of Berlin.

a side-to-side movement, much like that used by Europeans and Americans, but the reverse of Asian and Asian-Indian usage.

Hans points to directions with his nose: up, down, right, and left, although you may notice that, like a theater performer, he gives directions from your viewpoint rather than his own. That is, your right is his left. He can walk in the direction of an object if the answer to a question demands showing an object or if it is a command (for example, "Show me something red.") Von Osten has translated the numbers that Hans taps out into the alphabet, tones, and the names of playing cards. For example, with playing cards, one tap is an ace, two a king, and so forth. Hans can count from one to one hundred and from first to tenth if asked to use ordinal numbers. A shake of the head is naught or zero. When the answer to a question is a small number, he taps slowly, but when the answer is higher, he taps quickly, suggesting that he knows the answer before he starts and uses his time efficiently. When Hans has finished explaining an answer by his right-hoofed tappings, he gives a single, emphatic tap with his left foot.

RICHMOND, VIRGINIA, 1924–1952

Having now met Hans, let us leap ahead in time to visit the area around Richmond, Virginia, first in 1927 and then, later, in 1952. This horse is Lady Wonder, and her owner and companion is Mrs. C. D. Fonda. Lady is, on occasion, telepathic; she is able to "mind read." At first, she used her muzzle to manipulate blocks with letters on them to spell out English words. In later life, she used a "horse" typewriter, a device on which she pushed keys with her nuzzle, again to spell out words. She does so in Figure 5.2.

Place the twenty-six blocks on a table in two rows. On a pad, have two persons write words, known only to them. Ask Mrs. Fonda to ask Lady to spell the words. Lady selects the blocks that compose the following words selected by two other persons:

bed, kid, Mesopotamia, Carolina, Hindustan

You note that the relationship between Lady Wonder and Mrs. Fonda is of a different sort from that between von Osten and Hans. Mrs. Fonda uses a whip and uses it to prod Lady in the way that people interested in performance by animals often do.

If you will return to the early 1950s you will discover that Lady's telepathy now enables her to answer questions about lost objects and people. A missing child is said by Lady to be at "Pittsfield Water Wheel." A policeman, pondering this interpretation, wonders if it might be the "Field and Wilde Water Pit," a local quarry. It is a prediction both grim and true. When the quarry was dragged, the body of the missing child was found.[2]

On December 7, 1952, two children, a boy aged three and a girl of

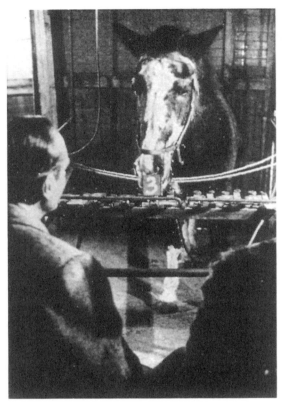

Figure 5.2 Lady Wonder, the horse who flourished in Virginia from 1920 to 1950. She is shown here using her "typewriter." She was able to spell out clues as to the location of missing children, and told visitors facts about themselves, such as their birthdate. Early human investigators of her powers expressed the belief that Lady Wonder was highly intuitive; professional magicians spotted an animal trained to do a standard magic trick.

six, were seen playing shortly before noon in the boy's backyard in Naperville, Illinois. They were not seen again. The search for them occupied the community: ponds were dragged, sewers investigated, old iceboxes overturned, bloodhounds engaged along with the boy's kitten (who led searchers to a nearby quarry which was then emptied of 70 million gallons of water). The search was in vain.[3]

The boy's mother decided to consult Lady Wonder, whose previous achievement in pointing the location of a missing child was well known. The facts were explained to Lady. Nuzzling her typewriter machine, Lady said that the children were dead and that they could be found together at "tree" and "water."

"What water?" the mother asked.

She was told to write on a piece of paper the name of any body of water near her home. She wrote "DuPage River," and the horse spelled out "river." Lady Wonder also predicted that the children would be found on a Sunday.

A *Newsweek* magazine account of 1953 read in part: "58 days after the disappearance, a workman walking over the frozen DuPage spotted a red coat under the ice. Police recovered (the girl's) body. The boy was found under the ice 50 feet downstream.

"The horse was right about everything but the day they would be found, (the mother) said."[4]

Lady's career as a telepathic horse was indeed boosted by these two predictions of the locations of missing children. Evidently, she performed, if that is the correct word, for thirty years on her Virginia farm. Although a small fee was charged visitors who wished to make inquiry of Lady, it cannot be said that Lady produced much money for Mrs. Fonda. Here we will examine how Lady (or Mrs. Fonda) came to perform these exceptional feats. Many persons of professional standing investigated the horse (and Mrs. Fonda) and came away convinced of Lady's talent, intelligence, and ability. Experiments were performed and alternative explanations examined. Lady's achievements, after all, were known a generation or more after those of Clever Hans's, so we might expect whatever was learned about Hans's abilities and training to be used in evaluating Lady.

Among those who investigated Lady were the Duke University botanist-psychologist J. B. Rhine and his colleague and wife, Louisa Rhine. Their initial analysis concluded that Lady was telepathic, but as noted earlier, a second investigation during that same year was unsatisfactory: Lady now showed no ability to answer questions correctly. The Rhines concluded, logically, that Lady had lost her telepathic ability at some period between the visits.

THE CARROT AND THE WHIP

Hans's chief social relationship was with human beings, especially with von Osten. Apparently, von Osten was of a kind and gentle disposition, preferring use of the carrot to the whip. He did not whip the horse, at least not frequently, and never during the sessions when Hans was "thinking." Von Osten gave Hans vegetables during these sessions. The precise relationship between the two is of some importance to our analysis, for it is evident that some training of Hans took place. The nature of that training is of more than passing interest, for if Hans was a clever horse, as he indeed was, then von Osten was a clever teacher. What Hans knew was how to tap his hoof, thereby indicating to human observers his answer to questions they posed.

Although history does not tell us much about von Osten, we do know that he taught mathematics to schoolchildren, retired from that occupation perhaps with some bitterness, and died shortly after the 1909 publication of Oskar Pfungst's analysis of Clever Hans and von Osten. We also know that Hans then became a teacher in his own right, but we must save that interesting reversal of profession until the next chapter.

Hans's abilities came to the attention of the people of Berlin in the first years of the century chiefly through newspaper articles. Belief in von Osten's honesty and integrity was enhanced by the fact that he never charged for demonstrations of Hans's abilities and that he, von Osten, welcomed all investigators. Those who knew both horse and man wrote that von Osten seemed mystified, but proud, of Hans's accomplishments. Those who worked with the pair seemed to regard von Osten as a little rustic, if not simple, not of high birth or education, but mannerly and thoughtful, if not plodding. These few descriptions of von Osten seem slightly condescending, if not arrogant, when we recall that von Osten, if a bumpkin, taught mathematics for some years.

Hans, the horse, was not particularly docile, nor was he the first or only horse to show high ability under von Osten's care. Hans was actually Hans II: earlier, von Osten had educated or trained a horse named Hans I. Hans I is said to have reached levels of ability that were only a little short of what Hans II achieved, but Hans I died when young.[5] Hans II was said to be inattentive and occasionally bad tempered, but he was receptive to von Osten's encouragements and attention. Once he got into position to tap his foot (with von Osten usually nearby), there was no reluctance to tap answers. Some people noted that the horse then seemed to move into a slightly trancelike state or, perhaps, a state of pure occupation with the task. It is suggestive that the same description of a trancelike state[6] was made about Lady, but none of the persons describing Lady appeared to be aware of the similar descriptions of Hans.

Among the many individuals who examined Hans, only Oskar Pfungst, a student of Professor Carl Stumpf, the professor of psychology at the University of Berlin, and a central figure in the story now unravelling, seems to have determined the relationship between the human and horse. Other watchers and investigators concentrated on documenting Hans's abilities, testing mainly to determine Hans's accuracy and the limits of his abilities. Pfungst alone seems to have attended to questions about the relationship of human and animal, and Pfungst alone appears to have asked von Osten how Hans was trained to tap. The lesson for us and our time is that investigators tend to bring to the investigation their own beliefs about how we learn. These beliefs shape the questions asked and the explanations posed. Those who assume that Hans was indeed a clever horse will test the thresholds and limits of these abilities. Their interest will be in demonstrating what Hans did and did not learn. Those, like Oscar Pfungst, will look instead, or in addition, to the relationship and to the question of how learning has occurred.

Articles about Hans first appeared in journals directed to horse trainers and teachers of human pupils. One such reader was the horse trainer at Elbersfeld, Karl Krall,[7] who then spent time in Berlin watching Hans and von Osten and, upon von Osten's death, acquired Hans. Few know that Krall successfully trained two other horses, Zarif and Muhamed, to perform tapping feats similar to those of Hans. Hans is surely the most famed, but he is not the only horse who showed his cleverness. Hans I, Zarif, and Muhamed are mostly unknown, I think, because Krall's book became rare after World War I and remains available only in the original and unreprinted German-language edition.[8]

By 1904, Hans, the clever horse, was a popular topic of the day for Berliners. His image appeared on postcards and liquor labels, and his name was honored in vaudeville at the theater. A commission of respected persons was organized to investigate the story of Hans. The members included two teachers, two zoo directors (one of whom, Oskar Heinroth [1871–1945], was an inventive and respected researcher on the behavior of animals), two military majors, a circus manager, a veterinarian, a count who was also a respected horseman, and two academics—Professor Konrad Nagel, the head of the Physiological Institute at the University of Berlin, and Professor Carl Stumpf, then director of the Psychological Institute at the University of Berlin. The group was called the September Commission.

Among those who observed Hans but who were not members of the commission were Professor Hans Schweinfurth, the explorer of Africa, whose surname is honored in the chimpanzee species (*Pan Schweinfurthii*); Dr. Helmut Schaff, director of the impressive zoo at Hanover; the zoologist Professor Kurt Mobius; and Hr. C. G. Schillings, another African explorer and zoologist. When Schillings first saw Hans in July 1904, he was so impressed that he remained in Berlin to test Hans and von Osten for trickery. Finding none, he became a leading advertiser and proponent of Hans's abilities. Indeed, all these gentlemen came away convinced of Hans's mental abilities.

THE SEPTEMBER COMMISSION

The September Commission went to see Hans and von Osten in order to determine whether the horse performed these astonishing acts by trickery, by intelligence, or by some combination of both. They tested Hans by having someone other than von Osten ask questions and, in other cases, by having the correct answers unknown to the questioner. They determined that von Osten's method for training Hans to show his ability was "like that used in the school system." A majority of the commission reported that they knew of evidence that Hans, in previous observations, had been able to answer questions correctly without von Osten's being present. This is an important observation, for the account of Hans as told by contemporary writers is often wrong on this critical point. Both von Osten and the Commission were aware of the possibility that he gave cues. The Commission found no evidence that signals or signs, intentional or purposeful, were used.

Some compromises were necessary in writing the report, for not everyone agreed with every word. Words were changed, meanings qualified by lengthy prepositional phrases, and a document was signed on September 12, 1904, somewhat hurriedly. The document describes the observations, the sources used and the results, and concludes that although (the report said) no trickery was found, the Commission could not pass judgment on how clever Clever Hans was, "since . . . there was a great possibility that other factors were involved which ought to be investigated carefully."[9] The Commission decided that *Clever Hans was himself providing answers to questions involving counting, music, and locating nearby objects; that this was done without trickery on the part of human beings; and, yet, that it was not yet possible to assess Hans's degree of cleverness or intelligence.*

One member of the Commission, Carl Stumpf, was a well-known Berlin professor. His most recent experiments had investigated altered states of awareness, especially hypnosis, a state that was then of particular importance because of the implication that people were capable of acting and behaving under the motivation of "unconscious" suggestion. Remember the special words that Freud had for those who would overapply the concept of suggestion. Oskar Pfungst, Stumpf's student, was quite aware of the power of the concept of suggestion and unconscious inference, and his understanding of the notions guided part of his analysis of Hans's successes. In the first pages of his account, Pfungst cites the words of another author, writing in 1901 about animal abilities: "The astounding phenomenon of an animal apparently possessing human reason is to be attributed solely to suggestion."[10] Carl Stumpf, as is the custom of many professors who "teach" advanced students, turned the problem of Hans and von Osten to student Pfungst.

No investigator can approach a problem without hypotheses, for without prejudgments human beings would have no myths to explain as to why certain events, but not others, present problems to our understanding. A horse that neighs presents no special problem to my understanding; a horse that does arithmetic requires me to reestablish my myths about humans and animals alike. Hans presents a problem to human beings only because his behavior does not correspond to our imagined structure of human or animal nature. When we do not understand, we form hypotheses to make our understanding sensible.

THE MIND'S HYPOTHESES

Depending on my hypotheses regarding human nature, I shall have a like hypothesis about how Hans did what he did. For example, if I depend on learning as the explanation of how human beings come to be the way they are, I will think of Hans as a trained horse taking his cue from someone or something. The hypothesis itself will compel me toward searching for just such cues.

If I have another hypothesis which states that horses are far more intelligent than other people believe, the first question I ask is this: With what hypothesis and what values do you approach the phenomenon? Schillings, for example, is said to have first observed Hans at work when he, Schillings, was convinced that he could expose trickery. When he was not able to do so, he became the firmest of believers and rejected his own hypothesis as a false one. His testimony is made more convincing by his frankness in admitting his change of mind, and, as is often true of the convert, the reversing of belief brings with it greater certainty. Oskar Pfungst, who later conducted his experiments with Schillings in attendance, found that Schillings was among the few people who, under Pfungst's arranged conditions, was a sufficient, but not necessary, presence for Hans to be able to answer the questions posed.

Those who believe human behavior is largely determined genetically are more likely to be sympathetic to a hypothesis about Hans that credits the horse's inherited mentality. They are more likely to ask about Hans's parentage and presumed genetic stock, and to accept his feats as evidence of his unusual mental ability, as this belief is the source of the hypothesis. We shall see that Pfungst changed his approach as his investigation continues, for he used the test of one hypothesis to generate the postulation of the next.

Pfungst showed that Hans performed admirably only if certain people were in attendance. What do we make of this? If we are ready to grant that no human investigation is undertaken without the presence of hypotheses and values, then we may proceed to see that even a finding as seemingly exact as this one may have one or more explanations. Data: Schillings was among those people whose presence was required for Hans to perform accurately. From this information, we may suppose either that Schillings had learned how to provide cues, whether intentionally or unintentionally, or that no cues were involved, but that Hans required the presence of believing or sensitive observers. The hypothesis determines not only how we test and gather data and information, but also the direction of the alternative explanations and, hence, the conclusion.

The problem is that such conclusions are logically like that reached by J. B. Rhine when he revisited the horse Lady Wonder, found her ability gone, and concluded that this was because her telepathy had dissipated, probably temporarily. He could have just as well assumed that Occasion 2 was the true state of affairs and that Occasion 1 was the aberration. He did not offer this explanation, however, perhaps because he had already committed himself to the explanation of telepathy at the time of Occasion 1. Having done so, he then assumed it and treated its nonappearance as a loss.

OSKAR PFUNGST'S HYPOTHESIS

Pfungst decided to separate the problem into three experimental considerations. Beforehand, he decreed some experimental conditions and procedures. A tent was erected in the courtyard so that the experiments could occur without people's random movements influencing Hans's behavior. Here appears an experimental paradox, although Pfungst was not aware of it. A sure goal of experimentation is to reduce the number of causative variables until only one remains. Pfungst's goal was to eliminate the variables of no apparent importance, such as the sound of a bird or the smell of another animal, so that when he was able to isolate something that caused Hans to be accurate, or inaccurate, he, Pfungst, would be able to name the single causative agent. To the other extreme, if all such random variables were eliminated, Hans would be living in a vapid environment, one without the varieties of stimulation that real horses and real people expect. To deprive Hans of the cacophony of stimuli that make life and experience meaningful and familiar was quite likely to interfere with whatever behavior we might normally produce. The issue is not insignificant, for it forms the alternatives promoted by those who believe that captive animals are no animals at all and those who insist on bringing environmental variables into the controlled experiment.

Deprivation of all stimuli that excite the senses is unreal and impossible. How the experimenter adjusts the two extremes calls for delicate designs and plans. The decisions demonstrate the art of methodology for the behavioral scientist, for the decisions require some notion of which variables are likely to be relevant and which irrelevant. Thus, the hypotheses the investigator brings to the experiment may account for the degree of controlled sensory deprivation. It is not that the task of setting the degree can't be done or shouldn't be tried, but that we need to understand the tacit hypotheses if we hope to evaluate the investigator's results.[11] The question of how much we need the natural occurrence of an event and how much we must subject it to laboratory control in order to understand it remains a longstanding issue in the application of science to animal mental life.

Just as J. B. Rhine brought with him to his visits to Lady Wonder several mental hypotheses of his own, so Pfungst took with himself his more skeptical hypothesis—namely, that the horse was trained, purposefully or accidentally, to tap in response to some sign, intentional or inadvertent, from a human being. Such expectations and hypotheses underlie all human behavior, for we are all natural investigators of the phenomena we sense. To understand why Pfungst asked the questions he did, and therefore why he performed the particular experiments he conducted, *we* need to think about the hypotheses that guided *him*. For his explanations can only be applied to the data he collected and, therefore, are controlled by the hypotheses *he* set. Let us attend to the details to see if we can uncover the investigator's thinking.

First, Pfungst removed both von Osten and Schillings from the tented courtyard. Pfungst writes:

"Now one would have thought that the horse would respond to any moderately efficient examiner. But as a matter of fact it was found that the horse would not react at all in the case of the greater number of persons. Again, in the case of others he would respond once or twice but then cease. All told, Hans responded more or less readily to forty persons, but it was only when he worked with von Osten or with Schillings that his responses were dependable. *"For this reason I undertook to befriend the horse, and by happy chance it came to pass in a short time he responded as readily to my questions as to those of the two gentlemen."* (italics, mine, for emphasis.) [12]

This comment gives us a lot to think about. First, Hans responded to great numbers of people but was dependable only when in the company of the owner, von Osten, or the skeptic turned convert, Schillings. Second, Hans came to respond to experimenter Pfungst. From this information we can deduce that Hans had learned to respond to some human cue, even when given by strangers.

Based on these data Pfungst concluded that Hans did not think independently of human beings. To the modern mind, the notion that a horse calculates independently does not seem to be a likely hypothesis to test. However, our disbelief is but evidence that we too carry with us hypotheses, or the lack of them, as part of our perceptions as we interpret our world. We might grant horses and other animals the possibility of limited counting and recognizing, but we are unlikely to ascribe to them sensitivity to advanced harmonics and musicality. So *we* would not examine Hans from the viewpoint that he possessed independent thought. Pfungst, however, arranged his experimental designs with the idea in mind that Hans *might be* capable of independent thought. Why not? The September Commission suggested such a possibility, and Pfungst's professional adviser was a member of the Commission.

Pfungst's conclusion is surely fair in terms of his hypotheses and the data, but it is not the only explanation consistent with the findings. Moreover, it is probably not the one that the contemporary mind would select. The same data, which to Pfungst deny independent thought, to us suggest that Hans had learned to associate a stimulus or cue with his response of tapping. Today, we understand that animals learn by reinforcement of correct responses, by shaping responses to cues, and by the associative techniques used to control behavior. Pfungst, along with educators of his time, knew, with like assurance no doubt, that animals could be encouraged to demonstrate novel, humanlike behavior, but they did not know the principles by which animals did so. Educators of children were the first to see the importance of Hans's abilities. They emphasized that if we could understand how Hans performed, we might discover something important about how to teach children as well.

The notion has persisted that animals are more like human children in mental ability and that children can therefore learn better when techniques known to be appropriate for animals are used. Some might call this "training" rather than education. In our time the notion is applied to human and ape communication (see Chapters 10 to 12). Our expectations have shifted since Pfungst's and Hans's time, both in terms of the kind and degree of thinking we attribute to animals and the nature of the species we think capable of certain kinds of thought. Our own myths and models of the animal and child mind have changed shape since Pfungst's time.

Pfungst now describes his experiments, of which the following is an informative model:

"During the course of these experiments Hans wore his accustomed trappings, i.e., a girdle, light headgear and snaffle, and he either stood alone, untied, or was held loosely by the bridle either by the questioner or (though only in a few instances) by his attendant. The questioner always stood to the right of the horse, as Hr. von Osten had been accustomed to do. As a reward for correct responses Hans received from the questioner—and from him only—a bit of bread or carrot, and at times also a square of sugar. Never was a whip applied." [13]

Here are the summary data from an experiment in which cards with numbers on them were shown Hans. Von Osten was the questioner. No other human present could see the numbers. Hans was asked to tap the numbers. At times, von Osten knew the numbers *(Wissentlich)* and at times he did not *(Unwissentlich)*.

Method [14]	The Number Shown	The Number Tapped
Without knowledge	8	14
With knowledge	8	8
Without	4	8
With	4	4
Without	7	9
With	7	7
Without	10	17
With	10	10
Without	3	9
With	3	3

Pfungst concluded: *"Whenever the questioner knew the solution, nearly all of the horse's answers were correct, but when the answer was unknown to the questioner, the horse's responses were, with only a few exceptions, quite unsuccessful. Since the few exceptional cases must be regarded as fortuitous, the conclusion is warranted that the horse was unable to read the numerals without assistance"* (italics added). [15]

Pfungst used like designs to test Hans's abilities to recognize names and pictures and to perform the other tasks for which the horse was famed. The method is therefore of great interest to us. While Pfungst concluded from these data that the notion that Hans could read numerals without assistance was not warranted, we contemporary readers can understand more from the data than Pfungst did. The reason is that we have available inferential statistics, a tool that was not available to Pfungst. Pfungst makes, as does Freud, general conclusions from data, and the highest form of statistical analysis used is the taking of percentages. Percentages may imply but they do not permit inference. Modern readers can use their knowledge of methodology and statistical inference to test hypotheses and to find patterns in the data that inspection alone may overlook.

Look, first, at the method. "Without knowledge" trials were alternated with "With knowledge," and the same number shown in the "Without knowledge" condition was then reshown in the "With knowledge" (i.e., 8, 8, 4, 4, 7, 7, 10, 10, 3, 3). The methodological "check" that repetition is needed so that every number used in the "Without knowledge" condition also appears in the "With knowledge" is sound experimental design. It is a form of counterbalancing done to be sure that the number itself (8, 4) is not solely accountable for the results. But when the two conditions and their numbers are merely alternated, the possibility exists that Hans learned the alternation. After all, Hans, no matter how he did what he did, was demonstrably and undeniably a clever horse. A contemporary experimenter working with the understanding of methodology and inference not available to Pfungst would not alternate the trials. Rather, he or she would prepare a random pattern in which, like the original, each number would appear in each knowledge-condition equally often, but in which, unlike the original, these conditions would be presented in a prearranged order that eliminated the possibility of Hans responding to the pattern.

When we examine the location and quality of errors, we find the following discrepancies in the number shown and the number of taps. "With knowledge," 0, 0, 0, 0, 0; that is, Hans was correct every time. "Without knowledge", +6, +4, +2, +7, +6, letting + indicate more taps than the number requested.

While Hans was correct when von Osten knew the answer, he was never correct when von Osten did not know it. Moreover, he always overtapped but never undertapped. (This finding seems to be a trivial one, but let us think about the meaning, for it is often the seemingly trivial observations that lead to an encompassing explanation. Often, it is such clues that head our thinking and hypothesis-forming in the right direction.) If Hans was merely in error, the number of overtaps should have approximated the number of undertaps, but this was not the case. Hans *always* overtapped when he was *incorrect*. Does this oddity tell us something about Hans's method? Pfungst became alert to the possible significance of the overtapping. His explanation accords with the explanation of how Hans accomplished these tasks, without clearly seeing why this was so.

The investigator now moved forward to an intriguing set of experiments in which he attempted to determine which sensory system Hans was using. The natural tendency on such matters is to test visual properties first, for we, as human beings, parochially assume that visual cues are the most likely to be used. We now know better: we know that horses are attentive to slight motions, to verbal commands, and to cutaneous sensations, that is, those that move or stimulate the skin. We know that many animals have far more acute sensory systems, especially auditory ones, than our own, but, limited as we are to our own sensory systems, we assume that what we use is the norm, if not superior. Pfungst rose above this restricted view of life and checked the possibility that Hans used auditory cues, among others, when he signalled the correct answers. Hans showed no particular ability to answer verbal questions posed by von Osten speaking into Hans's ears. Other experiments suggested that Hans was not adept at tapping correct answers, or tapping at all, unless he was able to use his eyes.

HANS'S SENSES

"We began," Pfungst writes, "by examining the sense of vision, and in the following manner. Blinders were applied, and it is worthy of mention that Hans made no attempt to resist. The questioner stood to the right of the horse, so that the animal knew him to be present and could hear, but not see him. Hans was requested to tap a certain number. . . . Hans would always make the most strenuous efforts to get a view of the questioner. . . . I am using, in the following exposition, besides the two categories of 'not seen' and 'seen,' a third which I have called 'undecided.' A total of 102 tests was made in which the large blinders were used. In 35 of these, the experimenter certainly was 'not seen,' in 56 cases he was 'seen' and the remaining 11 are 'undecided'." [16]

Using our modern knowledge of how to examine data, let's make a table of these findings: [17]

Condition	Number of Times	Percentage Correct
Experimenter not seen	35	6
Experimenter seen	56	89
Undecided	11	18

Working without the benefit of statistical inference, Pfungst concluded: "It is evident therefore, that the horse requires certain visual stimuli or signs in order to make correct responses." [18] Techniques of statistical inference now available would permit us to calculate the probability of the percentages falling into the three conditions as they do. If the conditions didn't matter at all—if Hans was able to solve the problems without the position of the observer being of any importance—we would expect the number of correct answers to be distributed evenly among the categories. But the percentages observed were approximately 6, 18, and 89. The probability of these percentages occurring is less than one time in a hundred. We, therefore, have reason to suspect that the condition affected the results—namely, that the "experimenter seen" condition yielded correct responses far beyond the proportion we should expect. The inferential analysis does not prove that the condition led to the effect, for other kinds of experiments would be required to establish the logical relationship. But the analysis shows a highly unusual pattern of the distribution of percentages. The pattern is sufficiently unusual in terms of probability to compel us to look for an explanation.

"I would add that the horse—in so far as it was at all possible to decide—never looked at the persons or the objects which he was to count, or at the words which he was to read, yet he nevertheless gave the proper responses."[19] We human beings would swear, in other words, that Hans was not looking at the question or question-card. There are at least three explanations, each of which highlights a methodological lesson to be acquired from Hans. First, human beings are not necessarily competent judges of where or to whom or what other animals are attending; second, Hans may not have needed to look *at* the source of the question in order to locate the cues, if any, that prompt tapping; and, third, the human being's idea of the source of the question (Here, see the card!) may have nothing to do with the source of the answer that is effective in prompting Hans.

Pfungst made progress by elimination: he eliminated sound, for example, as the way in which Hans learned, and he discovered certain correlations—namely, that the presence of von Osten or Schillings, or, eventually, Oskar Pfungst himself, was necessary for Hans to answer properly. Pfungst's work progressed to an hypothesis that was not only consistent with his eliminations of certain types of cues and correlations, but one that predicted other situations in which Hans should be both successful and unsuccessful.

"Investigations of the other senses became needless, for I had, in the meantime, succeeded in discovering the essential and effective signs in the course of my observation of Hr. von Osten. These signs are minimal movements of the head on the part of the experimenter. As soon as the experimenter had given a problem to the horse, he, involuntarily, bent his head and trunk slightly forward and the horse would then put the right foot forward and begin to tap, without, however, returning it each time to its original position. As soon as the desired number of taps was given, the questioner would make a slight upward jerk of the head. Thereupon the horse would immediately swing his foot in a wide circle, bringing it back to its original position. . . . Now after Hans had ceased tapping, the questioner would raise his head and trunk to their normal position"[20] (my italics).

Pfungst now believed that there was a cue, that the horse responded to the cue and not to the card or source of the question, that the cue was most easily grasped when given by von Osten, that the cue could be given by others, and that the cue did not have to be planned by them. So Pfungst readjusted his hypotheses. To him, we assume, the adjustment was merely a continuation of his own line of thinking: to us, we see a stepping aside, a reinterpretation, a willingness to adjust a hypothesis.

A test of his hypothesis, that human beings were giving Hans cues from their posture, requires a different method of experimentation from that described when Pfungst first undertook the investigation. Nothing he did to this point tested this hypothesis. No data from his tests suggested that this idea should be considered anything other than a hypothesis to be tested. What Pfungst thought he saw as reported in the italicized section was an observation to be examined, not one observed or verified.

That Pfungst had a new hypothesis does not mean that he had completed work on the old ones. But he had learned enough from the old ones to see that the new one might be a better bet. Not everyone would agree that experimentation by hunch is the best way.

PFUNGST WATCHES VON OSTEN

Pfungst did not know which movement of the person was the necessary and efficient one, any more than he knew that any single one was the cause. He therefore turned to that most elemental and necessary of scientific investigation—observation. He watched von Osten. This is what Pfungst tells us:

"With regard to the regular occurrence of the movements noticed in the case of Hr. von Osten, I was, after some practice, able to note carefully their particular characteristics. This was rather difficult, not only on account of their extreme minuteness, but also because that very vivacious gentleman made sundry accompanying movements and was constantly moving back and forth. . . . It was much easier to observe these movements in the case of Hr. Schillings. . . . usually he would raise the entire trunk a trifle, so that the movements could be noticed from behind. Beside these, I had an opportunity to observe the Count zu Castell, Mr. Hahn and the Count Matuschka. All three made the same movements, though somewhat more minutely than Hr. Schillings, yet none was as slight as those of Hr. von Osten. I further noted that Count Matuschka and Hr. Schillings often showed a tendency to accompany every tap of the horse with a slight nod of the head, the last being accompanied by a more pronounced nod and then followed by the upward jerk of the head, in other words, they beat time with the horse."[21] [My italics.]

But, note, these observations prove nothing alone: they too lead to a new hypothesis, one that requires experimentation and testing. Pfungst now designed an experiment in which different observers (von Osten, Schillings, and Pfungst himself) tested Hans while being observed themselves by four persons (two of whom were Pfungst and Stumpf). The purpose was to determine whether Hans's movements preceded or followed the cue, but the results say a lot about the influences of the different persons. The investigators were asked to give the signal, and the observers timed the duration—the reaction time—between the signal and Hans's response to it by moving his foot circularly. In short, instead of observers watching Hans, observers were now watching investigators watching Hans.

The suggestive differences among observers remained uncontrolled for in the experimental design and hence unproven in a tidy fashion. Pfungst's idea was a hypothesis that merely derived from and supported the chief hypothesis: The data did not *test* the hypothesis clearly, much less unequivocally. That Hans's reaction times were random, or that he responded well to one person and poorly to another, does not show the falsity or truth of the hypothesis that he responded to human cues.

In the behavioral sciences, hypotheses that are really logical statements, circular definitions, or descriptions are all too often taken as satisfactory hypotheses. Because they do not admit to falsification through empirical means, they are not testable hypotheses. The "proving" of them is done merely by restating them, not by falsification. If a hypothesis cannot be shown to be false, it also cannot be shown to be true.

Hans showed his most successful reaction times to Pfungst, then to Schillings, and finally to von Osten. Now Pfungst struggled to explain his results, chiefly because the experiment he designed was intended to solve another problem. He now retracted by attempting to account for the results *individually,* arguing that Schillings's results occurred because Schillings tested first or that von Osten was unsuccessful because his cues were more subtle than Schillings's and Pfungst's. These explanations, though true, are of little explanatory value. They are examples of the backsliding often found in science when applied to behavior; yet they require experimental testing if we are to assess their probity. Worse, the many retellings of how Pfungst discovered that von Osten was giving unconscious cues miss the truth of the matter. They give Pfungst credit he does not quite deserve (he did not, after all, test that hypothesis), and they thereby fail to give him credit for what he did (which was to organize a logical set of deductive experiments regarding Hans's senses).

The problem, of course, lies not in the data but in the conditions under which they were collected. Data, however sound, cannot be interpreted from experiments of poor design, and Pfungst's experimental design was indeed a poor one, for it tested no clear hypothesis. Because the hypothesis to be tested was not stated clearly, the advantage of having observers watching investigators watching Hans was lost. Nothing of consequence came from this engaging study, and Pfungst was at a loss to know why. The power of statistical deduction coupled with experimental design was not yet available.

Pfungst's attention was drawn to the individual, especially to what the individual might be doing that served Hans as a cue to tap and to stop tapping. He noted that "if the questioner retained the erect position he elicited no response from the horse, say what he would. If, however, he stooped over slightly, Hans would immediately begin to tap, whether or not he had been asked a question. It seems almost ridiculous that this should never have been noticed before, but it is easily understood, for as soon as the questioner gave the problem he bent forward—be it ever so slightly—in order to observe the horse's hoof the more closely, for the foot was the horse's organ of speech. Hans would invariably begin to tap when I stooped to jot down some note I wished to make. Even to lower the head a little was sufficient to elicit a response, even though the body itself might remain completely erect."[22] Pfungst conducted a small experiment in which he stooped, or did not do so, thirty times. Hans was "right" twenty-nine times. Note that no controlled experiment was done here either, except for the counting of stoops. Once again, we have an hypothesis based on observation, but no attempt was made to formulate it in a way by which it could be proven wrong.

Undaunted, Pfungst now analyzed Hans's steps, which he did by drawing the patterns of the horse's hoof motion. He found that Hans tapped with a continuous motion that was completed—unless a sign to stop tapping occurred before he started the circle. Here is one reason why Hans overshot the number of taps and why he erred not in too few taps, but by too many. One aspect, of course, is that observers were likely to wait until Hans had tapped the correct number, rather than to consider a test to have ended when Hans had not reached the number of the answer (whereas when the correct number was reached, the trial, we suppose, was terminated). A second reason is that the movement that comprised the tap was a long one; once begun, it continued, and was a full and complete motion, not a tap alone. Figure 5.3 shows one of Pfungst's drawings of Hans's motion.

Hans raised his foot to position *b* from *a* in seeming preparation for a signal. (How, we may ask, did he come to anticipate the possibility of a signal? What nature of learning leads one to anticipate?) At or around point *b,* a cue was observed: Hans now completed the swing of the foot from *b* to *d.* Note that once this arc had begun, it had to end: otherwise Hans would merely keep his hoof in the air. The arc describes a full and complete behavior, one that is seemingly unreducible. Once point *b* is reached, point *d* must be reached as well.

Certain aspects of learning also require attention—for example, the seeming anticipation of the cue. A second consideration is the nature of the unit of behavior being observed. With Hans, for example, we find that the irreducible behavior was a rather large arc of the foot, not a discrete tap as we may assume from our concept of "tap."

Pfungst is inching toward the hypothesis that the human lowered the trunk of the body to cue Hans to continue tapping, while making the human trunk erect cues Hans to stop tapping. Unable to find a way to test the hypothesis directly, he looked for indirect support for the hypothesis, for evidence that, though not directly testing the hypothesis, was nonetheless not inconsistent with it. It was a false argument, though

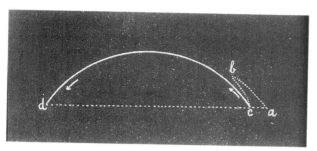

Figure 5.3 Oskar Pfungst, then a graduate student at the Psychological Clinic at the University of Berlin, investigated Hans II's abilities. As Hans answered questions by the number of taps of his hoof, Pfungust investigated certain qualities of the tap. "Hans would invariably begin to tap when I stooped to jot down some note I wished to make. Even to lower the head a little was sufficient to elicit a response, even though the body itself might remain completely erect. Hans would continue to tap until the questioner again resumed a completely erect posture. If, for instance, I stooped forward after having told the horse to tap 123, and if I purposely remained in this position until I had counted 20, he would, without any hesitation, tap 20." (Pfungst, p. 57). Here, Pfungst shows the parts to the tap (a–d), reading from right to left. a–b is seen by human observers as a slight hesitation.

understandable for Pfungst's time. That evidence is *not inconsistent* does not demonstrate that the hypothesis is the right one.

The kinds of experiments ranged from the arithmetical to those in which Hans indicated the location of objects, such as "blue cloth." With this sort of problem, Pfungst demonstrated that Hans was responding to more gross cues. For example, when Hans approached the "wrong" cloth, it was necessary only for von Osten to say "No" or "The Blue" for Hans to move on to an adjacent cloth. Hans appears to have learned this principle; namely, when the human speaks, try another, but similar stimulus, and do so until the reward of carrot or sweet is given.

Horses, like people, find objects in the last place they search.

WHY HANS WAS CLEVER

Two Cheers for Clever Hans and Clever Pfungst. Pfungst showed that Hans was indeed a clever horse, if not in the way that von Osten, Schillings, and the September Commission thought. For his part Pfungst can be seen to be a clever investigator. The horse learned to respond to delicate, but human-originated, cues. He did so with remarkable attention to human movements and by ignoring the numerous possible distractions in the world. Hans was able to transfer to a second person what he had learned about the cues from a first.

Pfungst's experiments and observations did not make Hans less clever, but clever in ways that Pfungst did not comprehend. Pfungst's experiments and analysis magnify the wonder of Hans's achievements. Some may say, "Ah, here is the evidence that the beast is but a machine, one made to tap out so long as the human dictates." Others may say, "Ah, here then is evidence in favor of education that stresses training of small steps, for seemingly small learnings add to big abilities. By training the small pieces, such as the tap itself, the horse comes to show what appears to be independent thought—here is a major principle of education, for here is the way that we should teach pupils." Yet others may state, "The true wonder of Hans is not that he learned to tap or nuzzle to human cues, but that he knew how to make the singular general—how to apply a principle. Hans was able to learn to accept cues from people other than Hr. von Osten; he learned the principle to try the next cloth when one cloth was indicated as incorrect. These abilities show that the learned aspect is but a small component of what the mind is able to accomplish."

It is a pity that Pfungst did not teach Hans the cues in an explicit way, rather than merely deducing their presence. Such an experiment would have provided the test of the hypothesis and given us a firm understanding of what Hans had learned.[23]

TO THE PSYCHOLOGICAL LABORATORY:
HUMAN SUGGESTION

In the laboratory, Pfungst used the principle uncovered in his work with Hans and von Osten to see if human beings could cue other human beings. Pfungst's reasoning was that, if "unconscious" cueing could be shown between human and animal, such cueing was also possible between one human being and another. "Suggestion" was the major explanatory principle used at those times to explain inexplicable behavior, and Freud had introduced the notion of the unconscious as the mechanism by which suggestion occurs. J. B. Rhine would explain like phenomena between the horse Lady and a human being as telepathy, and, later, that between human beings as extrasensory perception. Freud's objection to suggestion as an explanation, it may be recalled, was that it was but a term, a term that *explained* nothing. If, however, Pfungst could show the mechanism by which "suggestion" occurred, a major principle of psychology would be established.

Pfungst explains:

"The tests which are to be briefly reported here, were begun in November, 1904, and were carried out at the Psychological Institute of the University of Berlin. The purpose was twofold: first, to discover whether the expressive movements noted in Mr. von Osten, Mr. Schillings, and others, were to be regarded as typical and to be found in the majority of individuals,—and, secondly, to ascertain in how far the physical processes which I had noted in my own case and which I believed to lie at the bottom of these movements [he refers to the trunk-bending cueing between von Osten and Hans] were paralleled in, and confirmed by, the introspections of others." [24]

The general plan was for Pfungst to think of a number or name. Here is an example of what was done in the Berlin laboratory, in this instance between Pfungst and a person he calls Mr. von A.:

"I had one of the subjects (von A.) think of 'left' and 'right' in any order he chose. [The command was purposely given only in a general way: 'Think of "right" or "left." '] We had agreed that I was to try to guess the mental content of the subject's mind, but I was not to utter a word. Instead, I was to indicate 'right' in every case by an arm movement downward, and 'left' by a movement upward. To the subject I gave a fictitious but plausible reason for this. The behavior of the subject took the following course: that the first three test he moved his eyes to the right when he thought of 'right,' and to the left when he thought of 'left.' This was the normal expressive movement. In the fourth test, however, the thought 'left' was accompanied by an upward movement of his eyes. Two further tests again showed eye-movements to the right and left. In the seventh test with the idea 'left' the eyes moved first to the left and then immediately upward." [25]

However interesting it may be to know that one can interpret the movements of another, it does not show that Hans and von Osten used this technique. It is true that these two tests gave Pfungst the idea for a hypothesis, one he suggested both for horse and human and for the human subjects brought to the Psychological Laboratory, but demonstrations do not test hypotheses. We do not have experimental proof from Pfungst's work with Hans as to the nature of the cue.

Pfungst's researches on unconscious human communication and suggestion garnered little notice, either then or now. His reputation today rests solely on his work with Hans. Hans and Pfungst are an unread legend, a myth perhaps, as we will read in Chapters 10 to 12, wherein we find that some investigators misrepresent Hans's and Pfungst's achievements, sometimes arrogantly, by telling us that because their work on ape communication is ongoing and contemporary it is, thereby, too advanced to have made the errors made by Clever Hans's companions and investigators. For example, Pfungst is said by the lazy and inattentive to have exposed Clever Hans and Hr. von Osten by showing that unconscious cues were used. Neither Hans or Pfungst did what most contemporary writers said they did, while both achieved far more than these folk appear to know. Hans and Pfungst did not, alas, put an end to human self-deception about interspecies communication.

Pfungst exposed nothing, nor did he ever *demonstrate* that it was the leaning forward by people that cued Hans. He deduced this and showed that it was a possible means of communication with human beings, but he never attempted to demonstrate the efficiency of the cue of people leaning forward for horses. The critical experiment, one which shows that Hans was responding to people leaning forward and backward, and only to that, has yet to be done.

That Hans learned from von Osten and that von Osten worked with Hans by encouraging and correcting him are beyond doubt. Any parent or teacher will understand von Osten's pride when Hans began to answer correctly. Pfungst's significant contribution was to show that Hans did not have answers independent of his human companions and that he was able to identify small cues from people and to form generalities from what he had learned. Hans deserves to be remembered for his extraordinary abilities. But once it was proved that he did not do arithmetic, locating, or musical harmonics by himself, he was regarded as a no-account. If he was not a participant in a fraud, then surely he was just an animal-machine that responded when stimulated. Hans deserved a more accurate representation, one at least as laudatory as that enjoyed by the various chimps today who sign ideas and notions to human beings.

Von Osten died within the year of the publication of Pfungst's book. He had let no one else work again with Hans and, after Pfungst's book was published, he retreated with Hans, refusing to see the public or letting them see Hans. For Oskar Pfungst, whose fine researches with Hans were extended to investigate how human beings communicate to one another, even though they do not believe they are doing so, there was not even an academic degree as reward. In his book, his careful, detailed experimentation is sandwiched between an introduction and a supplement by his major professor, Carl Stumpf, the professor who seems concerned mostly with amending his own reputation.

What the modern reader learns from Clever Hans is that the beast was capable of the most impressive kind of learning; that human beings learn without their being aware of doing so; that human and animal may learn together, and by so doing may communicate, intentionally or not; and that to understand how this is done, we must turn our attention to how animals learn—that is, how they and we modify our behavior in response to aspects of our environment, whether these be the actions of others (such as von Osten's cues) or changes within our environment (shifts in the food supply, changes in the weather). What we should learn from Hans is the incredible sensitivity of the living being to its environment and of its abilities to change behavior in response to it.

The study of how animals modify their behavior will occupy us next, for it is just this kind of modification that gives living beings their unique qualities. Both Freud and Pfungst understood this to be so, and in their descriptions of their subject, both turned to a common theme: the contradictions of character.

FREUD, PFUNGST, AND CONTRADICTION

We began our analysis of Clever Hans as an introduction to behaviorism. Behaviorism can be practiced in many ways, but all these ways involve experiment. The story of Clever Hans emphasizes Pfungst's reliance on hypothesis, experiment, deduction, and generalization. Experimentation does not care for contradiction: Alternatives are to be tested and some discarded, not to be held alongside one another.

In contrast, by emphasizing the mythical quality of Little Hans's imagination, Freud explained that contradiction is an essential aspect of the mind; that it was not only possible, but humanlike, to love persons while wishing to be aggressive with them. Pfungst, who was also concerned with contradiction and its relation to the mind and behavior, spent the last part of his book not in describing Hans, but in setting out the arguments and psychology he used to judge that von Osten was not fraudulent but a teacher. (One can understand the potential for confusion.) He judged von Osten an honest man and ended his book by writing:

"To be sure, we must then reckon with curious inner contradictions in Hr. von Osten's character. But such contradictions are to be found, upon earnest analysis, in nearly every human character. And Hr. von Osten may say with the poet: *Ich bin kein ausgeklügelt Buch. Ich bin ein Mensch mit seinem Widerspruch!* "[26] (These are the words of the poet Schiller; very loosely: "I'm not some character in a well-wrought book; I am a man with his contradictions." Freud, by the way, uses the same quotation when describing aspects of the psychoanalytic method.)

As if to emphasize the contradiction, the two Hanses provide us with two seemingly contradictory ways of understanding ourselves and other living beings. The one way is the *construction* of myth, this being myth whose nature we grasp from the verbal evidence; the other is the *reduction* of behavior to its causative elements, the cue and the tap, a method applied, though not limited to, the nonverbal. The two methods would guide twentieth-century thought about human and animal nature. What Freud did for myth, Pfungst did for experimentation, by applying hypothesis-testing to questions of the nature of living beings.

I have told you of the future of von Osten and Pfungst. Other figures go their own way: Stumpf to be a professor for another thirty years; Schillings we know not where. But what of Hans?

His days as an experimental subject have only just begun and we are wise to withhold comment on the reliability and validity of experimentation and behaviorism until Hans's work is finished, as it is to be in the next chapter.

6   *Experimentation and the Experimenter: Clever Hans's Companions*

A M A J O R premise of this book is that observation and experimentation involving both animals and human beings alter the observer as well as the observed. Everyone known to us who came into Hans's life found that their relation to him and to one another changed. Hr. Schillings, who at first was a skeptic, became a true believer in the horse's ability to demonstrate humanlike intelligence. Professor Stumpf felt the need to defend his professional competence in many ways: by defending his work on the September Commission and by assigning the student Pfungst to clear up the mess created by the report of the Commission. Pfungst, we have described.

And what of Hans and von Osten? What became of them after the Pfungst/Stumpf report was issued? The report did exonerate von Osten of fraud, but in the process it destroyed Hans's popular reputation as a thinking horse. When the report was published, Stumpf's reputation was restored or, at least, the Commission Report was forgotten, and Pfungst carried on with his human research on suggestion. For Hans and von Osten, however, life would never be the same.

In this chapter, we learn something more about von Osten and his relationship with Hans and Pfungst, about what became of Hans after his master died and how Hans's new owner, Hr. Karl Krall, developed Hans's talents. The chapter is both a parallel and an extension of the previous chapter, sketching information regarding some of the people mentioned and telling us of Hans's new career. Its purpose is to show how human belief shapes our theories and myths about the mind and behavior. I do not write of fraud or gullibility, but of how our expectations compel us to stack seemingly pure research questions toward the answers that fit our perceptions, at first trivially, but eventually blatantly. In short, the meaning of an experiment is not completely known without knowledge of the investigator's motives and perceptions.

ABOUT HR. VON OSTEN

Wilhelm von Osten was born on November 30, 1838.[1] We know that his father kept a large estate, a fact that may account for the title "von" and the use of the title "Baron" by some writers of the time.[2] He studied at Köningsburg and Danzig, great universities then and now. He taught school in various places throughout what is modern-day Germany. In 1866, when he was twenty-eight years old, he appeared in Berlin where he established a residence at number 10 Griebenow Strasse. It appears that he owned the building, but he was content to live in two rooms. Some time around the year 1890 he was working with the horse Hans I, who, at this time, was twelve years old. Hans II, the horse whom we know as Clever Hans, makes his appearance in 1900. He is thought to have been then five years old.[3]

Pfungst's book appeared in March 1907. Von Osten read it and felt himself abused, exploited, and physically ill. But he directed his anger not at Pfungst, but at Hans who, von Osten believed, had somehow deceived him. Von Osten said that the horse's deceitful behavior had made him sick. And he had become very sick indeed with what the physician diagnosed as cancer of the liver.

Von Osten even placed a curse on his one-time companion. According to Karl Krall, who was with von Osten in the last days of his life, von Osten hoped that *"Hans spends the rest of his life pulling hearses!"*[4] Von Osten continued to accuse Hans of treachery while he took to his bed. Never forgiving Hans, and still blaming Hans's perfidy for the sickness, von Osten died on June 29, 1909, presumably of cancer. He was seventy years old.

Hans, it seems, had been promised to Karl Krall, a resident of Elbersfeld, said to be a jeweler with an interest in establishing communication with horses as a means of understanding their minds and demonstrating that animals are thinking beings, just as human beings are.[5] Krall made no pretense of being an uncommitted observer. He was convinced that animals think, and so he acquired Hans in order to continue the horse's training. In addition, he thought that Hans might serve as a teacher for other horses, thereby participating in animal–animal communication. Before we fault Krall for his commitment, we should remember that some very good science has been produced by those who are already committed to a point of view and argue for it. Galileo is a prime example. Krall persisted in his view, even though he was well aware that some people thought that trickery had been used to teach Hans. Some thought von Osten was a trickster; others thought he was innocent but foolish. Had Stumpf and the September Commission done their homework, they would have assessed these doubts, for Stumpf, at least, had already been involved in examining one such piece of trickery. Oddly, however, the September Commission makes no mention of this in its report, nor does it appear that Stumpf used what he had learned from this incident when examining Hans.

RENDICH AND NORA

Among the doubters was Signor Emilio Rendich, a painter. Sometime before 1905, before the appointment of the September Commission, even before the notoriety that came to von Osten and Hans II, and before Stumpf appointed Pfungst to investigate Hans's abilities, Rendich had watched Hans's performance and had come to this conclusion: that von Osten constantly watched Hans's hoofs, and that Hans constantly watched von Osten *leaning forward and backward.*[6]

Rendich argued that Hans had learned, not necessarily to tap, but to *stop* tapping. When in the appropriate situation—for this Hans, the horse, being in the courtyard—he learned to tap his foot until a sign was received; that sign, suggested Rendich, was von Osten's leaning. To demonstrate the plausibility of this view, Rendich undertook to train another animal, his dog Nora, to do the same thing. (I chose a picture of the Clever Nora as the frontispiece for this book for ironic reasons. While she learned to do the same acts that others thought evidence of outstanding intelligence when done by a horse, she is seemingly unknown to our researchers and storytellers.) Rendich and Nora demonstrated that Hans tapped until he received a cue signaling him to *stop* tapping, and that the cue was that of a familiar human being leaning forward and backward. Together, Nora and Rendich tested the hypothesis *directly* by showing that the technique could be learned by another human-animal pair. They did so both before the work of the Commission and before Pfungst's analysis, an analysis which, in any event, failed to test its hypothesis directly.

Professor Stumpf himself visited Nora before his appointment to the September Commission. The dog showed herself able to communicate in much the same way as Hans; however, Rendich made clear that she was unable to do so before he trained her and that he had concentrated on teaching the dog to respond, and to cease to respond, to the movement involved in his leaning forward.[7]

I also chose Nora for the frontispiece of this book because she serves as a reminder of how people create the psychologies of animals from their own images of themselves. We do not want to forget that, although Rendich offered a reasonable explanatory hypothesis regarding von Osten and Hans and tested it afresh with the assistance of another species, he and Nora also demonstrated that one aspect of the communication possible between human being and beast was a matter of training. Latter-day investigations of human and animal communication would do well to emulate Rendich's approach: to test the hypothesis directly rather than merely assume it to be true when a few alternatives have been eliminated.

HANS'S NEW LIFE

Krall had attended many of the sessions with Clever Hans in the courtyard in Berlin, as attested by photographs of the period in which Krall and often Schillings are shown to be near von Osten and Hans. We know from Pfungst's account that Schillings was present at the earlier demonstrations, while Krall is not mentioned at this time. Krall came to know von Osten so well that he was to be given charge of Hans. Krall is, in fact, our primary source regarding von Osten's life and the many demonstrations and tests in which Hans participated. Krall's book on the subject—a book in physical substance worthy of being a doorstop—exists only in its original 1912 German (script) edition. It is divided into two sections, each of approximately two hundred and fifty pages. The first section describes Hans's career at Elberfeld, but emphasizes Krall's attempts to communicate with two horses in his stable, Zarif and Muhamed. The second provides a careful, even tedious, accounting of demonstrations by Hans, an account of von Osten's life, and comments on Pfungst's analysis.

Krall recounts how von Osten undertook to test Hans's visual perception. Astonishingly, no one who had studied Hans, especially Pfungst, appears to have asked questions about Hans's visual abilities. Is the horse eye built in a similar fashion to the human eye? Does it focus at a different distance—is the horse near or far-sighted in our terms? Does the horse have a retina with color receptors? What is the nature of its depth of field? What is the visual world, the umwelt, of the horse? Is the horse capable of focusing on the materials presented by the people? At which point in the distance is accommodation at a maximum?

Establishing the horse's visual world would be among the first tasks on a modern investigator's agenda, whether behaviorist or phenomenologist, but the nature of the horse's visual thresholds was not yet an aspect of Stumpf's or Pfungst's experimental procedures. Somewhere around 1907 von Osten and Krall, seemingly later and separately, undertook psychophysical examinations of Hans's visual acuity by using the standard forms of measurement then available. They are not unlike the preliminary methods used today by optometrists.

According to Krall, von Osten first measured Hans's ability to distinguish the letter "E" as follows: the letter was placed horizontally on its spine and the spine was placed on top. Figure 6.1 shows Krall's picture of von Osten showing Hans the letter. Note that the distance the "E" is held from the horse's *nose* is approximately the distance that would be expected from the eye of a young human being. Hans is asked to tap his hoof to indicate whether the "E" is "closed" or "open." If open, presumably the horse sees the prongs on the spine of the letter "E" separately; if closed, the prongs are seen as blurred, and the "E" is closed. Von Osten concluded that Hans's length for accommodation was 2.62 times that of the human being—that is, two and one-half times as long. The data demonstrate that, compared to the human being, the horse sees most clearly at about two and one-half times the appropriate length for people. If my accommodation point is 10 inches from my eyes, that of the horse is approximately 25 inches. Stimuli closer and further away become progressively more blurred. Modern data suggest that the ac-

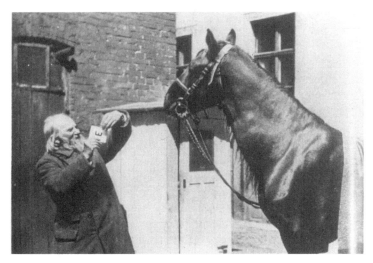

Figure 6.1   Von Osten took part in tests administered by his friend Karl Krall to determine Hans II's visual acuity. Herr von Osten uses the opticians' E to measure the distance required for accommodation of the lens. The results were inconclusive, although it now seems evident that the distance for a horse is far greater than that for a human. Hans II was, like all horses, far-sighted by human standards.

commodation point for horses in general, not for Hans in particular, is far greater than this number. The length for the human being, and probably for horses, would be variable, for age has a powerful effect on the ability of the muscles to accommodate the lens to distance.

Von Osten investigated Hans's visual astigmatism, if any, by using curved lines and asking Hans to tap whether they were straight or curved. He asked Hans to examine the well-known Müller-Lyer illusion (shown in Figure 6.2). We know that human beings and many, if not all, species ever tested, think the top line to be longer than the bottom line. All mammals tested on it see a like illusion. Hans, however, when asked, tapped, "No, neither line is longer,"[8] thereby giving a physically correct but perceptually wrong answer, at least for all other animals tested during this century. Either Hans's perceptions were distinctive, or whoever cued him thought that only people saw the illusion.

The ability to focus, distinguish, and count, which taken together are rather complex visual abilities, are tested by means of cards. For example, one card has 12 dots and the other 9. When placed 10 to 15 centimeters from his "face," Hans was able to count the 9 dots in one second and the 12 in two seconds. The difference in time suggests that Hans saw the cards as different from one another. Hans's vision at a distance was excellent, at least compared to human sight. When a neighbor was asked to hold up differing numbers of fingers while standing at the top of the courtyard next to an upstairs window, Hans correctly tapped that five

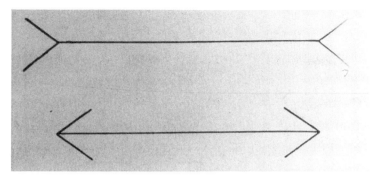

Figure 6.2 The Müller-Lyer illusion in which the bottom line appears to be shorter than the top line, even though the lines are of the same length. The causes of the illusion are complex. The illusion was shown to Hans II, and, by tapping "yes, equal length" or "no, unequal length," he reported seeing the illusion in the way human beings do.

fingers were shown. In a test of color vision, von Osten was pleasantly surprised that Hans discriminated a blue-green from a green, colors that the attending human beings themselves had difficulty in discriminating under conditions of low illumination. Hans demonstrated splendid visual ability in other tests, such as discriminating the number of candles lighted on a table. This assessment was a standard way of measuring the threshold for brightness more than for counting. It is often used to demonstrate the validity of Fechner's law relating the physical to the psychological world.

Our best guess is that Hans would see the images on many cards presented to him as only blurs, for as judged from photographs, the human observers appeared to be holding the cards at a distance from Hans's nose appropriate to the distance comfortable for a human person to see most clearly. From pictures of the crowds attending the demonstrations, or from those showing only a few people gathered around, it would seem that the distance kept by these observers from the horse was nearer to the maximal distance of focus for the horse than to the minimal. By human visual standards, Hans's thresholds were excellent indeed, and his ability to focus was, to a human, farsighted. These measurements of Hans's abilities—these particular ways in which the visual stimuli are presented—suit human standards and are judged against human abilities.[9]

AT ELBERFELD: MUHAMED AND ZARIF

At Krall's home in Elberfeld, Hans met with unfamiliar situations, requirements, people, and assignments. Here he met new companions, the horses Zarif and Muhamed, brought to Krall at the ages of two and two

and one-half years, respectively at the suggestion of the military.[10] Figure
6.3 shows the horses who would be Hans's companions for the remainder of his life.

At Elberfeld Krall commented on Hans's treatment by his former master. But Krall made some criticism of Von Osten as a master only after
von Osten's death and Hans had been given to Krall. While it was true,
writes Krall, that von Osten did not use the whip to "encourage" Hans
in public, he "probably" did so in the private instructional sessions. Nor
was Hans well cared for, continues Krall, as he seemingly warmed to the
topic: his stable was closed and dirty; he was rarely combed and cleaned;
the food offered him was old and rotting. Perhaps, he surmises, Hans's
notorious bad temper (he bit Pfungst often enough, and von Osten found
him moody and unpredictable on some days, so much so that on those
days Hans could not be expected to learn or to perform) was related to
von Osten's lack of care and concern for him.

Krall, like Freud in the case of Little Hans, announced clearly the
hypothesis he was supporting and testing, thereby leaving it to us to

Figure 6.3   Zarif and Muhamed were the stars of the Elberfeld stables where
Hans II resided after von Osten's death. At first, Krall expected Hans to teach
the other horses, but his work proved to be unnecessary. Zarif and Muhamed
were rapid learners, perhaps because Krall invented a new "writing board," so
organized that the horses could pick a letter by identifying its column and row,
using one hoof to signify the row and the other to signify the column. These
two horses are reported to have used the writing board to hold conversations
with one another.

judge the success of the work. Krall's hypothesis was not implicit; it was stated clearly. Animals are capable of self-initiated thought, he believed, and he set out to document it: At the same time, Krall grasped the impossibility of fully knowing the mind of another. He believed that one cannot make a conclusion about the minds of animals or people, for we cannot know beyond our own capacity for self-understanding. His task, he said, was to show the nature of such thought, and of that nature he would conclude: "I have proven [there to be] no cues—there is only one explanation that remains—the self *[die selb]*."[11]

Krall suggested several modifications and improvements on von Osten's approach. First, he provided a slanted platform on which the horse was to tap. The purpose was to help the human observer know when a full tap had been made, as opposed to a half-tap or a pawing of the ground. Second, a new code was devised (after a false start) to permit the horse to tap the letters of the alphabet. The newly devised "codeboard" is shown in Figure 6.4. The code required that the horse tap with both hooves. The right hoof determined the rows 1 to 6; the left the units 10, 20, 30, and so on. The letter "E" is therefore one right followed by one left tap.

The spelling of the German word "Mann" would therefore be as follows:

| Right hoof | !!!!! | ! | | !!!!!! | | !!!!!! |
| Left hoof | ! | | !! | !!!! | | !!!! |

	1	**2**	**3**	**4**	**5**	**6**
10	e ℭ	n ℜ	r ℛ	ſ ſ ẜ	m ℳ	c ℭ
20	a 𝔄	h 𝔥	l 𝔏	t 𝔗	ä 𝔄̈	ch
30	i ℑ	d 𝔇	g 𝔊	w 𝔚	j ℑ	ſch
40	o 𝔒	b 𝔅	f 𝔉	r ℜ	ö 𝔒̈	y 𝔜
50	u 𝔘	v 𝔙	z ℨ	p 𝔓	ü 𝔘̈	
60	ei	au	eu	r ℨ	q ℭ	

Figure 6.4 The "writing board" used by Zarif and Muhamed, among other horses in the Elberfeld stables. By tapping both hooves, the horse indicated the column and row of the letter wanted. Evidently, the horses did well at German script as well as French, for when the French-speaking Eduard Claparède visited, they replied in German to his questions posed in French.

Note that the shift of the tap from the left to the right hoof indicates the end of the code for each individual letter. I mention this because, if an observer hears only one hoof tapping it becomes the perceptual task of the observer to interpolate the end and beginning of letters, words, and sentences. This was required of von Osten's technique, although no commentator appears to have noted that it was the human being, not the horse, who was compelled to separate the sound of the taps from one another. What Hans tapped and what the human heard may, or may not, have been identical. As it is, the new procedure still required the observer to determine the end and beginning of words, although presumably not those of the letters. Von Osten's technique left ample room for the human observer to separate or organize the taps. Krall's code was an improvement in that the left tap should have signaled the onset of a separate letter. However, as left taps themselves required organization as to whether there was one or two or three in sequence, the separation to the human hearer remained imperfect.

The horses Muhamed and Zarif became adept at using the letter-board to answer questions put orally, mostly by Krall, but sometimes by a Dr. Schoeller. We learn that Schoeller did the teaching and demonstrating when Krall was away and later assisted in an important experiment in which Krall's questions and directions were conveyed to the horses by telephone. The horses' original learning of the letter-board required only two weeks, with two instructional sessions of two hours daily. Krall's description contains two curiosities:

First, when given a number such as 126, the horse (tapping on the arithmetic board) tapped out the 6 first, then the 2, and then the 1. Human children sometimes read numbers from right to left when learning numbers, but most who write this number as adults do so from left to right, as they conceptualize first the 100, then the 20, followed by the 6. But, in spoken German, the number "26" is pronounced not "twenty-six," as in English, but "six and twenty." Is the horse a natural German-speaker, or is it responding to the way in which a human speaking German expresses words?

Second, among the supporting evidence for the idea that the horse is "self-thinking" is the observation that the horse misspells, or spells with variants, in ways that approximate the sound of the spoken word. For example, Muhamed and Zarif spelled "pferd" (horse) as

bfert, bferd, faart, fared, pasrd, pfrde, vard, nechrd, tfert, sdfert, bfahd, bffet, fadrb, fdaerp.[12]

Are these misspellings? Or are they mishearings based on a human observer? Or a wanting to hear by the observer? Or simple "mistakes" related to the location of letters on the letter-board?

Consider, for example, the proximity of letters on the board to one another. The horses appeared to mistake "v" for "au," an "error" that may be related to the similarity of pronunciation. (An example, in English, would be saying the letter "f" without adding the vowel sound of the "e" before it; our natural way of saying the letter "f" is "ef," just as

in horse-German the "v" becomes "vau." Is the horse misspelling, mis-
hearing, or mistapping?) Notice that in the letter-board, wherever placed,
the au, v, and eu differ by one tap. Indeed, most letters differ from the
eight surrounding letters by one tap, thus offering many more possible
"misspellings" than one might suppose.

Zarif and Muhamed were able to recognize portraits of people. These
portraits were of the then Kaiser and Kaiserin, the philosopher Schopenhauer, and the poet and writer Schiller. Here is a sample of the dialogue
between Krall and the two horses when the pictures were shown. Krall's
quotations are words, of course, and those of the animals are the letters
tapped out, as decoded by Krall and his associates: [13]

Krall	*Replies of M (Muhamed) or Z (Zarif)*
Here are pictures of the Kaisers	—
A picture of the Kaiserin	M: keisr
That is not the Kaiser, it is the Kaiserin!	M: in
The picture of Schopenhauer	Z: ksobnd
False!	Z: ndauer
Muhamed? False!	M: p schobndvn M: r
Schiller picture False!	Z: schiln Z: schooo
Schiller! False!	Z: ilhz Z: lr
A picture of Stumpf Zarif!	Z: tnurrrz Z: st
Again!	Z: r
False! (Krall writes a "u")	Z: umf
Muhamed, What is his name?	M: sumf

Let us keep *our* eyes focused on the issue: To what is the horse responding? Is it responding to some knowledge that can be expressed
across species, and to human beings, by using the tap and the letter-
board? Or, to be skeptical, can it be that the horse has learned two sig-

nals, each with two variants: left and right, start and stop tapping. Then what is the signal? Leaning forward, as was shown with the dog Nora or the horse Clever Hans, or the movement of the human eyes, as is so often a significant and relevant stimulus for dogs, or some other movement, given intentionally or unintentionally?

Investigations into the mind of the horse continued. Krall and a retired school inspector, Dr. Grabow, selected from the pocket of Grabow a set of small cards with simple problems in addition on them (e.g., $2 + 3 = ?$). Krall went to one corner; Grabow showed the card. The horse then tapped out 2 and 3 and, in answer to the request to add them, tapped 5. Notice that Krall remained in the area, in the line of the peripheral sight of the horse.

In another test Krall had a box containing boards with arithmetic questions on them. The horse tapped an answer to one or two, then knocked over the box and was defiant. He was bored, Krall said, and when horses are bored or out of sorts they will not do the problems. That was often the problem with Hans when he was with von Osten, says Krall. Von Osten rarely varied the problems but had Hans do the same task day after day. Did this lead to Hans's learning a series of answers regardless of the questions asked?

Another time, Dr. Schoeller conducted a test in which Krall spoke to the horse by telephone. The evident purpose of the telephone was to do away with the possibility of visual cueing. Telephones in those days had a separate voice-piece and ear-piece, so Dr. Schoeller held the ear-piece to the horse's ear as Krall's voice came over the line. The following conversation took place:

> *Schoeller:* "Who is that [speaking]?"
> *Muhamed:* krl [Krall?]
> *Schoeller:* Who?
> *M:* vatr [Vater?]
> *Schoeller:* "What did he say?"
> *M:* m gbn

Schoeller reads this as "mohren geben."[14] (Give me a carrot.)

Of course, since Schoeller was near the animal, the horse's seeming recognition of Krall's voice may not be astonishing. Moreover, Schoeller showed more than a little imagination in his crediting the horse with replying in translatable words.

Another time Krall asked Muhamed, "Why is Zarif lazy? "Because he doesn't want to know," answers Muhamed. "What should I do with him?" Answers Zarif, "Be nice to him." Later, Krall asked, "What should I do with him?" Zarif answered, "Beat him." Evidently, all was not well between Zarif and Muhamed.

Some tests fail, says Krall, because the experimenter takes away the soul, the relationship between animal and human being that is the essence of the means of communicating. This is why, he suggests, tests in the laboratory fail. The laboratory is sterile, and the experimenter cannot make contact with the animal mind, nor the animal with the human mind. Here Krall introduces an issue that concern all who would understand animals. It is an issue that is of vital concern to us throughout this book: Can animals in confinement be expected to show their mental abilities? Can Kaspar Hauser, raised in a prison, or the wolf-girls, raised by a nonhuman without soul, only be brought forth in an environment of human beings? When an ape does not communicate, can it be because the ape is behaving under controlled conditions, where control itself diminishes the animal mind? Or is the control necessary, even vital, if we are not to be fooled, as was von Osten, as to the cause of the seemingly impressive behavior?

Krall, who was not to be alone in his view, believes that the experimental approach with its controls and countings destroyed the quality of true and meaningful communication, and thereby presented the abilities of the animal in the weakest of lights. Krall appears to have believed that Hans, Muhamed, and Zarif had profited from instruction, not that they had been trained. To support his position that animals show self-initiated thought, he turned to the evidence available on dogs and chimpanzees. Before we leave the horses and their response to this special kind of experimentalism, let us consider the visit to Krall's horse-farm of Eduard Claparède, the distinguished educational psychologist. What could be more natural than someone interested in the education of children showing an interest in the education and intelligence of the horses of Elberfeld? His descriptions of his visits to Elberfeld give us valuable insights into the conduct of experimentation and the nature of the people at Elberfeld, for he was an outsider commenting on Krall and his colleagues.

CLAPARÈDE'S VISITS

Eduard Claparède (1873–1940) was the foremost French-speaking Swiss psychologist of his day and thereby one of the major Continental figures in psychology. His work remains well known, especially in Geneva, where he held an academic post. He was to Geneva what Stumpf was to Berlin and to French psychology what Stumpf was to German-speaking psychology. Claparède read both Krall's and Pfungst's books, and in 1912 he described the clever horses for the benefit of the French-speaking public.[15] The first paper he published on the matter is chiefly a review of what Pfungst found and of the information to be found in Krall's book. Here Claparède focused on the new alphabet code-table that required tapping by each hoof (the right for the units and the left for the tens) and on the curious but perhaps instructive spelling mistakes (ferd or pferd). He also described two other horses in Krall's stable, Hanschen and Berto, who were little mentioned by Krall.

Hanschen did not show outstanding ability, but there may be a reason. His consistent "error" was to tap 53 when 35 was the correct reply, or 27 when the answer was 72. Remember that in German such a number is expressed as "Three and fifty" but in English as "fifty-three." Did

Hanschen merely read backwards, as human children sometimes do? Or, if he had only learned to tap and to stop tapping at a signal from a human person, was he responding to cues from a German speaker who thought of 53 as "three and fifty" or an English speaker who spoke of "fifty and three?"

If I, the trainer, whether knowingly or innocently, think the correct answer to be 28, and if the horse misses my cue during the first tap and moves beyond the 2 by continuing to tap, might I not encourage the animal to continue to the second number (assuming that it is larger) and then re-cue the animal to tap the first number correctly as the second? In this way, human beings may think the horse is merely reversing, rather than being in total error. And, if this system is prompted, knowingly or not by the human, might it account for the "mistaken" letters that appear when the horse taps to the alphabet-board where, as we have noted, each letter is surrounded by a large number of letters each of which differs from the one only by a single tap?

Berto provides a potentially more interesting set of data, for Berto was congenitally blind.[16] Clearly, a horse without sight is a horse that cannot be cued by the visual movements of others, although slight sounds, sighs, and rustles of cloth might be sufficient. Tests made of Berto's visual acuity support the view that he could not see. Yet Berto was able to do simple calculations, such as $5 + 1$ and $5 + 2$, though nothing more complex. He may have learned this ability through questioning by the use of a code represented by fingers touching his coat, much as did Laura Bridgmen, who was taught to use tactile senses to discriminate and recognize letters of the alphabet. We have no more information on how the tactile sense was used to teach or to communicate with Berto. We are left with the knowledge only that he was trained by this skin sense through which he could give single-tap answers to arithmetical problems.

Caparède suggests several explanations for what he had read about Krall's horses: the animals were of superior intelligence; they were responding to the unconscious signals from human beings; they were telepathic (a characteristic attributed to the U.S. horse, Lady, it will be recalled); and they were taught to use some device to aid memory.

As for the first, Claparède quotes from a letter to the French daily, *Le Matin,* seemingly published in 1910 or 1911. The letter is signed *"artiste qui s'est specialisee dans la presentation d'animaux savants."* All we know of the writer is that it was a woman and an "artist who specializes in the presentation of animal savants." She wrote to say that, while working in Berlin in 1909, she wanted to prepare a music-hall production to show horse intelligence. She therefore used telemetric means, that is, a radio receiver placed under the bridle of the horse and controlled by the human being by means of a transmitter kept in the pocket. When the horse was asked to locate an object, the trainer waited until the horse was near the object selected, then signaled the animal, leading it closer and closer to the "correct" object. She was convinced, she writes, that Krall of Elberfeld used something like this technique, so that the horse was merely the recipient of "wireless" (radio) communication. This suggestion of a telephonic device appears elsewhere, but apparently no one checked on this matter—-or at least no observations on it remain available to us.[17]

The idea that signals were given, whether knowingly or not, was an attractive one, writes Claparède. After all, the horse might be sensitive to certain auditory frequencies that human beings cannot bear. It would appear that here Claparède is referring to what we now call ultrasonic sounds—namely, those sounds that lie above the auditory threshold of human beings. Claparède was aware that some dogs could hear tones and whistles at frequencies higher than those useful to the human ear. He therefore wondered whether some sounds might be made, perhaps by a human being, that lie above the threshold of listening human beings but that are within the range of the horse. This idea, too, does not seem to have been examined.

As for telepathy, Claparède recognized that telepathy would refer to a situation in which the horse was able to "receive" what the human being was thinking, rather than what was being heard or seen. Claparède thought this explanation to be unlikely and unprovable, except by those who claimed that, when all other hypotheses were shown to be wrong, telepathy remained the explanation. (Telepathy was to Claparède what suggestion was to Freud.) That the horse, or these horses at least, might have superior intellect was possible, writes Claparède. He mentioned a number of unusual descriptions of animals, dogs, horses, and pigs, who were reported to have shown unusual mental ability, but proof was lacking, he concluded. Memorization, the next explanation, was possible. This technique would explain the nature of some of the "mistakes," but data were also lacking. Nonetheless, Claparède observed that if Krall had indeed found a way to permit animals to communicate with people, Krall's name should be "put above that of Darwin."

Remember that Claparède had not yet met the horses or Krall, nor had he studied the abilities of either firsthand. It is time for a visit, a seance, or rather a set of four seances. The dates are August 30 and 31, 1911. Let us join them as we may see how an outsider judges the situations described by Krall. This view is not unimportant, for to date we have no independent evaluations of the events: Claparède is the Pfungst of Krall's horses. What, we may ask, has come of Hans? It seems that after he is used in an attempt to teach the other horses, he becomes background. At least, Krall has no more to say about him.

At six in the evening, Claparède met Zarif, the horse, who demonstrated his mathematical abilities with Krall in attendance. Later we learn, incidentally, that a stableboy was also present, but since he was of little account, in Krall and Claparède's view, he and his activities are mentioned only in passing.

The text is not clear on this point, but it appears that Claparède set the problem in French, and Krall then restated it for Zarif in German.

Both Hanschen and Muhamed ("the genius of the troop," writes Claparède) did a few square roots. For example, to the problem

$$\sqrt{1296 - 81} \times \sqrt{144 - 49} = ?$$

Muhamed answered 115 which is, of course, false. Krall, who spoke to Muhamed in the familiar, "du," asked for a reconsideration. Muhamed tapped 25, then 125: close, for some of the individual numbers are correct, but the correct answer is 135.

Turning to spelling and naming, Muhamed spelled Claparède's name as *Klapard* (not bad at all, considering Muhamed's seeming grasp of German but not of French). On the second try, Muhamed suggested adding the e *(Klapared)*. The name of Monsieur Tanski, who was also around somewhere, was spelled *Teauske* by Muhamed.

By the time of the third seance, Muhamed was correctly answering such problems as

$$\sqrt{614656} = ?$$

Yet he was giving a set of wrong responses to

$$\sqrt{9834496} = ?$$

The answers he gave were 43, 73, 267, 34, 74, 84, among others. (Something may be learned by your tapping these numbers to a stranger who is entrusted with the decoding of them. In this way you approximate what the horse does, and how the human hears it.)

Zarif showed himself to be the better linguist. Consider this exchange:

> *Krall:* Where is Zarif? (The question is reported in French, but as the account is taken from Claparède, we do not know whether the *original* question put by Krall was in German [and here translated] or in French.)
>
> *Zarif:* iig

Claparède read this as *Ich,* or *moi.*

Later, Zarif showed himself competent in French, at least if only numbers were to be tapped:

Krall asked *(in French):* 12 + 11?

Zarif responded: 23

Claparède wrote in his notes: *"très bonne."*

At the conclusion of the two days of seances, Claparède reviewed the ways by which the horses might be using their intellects. He decided that the truth required a mixture of them; that not one, but some combination, was used. A number of French physicians and professors agreed, and Claparède took their written assurances as evidence that he was headed in the right direction.

Six months passed, during which Claparède's first paper was published. Now it is time for a second visit. This visit occurred on March 26, 27, and 28, 1913. In attendance was Dr. Modzelewski, who, fortunately, provided an appendix of data of much more interest—chiefly because Krall was absent on the first day. Here are the data for March 26, a day when Krall was away: [18]

Horse	Time of day	Mistakes	Right, with help	Right
Hanschen	Morning	42	5	5
	Afternoon	47	6	2
Berto (blind)	Morning	23	5	2
		24	10	13
Muhamed	Morning	0	2	0
	Afternoon	46	6	4
Zarif	Afternoon	9	2	3
		200	36	29

The table refers to general arithmetic problems—for example, how much is 2 and 6? Note that some horses, in some sessions, received many more problems than did other pupils. One can guess that the more problems a horse answered, the more he was likely to get, while recalcitrant horses, such as Zarif, appear to have tired their questioners and so had fewer tasks to answer.

Upon Krall's return the next day, the horses did far better and returned to their state of *"très bonne."* Modzelewski had no doubt as to the meaning of this, but Modzelewski was not heard of again. Unfortunately, we do not know the location of the stableboy during all of this. After Claparède left, Krall reported that the very next day a retired army major came to see the horses, and that the horses performed for him admirably. This was evidence, wrote Krall to Claparède, that one's attitude toward the animal made all the difference in being able to communicate. The major was sympathetic, reported Krall; hence, the horses were helpful. Who can deny that any communication is enhanced by empathy? The rebuke is clear, but whether it was of Modzelewski or Claparède is not clear.

Our knowledge of these horses, as well as of Krall and his associates, now ends, not because nothing further occurred, but because this part of the world would soon be at war and would take with it almost all else into World War I. Draperies cover whatever happened at Elberfeld in the 1920s and 1930s, hiding all except for a few, unexceptional views. Krall became well known for his ability to train horses to do precision movements, especially those done by horses in time to the movements of other horses.

Krall's success as a horse trainer should not compel us to think of him as having done nothing more with Muhamed, Zarif, and Hans than training them in some way that merely showed that the animals were thoughtful and calculating. By training the horses—much as people today train chimpanzees and gorillas to use the movements involved in sign language—he was giving the animal and person a way to communicate.

SOULS AND MINDS OF THE ANIMALS

Krall discussed the ways in which various animals have been able to communicate with human beings. His definition of communication was surely less rigid than what is suitable today, for he included the parrot's imitation of speech as a primary example. Parrots and other birds, such as the mynahs, emit sounds that many human listeners can hear as human speech. Nonetheless, Krall began with the belief that, insofar as we can know, animals have minds that function in ways demonstrable to the human being. In asking Muhamed to comment on his fellow horse Zarif, Krall took us an additional step, but it is a step measurable in light-years. By so doing, he presumed that a human being could "tap in" on the communications between and among animals, rather as a third party overhearing a conversation.

HANS AND LADY WONDER

As additional evidence supporting his belief in the self-initiated thoughts of animals, Krall brought to our attention a number of cases of animal intelligence that are not well known to modern readers. Two of these, those of Don the dog and of an unnamed chimpanzee, deserve our attention, not merely for their oddity, or for their pioneer status, but as the beginnings of investigative movements that would dominate the study of animal mentality in the remainder of the century.

However much observers were moved to find uncanny explanations for Lady Wonder's abilities, the explanation of her successes would appear to be of a different nature from that suitable to describe Hans. Hans appears to have been a shrewd horse who learned to respond to specific stimuli from people. There is no reason to suppose that the people were aware of the cues they were giving, nor is there reason to think that the horse was trained with the expectation of fraudulent notoriety. Lady Wonder's abilities, to the contrary, appear to be the product of a standard "magic" trick, that of pencil-writing. In this trick, the magician is able to state accurately information about members of the audience. A confederate gathers the information in any number of ways, such as asking questions beforehand, examining credentials, or gathering information from third parties. The confederate is seen to be writing, as if to take notes. In fact, the pencil is moved in shapes and patterns that indicate to the magician information.

If Lady Wonder had been trained to move her muzzle to several designs made by the pencil in movement, the human controlling the pencil could control the movements made by the horse. The horse, of course, needs to be trained to respond to different patterns, but, as we have learned from Hans, horses are well equipped to learn minute movements.

And, how did Lady Wonder know the location of the dead children? Note the actual responses she made to the question, and you will see that there is ample room for human interpretation, just as the horses' conversation regarding the Kaiser, Kaiserin, Schopenhauer, and Schiller give ample room for human interpretation. I presume that the policeman was thinking of the quarry in any event, and, finding some similarity between the horse's choice of letters and the site, moved to drag it. Lady Wonder's success was really a projection on the part of the human hearer, and the successful result, of course, made a reputation for Lady Wonder. The questions asked of the horse regarding the children to be found in the DuPage River included mention of the river and, to my reading, the questions posed merely investigated the intuitions of the mother. Alas, the intuitions were correct.

It is sometimes argued that fraud is unlikely in the case of either Hans or Lady Wonder because no or little financial gain was involved. Human beings respond to other motives, power, authority, and fame being among them. That Hans was the toast of Berlin, and Lady Wonder was featured in a major U.S. magazine, is motive enough for some people. Both von Osten and Krall, I think, are guilty only of wanting to believe in the qualities of another. Von Osten came to admire and respect his companion, while Krall appears to have been eager to show the humanness of animals—both are noble motives.

DOG AND CHIMP

Don, the dog, belonged to the Hermann Ebers family. He is shown in Figure 6.5 negotiating with Fraulein Ebers, who is holding a piece of "cake." The word she used for "cake" was *Kuchen,* and surprisingly Don could speak this word using a voice that permitted human beings in attendance to verify that they heard the voice saying "Kuchen." Don used other words, including his name, *Don, Haben, Hunger, Ruhe, Ja,* and *Nein.* When there was a noise in the street, Don would reportedly go to a window and say *"Ruhe."* When he said, "Hunger," he was presumed to be hungry and was fed some morsel.

Don was somewhat selective regarding the people to whom he spoke. To the human observer, his pronunciation differed from day to day: on some days it was excellent, and on others it was throaty—what the Germans call "wolf's throat." His name, "Don," sounded like "ong" to most

Figure 6.5   Fräulein Ebers and her dog, Don. Don speaks six words, all nouns, including his name. The speech was said to be intelligible to nonfamily members. His name, Don, sounded like "ong" to some listeners. Why he could not speak when it was raining remains a mystery.

listeners, probably because Don did not find it easy or convenient to bark the sound of the "D". For an unknown reason, when it rained, the quality of his speech suffered.

Don was not the only dog known to Krall who was able to speak, or at least to be heard by human beings to speak a language familiar to human listeners. No less a respected person than Alexander Graham Bell, interested in understanding how the throat muscles produce sound (Bell's invention of the telephone was interwoven with his interest in teaching the deaf), shaped, by his hand, the throat muscles of his dog to say "Mama" and "How are you." But Bell appears to have claimed no more for the dog than that he, Bell, had trained the dog to move throat muscles in response to a command. In later years, especially in our own times, we shall read that persons attempting communication with the great apes also "shaped" the throat and mouth to help the animal produce the effect desired.

Our colleague Oskar Pfungst returns for a curtain-call. Now a recognized authority on animal communication, he was called to investigate Don. His technique was to record the "speech" of the dog on the phonograph and to determine if members of the family could now understand the "words" when they were heard outside of a context. They were unsuccessful.[19]

In anticipation of what would become a major scientific industry, Richard Lynch Garner, with access to six chimpanzees, was reported by Krall to have taught the animals to speak. Krall also reported that two of these chimpanzees, Cicero and Demosthenes, were taught how to use blocks by Garner, just as children do in kindergarten. (Garner is shown teaching a child and chimpanzee together in Figure 6.6.)

It is not evident where Krall learned of Garner, but Krall seems to have accepted uncritically the testimony that Garner taught chimpanzees to communicate with him. The suggestion is that they "spoke."[20] As we will read, Garner, who at first was a thoughtful investigator, came to believe that he had, in fact, accomplished this goal. Krall's book claims that Garner was successful in teaching one chimpanzee to say "red ball" to the sight of a red ball. Krall wrote that the ape was able to speak in continuous sentences, but there is no evidence for this claim. My own view is that Krall was taken in by Garner, whose exaggerated claims for his work became evident to the English-speaking world at the time of World War I.

While Krall can be faulted for accepting reports uncritically, for "wanting to believe," he was certainly not the first to be deceived, nor will he be the last. That being said, we should recognize that Krall's work with Clever Hans and other horses is informative for reasons that the judgment of history has seemingly missed. The experiments performed were themselves critical in this sense: no one else thought to perform threshold measurements using psychophysical methods; no one else saw that any explanation depended on establishing Hans's visual world. Krall grasped the problem with Hans's method of tapping (that it left to the hearer the task of sorting out beginnings and endings) and invented a

Figure 6.6   Richard Lynch Garner teaching a chimpanzee to turn alphabet blocks. The function of the child is unclear. Garner was by now well-known for having lived in a cage in Gabon in the 1890s while conducting observations of chimpanzees and gorillas. His book-length reports sold well and established him in the forefront of those studying speech in monkeys and people.

method that permitted the horse, not the person, to signal such. He grasped the need to test alternative theories, such as telepathy and telephony.

In some ways Krall's sense of experimental method was superior to Pfungst's. Pfungst understood the need to eliminate alternative explanations, but did not seem to sense that the number of explanations is infinite. He never appears to have understood that his hypothesis required testing. Rather, he quit work with Hans once he developed this particular hypothesis—that men leaned and, by so doing, signaled Hans—letting the goal of complete evidence escape him. Krall understood the importance of testing a hypothesis, and he did so thoughtfully when von Osten was alive and perhaps, later, at Elberfeld as well. But his belief that animals and people communicate overwhelmed his critical skills: he began to marshall evidence, such as that from Fraulein Ebers and Richard Lynch Garner, to support his belief. Skepticism and critical thinking were lost in favor of gathering one more supportive case.

EXPERIMENTALISM AND BEHAVIORISM: RELIABILITY AND VALIDITY

The experimental method of forming hypotheses, performing controlled experiments, and deducing causal relationships is truly scientific in the sense that it tries to remove the subjective, the inconsequential, by finding the true links between observed behaviors or events. It may or may not work as well with planets as with people and dogs, but it can be powerful in what it reveals. On reflection, we can see that this method, however compelling, is no more or less of a myth about ourselves and our world than any other. Its charm and success lie in its ability to reduce causal relationships to any degree of causality and prediction we wish. It is a method that permits us to assess reliability to a fine degree. It is for this reason, one supposes, that it is often thought to be *the* scientific method. If one is interested in problems that can be solved by deduction, we have no more reliable and valid a method.

It may be misapplied, however, and the examples described here describe that special misuse of the method. The method can never be more accurate than the hypothesis being tested. A sound hypothesis must be capable of rejection or verification. Time and time again, we saw Pfungst and Krall floating hypotheses incapable of either. From time to time, both abandoned the logic of their hypotheses, never testing them but moving on to some other murky hypothesis. Such is a misapplication of the hypothetico-deductive technique, for to work properly the hypotheses must lead to verification or rejection as well as new, more defined hypotheses.

The power of the method is its danger. When used well, the logic is reliable and unassailable. But when used poorly, as Pfungst did in his innocence of the philosophy of experimental design, the procedure has only the trappings of power. The results are believed and repeated, just as writers for almost a century have regarded Pfungst as having proved an explanation that was, in fact, only a hypothesis, and one that remained untested at that. Judging the value of the hypothetico-deductive method in particular situations requires care. The use of the method alone proves nothing and, as can be seen, is sometimes deceptive.

Some forms of behaviorism rely on the hypothetico-deductive method; some do not. The behavioral approaches have in common a reliance on observable and reliably measured behavior and an insistence that the concept of "mind" is not itself a behavior. The reliance of some kinds of behaviorism on experimentation is a characteristic, but one can conceive of a behaviorism that uses methods other than the hypothetico-deductive one. The question of the validity of experimentalism and behaviorism, I think, depends, like that of psychoanalysis, on what myths one finds acceptable. There is a scientific myth, just as there is a psychoanalytic myth, and I can see no criteria to help us select one as "more right" than another, although one may be found to explain more kinds of events.

The Hanses, both Little and Clever, and all who knew them are, like Dash, now gone from our direct view, but they leave a legacy of methodology that continues to guide us. They teach us that the perceptions and the motives of the experimenter both determine the range of the outcomes and, ironically, alter the minds and behavior of all the agents involved in the experiment.

7 *The Psychologies of Perceiving: Phenomenology and Ethology*

WHEN WE think about the character of other animals or other people, it is behavior that we classify and speak of, not perceptions. Thus, whether we think of Dash, the Hanses, or the wolf-girls, it is their behavior that we categorize and remark on, but it is their perceptions, their motives, their ideas, that attract and mystify us. Those perceptions we infer from the behavior of the subject of our observations. Perception guides us in another way, for it is the perceptions and motives of the experimenter that determine the range of the outcomes of the experiment while, somewhat ironically, altering the minds and behavior of all the agents involved in the experiment.

Even the most commonplace observations demonstrate the point:

"Or take those who live alone with a dog. They speak to him all day long; first they try to understand the dog, then they swear the dog understands them, he's shy, he's jealous, he's hypersensitive; next they're teasing him, making scenes, until they are sure he's become just like them, human, and they're proud of it, but the fact is that they have become just like him: they have become canine."[1]

The events described in this chapter have several themes, and if we are not attentive we may think them unrelated. The unfortunate situation of the child Laura Bridgman at first appears to have nothing to do with the tasks set for the dogs Van and Roger, but the relation conveys to us the distinctive philosophy of our times that promises understanding of animal and human life. This philosophy is phenomenology.

We are not as familiar with this philosophy as we are with psychoanalysis and behaviorism. In this chapter, I show this third approach that has been woven in and out of the tapestry of our understanding. To make evident its presence in the design, again, I rely on plucking from the fabric certain stories that, taken together, illustrate the nature of the contribution.

LAURA BRIDGMAN

Laura Bridgman was born in the United States in 1829. As a result of a disease she contracted when she was a small child, she was blind and deaf. Her contact with the outside world was limited to cutaneous sensations, such as being tapped on the back. This limited contact with the outside world makes her of interest, for while her situation was much like the feral children, in that she had minimal contact with civilization, it was also distinctly nonferal in that she was raised among her family.

Her situation was widely known. The Smithsonian Institute investigated her case, and Darwin, in his 1872 book on emotions in human beings and animals, was interested in knowing whether she showed the facial and postural emotions that naturally accrue to human beings. She was also studied and trained by a hard-working staff led by a leading educational and medical reformer of the day. Although her circumstances did not command the public interest that was shown in Peter, Kaspar Hauser, or the wolf-girls, it did attract the attention of researchers interested in exploring how we come to understand the world outside ourselves. Her situation led these thinkers to consider whether it was possible to make contact with the silent minds of people and animals by training them to communicate.

Among these thinkers was a friend and neighbor of Darwin's, John Lubbock (1834–1913). Lubbock was among the most thoughtful and liberal of that special stock of British natural scientists of the nineteenth century. His contributions, especially those concerning insect behavior and the scientific education of the young, are appealing and topical a century later for their clarity of reasoning and expression. In his most comprehensive book, *On the Senses, Instincts, and Intelligence of Animals with Special References to Insects,*[2] Lubbock concludes with a chapter titled "Education of the Deaf and Dumb; On the intelligence of the dog." The reason for the juxtaposition is not obvious, but, as we are about to read, Lubbock has a point to make about how we might contact the minds of animal species:

"Considering the long ages during which man and the other animals have shared this beautiful world, it is surely remarkable how little we know about them. . . . As to the intelligence of the dog, a great many people, indeed, seem to me to entertain two entirely opposite and contradictory opinions. I often hear it said that the dog, for instance, is very wise and clever. But when I ask whether a dog can realize that two and two make four, which is a very simple arithmetical calculation, I generally find much doubt expressed. . . . That the dog is a loyal, true, and affectionate friend must be gratefully admitted, but when we come to consider the psychical nature of the animal, the limits of our ability are almost immediately reached. I have elsewhere suggested that this arises in great measure from the fact that hitherto we have tried to teach animals, rather than to learn from them—to convey our ideas to them, rather than to devise any language or code of signals by means of which they might communicate theirs to us."[3]

"We have tried to teach *animals, rather than to learn from them."* The thinking expressed in this introductory passage is prescient; there can be no other word for it. In the century since it was written, we have come to understand a little more about the interactions of different species, about, along with the seeing of it, the "hearing of this beautiful world." Lubbock points out that we have tried to understand the animal mind by *teaching* the animal, and by so doing, asking it "tell me what you can do" rather than "tell me what you know."

Lubbock suspected that, by training an animal in order to assess the limits of its abilities and intelligence, we might not offer the animal the opportunity to teach us what it knows. Many would agree with this notion, that by training and conditioning animals to perform as a means to measure their talent, we take from them what they know and do best— namely, how to react spontaneously and constructively in response to the aspects of the environment that are salient to them rather than to us.

Comparison of the analysis of Clever Hans with that of Little Hans illustrates the distinction. Some observers suspected that Clever Hans had been trained ("conditioned" to use the modern term) not to think, but to connect an action with a signal. Other people thought that Hans had a mind able to compute and to be rational, and that the training was merely a technique to help Hans and people communicate. Those who accepted the first hypothesis looked for evidence of teaching by training; those who accepted the second accepted the presence of training, but saw it not as the explanation but as the means by which Hans could make known his mental activity. This distinction will vex us, as it has confounded all twentieth-century attempts to understand the animal mind, for the same issue reemerges when we think about investigations that purport to teach chimpanzees and gorillas to communicate with people. Lubbock presented the question—indeed, the wish—that a means might be found by which animals and human beings might communicate. He wondered what kind of method could establish this bridge:

"It occurred to me [Lubbock continues] whether some such system as that followed with deaf mutes, and especially by Dr. Howe with Laura Bridgman, might not prove very instructive if adapted to the case of dogs." [Dr. Howe was the director of the Massachusetts Asylum for the Blind, an asylum that owed the idea for its being to Itard, teacher of Victor of Aveyron.] Laura Bridgman was "deaf, dumb, and blind, almost without the power of smell and taste. . . . Until brought under Dr. Howe's skillful treatment and care, her physical defects excluded her from all social intercourse."[4]

"Laura Bridgman [continues Lubbock] was born of intelligent and respectable parents, in Hanover, New Hampshire, U.S., in December, 1829. She is said to have been a sprightly, pretty infant, but subject to fits, and altogether very fragile. At two years old she was fairly forward, had mastered the difference between A and B, and, indeed, is said to have displayed a considerable degree of intelligence. She then became suddenly ill, and had to be kept in a darkened room for five months. When she recovered she was blind, deaf, and had nearly lost the power both of smell and taste."[5] So writes Lubbock, who now turns to Dr. Howe's description:

"What a situation was hers! The darkness and silence of the tomb were around her; no mother's smile gladdened her heart, or 'called forth an answering smile'; no father's voice taught her to imitate his sounds. To her, brothers and sisters were but forms of matter, which resisted her touch, but which differed not from the furniture of the house, save in warmth and in the power of locomotion, and in these respects not even from the dog or cat." "Her mind," however, comments Lubbock at this point in Howe's description, "was unaffected, and the sense of touch remained." Howe continues: "As soon as she was able to walk Laura began to explore the room, and then the house; she became familiar with the form, density, weight, and heart of every article she could lay her hands on.

"She followed her mother, felt her hands and arms, as she was occupied about the house, and her disposition to imitate led her to repeat everything herself. She even learnt to sew a little, and to knit. Her affections, too, began to expand, and seemed to be lavished upon the members of her family with peculiar force.

"The means of communication with her, however, were very limited. She could only be told to go to a place by being pushed or to come to one by a sign of drawing her. Patting her gently on the head signified approbation; on the back, the contrary."

Lubbock clarifies: "The power of communication was thus most limited, and her character began to suffer, when fortunately Dr. Howe heard of her, and, in October 1837, received her into the institution."

"For a while she was much bewildered, till she became acquainted with her new locality, and somewhat familiar with the inmates; the attempt was made to give her knowledge of arbitrary signs, by which she could communicate thoughts with others.

"The first experiments were made by taking the articles in common use, such as knives, forks, spoons, keys, etc., and pasting upon them labels, with their names embossed in raised letters. These she felt carefully, and soon, of course, distinguished that the crooked lines s-p-o-o-n differed as much from the crooked lines k-e-y, as the spoon differed from the key in form. Then small detached labels with the same words printed upon them were put into her hands; she soon observed that they were the same as those pasted upon the articles. She showed her perception of this similarity by laying the label k-e-y upon the key, and the label s-p-o-o-n upon the spoon.

"Hitherto, the process had been mechanical, and the success about as great as that of teaching a very knowing dog a variety of tricks.

"The poor child sat in mute amazement, and patiently imitated everything her teacher did. But now her intellect began to work, the truth flashed upon her, and she perceived that there was a way by which she could herself make a sign of anything that was in her own mind, and show it to another mind. At once her countenance lighted up with human expression. It was no longer a mere insensitive animal; it was an immortal spirit, eagerly seizing upon a new link of union with other spirits. I could almost fix upon the moment when this truth dawned upon her mind, and spread its beams upon her countenance: I saw that the great obstacle was overcome, and that henceforth nothing but patient and persevering, but plain and straightforward, efforts were necessary.

"The result, thus far, is quickly related and easily conceived; but not so was the process, for many weeks of apparently unprofitable labour were spent before it was effected.

"The next step was to procure a set of metal types, with the different letters of the alphabet cast separately on their ends; also a board, in which were square holes, into which she could set the types, so that the letters could alone be felt above the surface.

"Thus, on any article being handed to her, as a pencil or a watch, she would select the component letters and arrange them on the board, and read them with apparent pleasure, assuring her teacher that she understood by taking all the letters of the word and putting them to her ear, or on the pencil.[6]

"Now," adds Lubbock, "it seemed to me that the ingenious method devised by Dr. Howe, and so successfully carried out in the case of Laura Bridgman, might be adapted to the case of dogs, and I have tried this in a small way with a poodle named Van."

John Lubbock's importance to the development of our thinking about silent minds has two faces: one turns toward the past, toward the same set of intellectualized questions that attracted the investigators of feral and imprisoned children. But he is not a human being of those times; concepts such as "rights" and "liberty" do not dominate his nineteenth-century mind. The second face is turned toward the future, for the concepts that interest him are those related to how human society can be made "better," how education can improve the human condition. His faith is that the world is destined to become "better," in the sense of more efficient. He is interested in nature, in electricity, magnetism, and insects, and in Dr. Howe and Laura because he sees in Howe's goals for the unfortunate Laura an example of how education can be made efficient and powerful.

He also sees that training a mind is not the same thing as educating it. In considering this distinction, he intuits that when we train a human or animal or perform a task, we are not necessarily determining what that being has learned. Performance may tell us what is learned, but what is learned is not necessarily what is performed. Were Lubbock alive today, he would remind us that a sign-making chimpanzee tells us what it can *perform;* what we want to know, however, is what is *learned.* This was the question that prompted Itard, Feuerbach, and Lubbock, and it continues to prompt us.

THE DOG, VAN

Lubbock was interested in what Van might know, and to determine this, Lubbock taught the dog some responses, which thereby permitted Lubbock to make contact with Van's thinking via Van's behavior. We have met up with this reasoning before, and we shall surely meet it again in our own times in the continuing attempt to reach the animal mind through conditioning the animal's behavior.

In describing Van and Lubbock's achievements, we are best instructed by Lubbock's description:

"I took two pieces of cardboard about ten inches by three, and on one of them printed in large letters the word

FOOD

leaving the other blank. I then placed the two cards over two saucers, and in the one under the 'food' card put a little bread and milk, which Van, after having his attention called to the card, was allowed to eat. This was repeated over and over again till he had had enough. In about ten days he began to distinguish between the two cards. I then put them on the floor and made him bring them to me, which he did readily enough. When he brought the plain card I simply threw it back, while when he brought the 'food' card I gave him a piece of bread, and in about a month he had pretty well learned to realize the difference. I then had some other cards printed with the words

OUT

TEA

BONE

WATER

and a certain number also with words to which I did not intend him to attach any significance, such as

NOUGHT

PLAIN

BALL

"Van soon learned that bringing a card was a request, and soon learned to distinguish between the plain and printed cards; it took him longer to realize the differences between words, but he gradually got to recognize several, such as 'food,' 'out,' 'bone,' 'tea,' etc. If he was asked whether he would like to go out for a walk, he would joyfully fish up the 'out' card, choosing it from several others, and bring it to me, or run with it in evident triumph to the door.

"I need hardly say that the cards were not always put in the same places. They were varied quite discriminately and in a great variety of positions. Nor could the dog recognize them by scent. They were all alike, and all continually handled by us. Still, I did not trust to that alone, but had a number printed for each word. When, for instance, he brought the card with 'food' on it, we did not put down the same identical card, but another bearing the same word; when he had brought that, a third, then a fourth, and so on . . . . For a single meal, therefore, eighteen or twenty cards would be used, so that he evidently is not guided by scent. No one who could see him look down a row of cards and pick up one that he wanted could, I think, doubt that in bringing a card he felt that he is making a request, and that he could not only distinguish one card from another but also associate the word and object.

"I used to have a card marked 'water' in my dressing-room, the door of which we used to pass in going to or from my sitting-room. Van was my constant companion, and passed the door when I was at home several times in the day. Generally he took no heed of the card. Hundreds, or I may say thousands, of times he passed it unnoticed. Sometimes, however, he would run in, pick it up, and bring it to me, when of course I gave him some water, and on such occasions I invariably found that he wanted to drink.

"I might also mention, in corroboration, that one morning he seemed unwell. A friend, being at breakfast with us, was anxious to see him bring his cards, and I therefore pressed him to do so. To my surprise he brought three dummy cards successively, one marked "ham," one 'bag,' and one 'brush.' I said reproachfully, "Oh Van! bring "food," or "tea"' on which he looked at me, went very slowly, and brought the 'tea' card. But when I brought some tea down as usual, he would not touch it. Generally he greatly enjoyed a cup of tea, and, indeed this was the only time I ever knew him to refuse it. [Presumably evidence of his illness.]

"A definite numerical statement always seems to me clearer and more satisfactory than a mere general assertion. [Again, Lubbock is prescient: this view will be that of our century's understanding of how science is carried out.] I will, therefore, give the actual particulars of certain days. Twelve cards were put on the floor, one marked 'food' and one 'tea.' The others had more or less similar words. Van was not pressed to bring cards, but simply left to do as he pleased.

Trial

1	Van brought "food" 4 times	"tea"	2 times
2	6		2
3	8		2
4	7		3
5	6		4

6	6	3 "nought" once
7	8	2
8	5	3
9	4	2
10	10	4 "door" once
11	10	3
12	6	3
Total 80		31

"Thus out of 113 times he [Van] brought food 80 times, tea 31 times, and the other 10 cards only twice. Moreover, the last time he was wrong he brought a card—namely, "door"—in which three letters out of four were the same as in "food.""

"This is, of course, only a beginning, but it is, I venture to think, suggestive, and might be carried further, though the limited wants and aspirations of the animal constitute a great difficulty. . . . My wife has a beautiful and charming collie, Patience, to whom we are much attached. This dog was often in the room when Van brought the "food" card and was rewarded with a piece of bread. She must have seen this a thousand times, and she begged in the usual manner, but never once did it occur to her to bring a card."[7]

Having judged the experiment in reading more or less successful, Lubbock continued his investigations by examining Van's arithmetic ability.

"I then endeavored to get some insight into the arithmetical condition of the dog's mind. On this subject I have been able to find but little in any of the standard works on the intelligence of animals. Considering, however, the very limited powers of savage men in this respect—that no Australian language, for instance, contains numerals even up to four, no Australian being able to count his own fingers even on one hand—we cannot be surprised if other animals have made but little progress. Still, it is curious that so little attention should be directed to this subject. . . . We tried our dogs by putting a piece of bread before them, and preventing them from touching it until we had counted seven. To prevent ourselves from unintentionally giving any indication, we used a metronome (the instrument used for marking time when practising the pianoforte) and to make the beats more evident we attached a slender rod to the pendulum. It certainly seemed as if our dogs knew when the moment of permission had arrived; but their movement of taking the bread was scarcely so definite as to place the matter beyond a doubt. Moreover, dogs are so very quick in sizing any indication given them, even unintentionally, that, on the whole, the attempt was not satisfactory to my mind. I was the more discouraged from continuing the experiment in this manner by an account Mr. Huggins gave me of a very intelligent dog belonging to him. A number of cards were placed on the ground, numbered respectively 1, 2, 3, and so on up to 120. A question was then asked: the square root of 9 or 16, or such a sum as $6 + 55 - 3$. Mr.

Huggins pointed consecutively to the cards, and the dog always barked when he came to the right one. Now, Mr. Huggins did not consciously give the dog any sign, yet so quick was the dog in seizing the slightest indication, that he was able to give the correct answer.

"Mr. Huggins writes, 'The mode of procedure is this, His master tells him to sit down, and shows him a piece of cake. He is then questioned, and barks his answers. Say he is asked what is the square root of 16, or of 9; he will bark four or three times, as the case may be. Or such a sum as $6 + 12 - \frac{3}{5}$ he will always answer correctly. The piece of cake is, of course, the mead of such cleverness. It must not be supposed that in these performances any sign is consciously made by his questioner. None whatsoever. We explain the performance by supposing that he reads in his master's expression when he has barked rightly; certainly, he never takes his eyes from his master's face."[8] Lubbock resumes: "This observation seems to me of great interest in connection with the so-called 'thought-reading.' No one, I suppose, will imagine that there was in this case any 'thought-reading' in the sense in which this word is generally used. Evidently, 'Kepler,' the dog, seized upon some slight indication unintentionally given by Mr. Huggins."

Huggins, let it be added, as was true of Emilio Rendich who demonstrated the Clever Hans effect with his dog, Nora, was a common-sense scientist. Both understood the need for control groups and for formulating hypotheses that lead to a clear test of their veracity. Using this criterion, Pfungst was not a common-sense scientist, for he failed to test his own hypotheses. Huggins's and Rendich's work was science at its most pure, but the attention they received was minimal. Perhaps this is the common fate of "pure" work, work that wants only to solve and to know, rather than to apply and to sell. Perhaps we are not much interested in demonstrations and experiments that explain, fail to ascribe mystical qualities to the mind, or give but ordinary explanations of seemingly unusual events. Lubbock, however, was interested in experimenting, since he recognized the need to quantify the behavior he observed. He was less skilled at working in the control conditions necessary to test his hypothesis, but with this failing, he has much company in our age.

LUBBOCK AND VAN

By contemporary standards, John Lubbock's observations, deductions, and generalizations about animal life, especially about the life of insects, are of excellent quality. He worked and wrote almost precisely a century ago, yet his work is sensitive to issues that are current today. He did not merely look at animal behavior, but he knew what to look for; he had an understanding far in advance of his times about what would become important and what would become trivial. He understood the idea of an experiment, of the need to set hypotheses, to measure, and to form general conclusions that would be found to be true in the specific case. Unlike Georges Romanes, who was only a few years older, and more like C. Lloyd Morgan, who was a few years younger, Lubbock saw the unique behavioral event not as proof of the unique quality of an animal, but as an example of some general principle worthy of investigation.

If we were to point to a watershed between the old way and the new—between the almost gullible collecting of examples of superior animal mentality by the immediately post–Darwin workers and the application of experimental techniques that seek to identify causative variables by systematically eliminating possible ones that turn out not to be so— we could find no better passage to mark the demarcation than Lubbock's description of his purposes and examinations of Van than this: *"We have tried to teach animals, rather than to learn from them."* The watershed that Lubbock and Van provide is between the belief that animals enjoy some kind of mentation, perhaps consciousness, some ability to reason and feel—and that with appropriate techniques, human beings can come to understand and appreciate the mentality of animal life—and the belief that would deny that mind is a useful concept because it is not observable or measurable directly, as is behavior.

Lubbock, with Van's assistance, shows the fullness of the first belief with an invigorating attempt to utilize the advantages of the second. Lubbock believed that Van was exceptional (as whose dog or cat is not?). Having read of Laura Bridgman's situation, he undertook a method, not so much to assess Van's talents, but to communicate, by determining if he could teach Van to read. If Van could learn to read, as Laura Bridgman learned to do by using nonverbal abilities, presumably human and animal would have much to say to one another. The fruition of this idea, an idea well reflected in the stories by Hugh Lofting about Dr. Dolittle, the fictional veterinarian who could talk to animals,[9] reappears in both the popular and scientific literature of our times. The teaching of American sign language to great apes (Chapters 10 to 12) is a contemporary restatement of the connection Lubbock perceived between the training given respectively to Laura Bridgman and Van in the hope that, if Van could learn to read, he would communicate his mentations to Lubbock. Itard's idea remains a powerful one.

When we read about Lubbock's work with Van, we need to appreciate Lubbock's grasp of the concept of *control* in experimental design. By control is meant nothing more complex than including in the procedure the means to assess alternative explanations that may account for the behavior we observe. Although the concept is itself not difficult, the use of controls *appropriate to the interpretation* is what separates the professional scientist from the true believer (i.e., what separates Pfungst's approach from Krall's). The trick is simply appropriateness. Think of Freud. When writing about Little Hans, he saw the unique case as an example. Little Hans's fear of horses was of value because from this unique example, we could deduce causation. The point of the "cure" lay in Little Hans's own ability to identify and express the cause of the fear. The psychoanalytic kind of data, however, does not permit alternative explanations, because once the data are collected (these being the words spoken) no other hypotheses can be examined. Freud once described the system as a case of "tails I win, heads you lose," for he was aware that the psychoanalytic method offered no true alternative explanations to be evaluated.

Although Lubbock's understanding of the need for control was rudimentary, he was ahead of his age in terms of both scientific technique and the notion that he might be able to train Van to communicate through the written word. But the control can only be as good as the hypothesis it is meant to test. Lubbock "controlled" for the human scent on the card by replacing the cards from time to time with new ones, but he did not take seriously enough the hypothesis that scent, not sight, is responsible for the animal's discrimination. When Van brought the card "door" rather than "food," it was not that Van was wrong, Lubbock thought. Rather, he took the act as proof that Van could read, because he mistook the "D" for the "F." When Van did not want the cup of tea, it was because he was ill. (If he did bring the card, it is presumed, it was *despite* illness.) Note how Lubbock's hypothesis guides his interpretation and how the control offered is, in fact, unrelated to the interpretation. We have met this type of thinking before, most obviously in Rhine's shifting analysis of the telepathic horse, Lady Wonder.

We must be troubled by questions unanswered even by Lubbock's thoughtful analysis. Where was Lubbock placing himself in regard to Van? What was he doing while Van searched? If Clever Hans was responding to bodily cues provided by the trainer, and later, by other human observers, could Van do the same? Was Lubbock looking toward the card he expected to have returned? Was Lubbock moving his body in some small way that was unknown to him but perfectly clear to the attentive dog? Could other human beings achieve the same success? Could Van collect cards for anyone who asked, or could he do so only for Lubbock?

Lubbock, being a skeptical and thoughtful scientist, understood the need to collect data, but he did not understand that the hypothesis posed by the observer also determines the nature of the questions asked, that the nature of the question should determine the nature of the controls, and that the nature of the controls should determine permissible interpretations. While Lubbock was training Van with the hope of their joint eventual correspondence, while Clever Hans was performing his arithmetical skills in Berlin, and while Little Hans and his father were providing Freud with the data that would lead to the psychoanalytic interpretation of human development, another dog, Roger, was showing himself better educated than Van. Roger could read not only English, but German and French as well. His owner had a grasp of experimental design and surely of the utility of alternative explanations. Roger, like Clever Hans, was observed and checked by a psychologist, but with one important difference: Roger's psychologist was trained in the principles of design and deduction in the behavioral sciences. The episode tells us much about changes that were taking place in the understanding of scientific method and experimental design. It also leads us, not unimportantly, into a consideration of what goes on in the mind of the dog. And it is to the mind of the dog that this chapter eventually turns its attention, for it is here that the importance of philosophies of perception can be seen most clearly.

EXPERIMENT AND CONTROL: ROGER AND ''B.B.E.''

In 1907 the *Century Magazine* published a report by "B.B.E." regarding his or her dog Roger under the title "A Record of the Performance of a Remarkable Dog."[10] A portrait of Roger is shown in Figure 7.1. As we read B.B.E's description of his or her relationship with Roger, and of Roger's talents and abilities, we immediately recognize that the author's understanding of experimental control surpasses that of any of the persons we have examined to this point. And yet B.B.E. was writing at the same time as the investigators of both Hanses, horse and child, and Lubbock's dog.

"Roger came to us three years ago," writes B.B.E., "a forlorn and

Figure 7.1   Roger's first two homes were abusive ones, but his third home was with a master who taught him how to recognize the suits and numbers of playing cards, how to spell his name, do arithmetic, and translate among English, French, and German. The dog's and master's accomplishments were reported and analyzed in magazines in the United States at the same time as Clever Hans's growing fame in Berlin.

hopeless-looking puppy. He was a full-blooded mongrel of the cocker spaniel persuasion. . . . He had been taken into two homes only to be turned out again as utterly 'impossible.' " Meanwhile he had evidently met with harsh treatment, and his spirit was entirely crushed. At sight of a broom or a stick he would tremble and run for a corner. It was only after three months of kindness that he began to recover his spirit and 'take notice.'

"About this time I began his education. I taught him to 'speak,' roll over, say his prayers, die, trust, sit up, wave his paws, and other similar tricks, until I considered him quite accomplished. In fact, he had done so well that I decided to see how much I could teach him. I had heard of dogs who had been taught to pick out cards, and I decided to make the attempt with Roger.

"As playing cards were the most available and convenient in size, I selected eight and laid them on the floor in front of Roger. Then I placed his foot on the ace of clubs and said, 'That is the ace of clubs, Roger—ace of clubs.' I repeated the name of the card four or five times and Roger looked very much bored. Then I gave him a piece of cooky, and he looked more cheerful. I kept this up for about ten minutes a day for two months until he had learned to find the ace in one position. Then I changed it about until at last he could find it in any position and I felt that I had gained a great victory. How great that victory really was I did not realize until a year later.

". . . After a month or so, however, I selected the ace of hearts, put his foot on it, and repeated the name of the card as I had done in the first instance. Then I gave him a piece of cooky and said, 'Show me the ace of hearts, Roger.' To my surprise, he did so at once. I put the card in different positions, and he found it every time. . . . In the course of time I taught him eight different playing cards, which he seemed to distinguish perfectly.

"Then I resolved to teach him to spell his name. I drew the letters of the alphabet on square bits of cardboard and laid them in front of him, the letters of his name mixed in with others, which he was not to use. I did not teach him the names of letters but simply said: 'Let me see you spell your name. Where is the first letter?' . . . I spent five or six lessons teaching him, but he learned very readily. . . . Then I taught him to spell his last name. He seemed to learn it as fast as I told him the letters, and did not confuse the two words in the least. More surprised than ever I taught him still another word, with the same result.

"Next I decided to try arithmetic. I taught him to add every combination of two as far as twelve. For instance, I would say, 'Show me six and two,' at the same time putting his foot on eight. He seemed never to forget after I had once told him. . . . Not knowing what to think, I took out the letters and said, 'Spell dog.' This was a word that he had never spelled before, and I gave him no clew whatever, yet he spelled it correctly and without hesitation. I said, 'Translate it into German, Roger,' and he spelled *'hund.'* Then I said, 'Spell it in French,' and he spelled *'chien.'* . . . Not once did I previously indicate the proper cards. He seemed to know them without even being told. I resolved to experiment a little so I took out the figures again and said, 'Show me two times three,' at the same time fixing my attention on the eight. He put his foot firmly on the eight. Here was the clew! *All this time when he seemed to be learning so rapidly, he had been simply getting the cards of which I thought"*[11] (Italic added).

On the one hand, B.B.E. used a control: he or she had himself or herself "think" of the wrong answer, and found that Roger responded to the wrong card as easily as to the right one. On the other, B.B.E. now believed that it was his or her own (B.B.E.'s) *thinking* that the dog understood. While the result of the simple control, the simple shift in the experimental presentation, shows that Roger was not himself solving the arithmetic and linguistic problem, B.B.E. took the result to mean that Roger was, in fact, attending to the thoughts of the human being; that while the dog may not have been a self-taught linguist and mathematician, he was a reader of human minds. From this anecdote we learn that merely establishing a control condition is not sufficient; it may lead to results that take us astray, unless we understand exactly for what interpretation the control is intended. The control used here led B.B.E. to believe that he or she was communicating with a telepathic dog, and B.B.E. proceeded to investigate how the dog engaged in human telepathy. "The great problem, of course, is, How does he know the cards of which I am thinking?" This is a problem, of course, only if we consider the animal's inability to respond to the correct card, when the human is thinking of the incorrect one, as having as its explanation, telepathy.

B.B.E. now examined the possible means of communication that could explain the apparent telepathy:

"At first, psychologists advanced the theory of motor suggestion. That is, they said that it was a form of muscle reading by which hypnotic subjects and so-called mind-readers often receive suggestions. For example, they said that as I thought of 'W,' I pronounced the letter to myself and unconsciously moved the muscles in my throat." We remember Pfungst's occupation with the notion of suggestion, and Freud's criticism of a psychology based on suggestion, and his favoring the idea of the importance of the unconscious. The theories of the day do, of course, influence, if not determine, how we phrase a problem, how we analyze it, and the kind of explanations we find plausible.

But B.B.E. rebuts this notion: "I had never taught Roger the names of the letters, and therefore if he did hear W, it would have no meaning whatever for him." Apparently, B.B.E. believed that, because Roger had attached no meaning to the signs, he was therefore unable to receive signals. Here, B.B.E. appears to assume that there must be "meaning" in order for communication to occur—the unstated assumption of the rebuttal is that meaning is an essential aspect of communication. This may or may not be so. Note how the assumption guides and controls the interpretation, blinding B.B.E. and us from the alternative explanations. For example:

"Another possible theory is that he [Roger] gets the picture which I have in my mind, but how he does it of course cannot be explained. I visualize very strongly and see things just as they are in their true colors. Some people think that I indicate the direction of the card by the direction of my glance." A sound hypothesis, we would say, considering what we have learned about how other animals of supposed unusual intelligence accomplished their feats. But B.B.E. objects, "This is not a sufficient explanation, because he will pick out the right card when my eyes are closed and even when I am not in sight." We are reminded that the control must be a satisfactory test of the hypothesis: if Roger can respond with the eyes closed, it may be that the eyes alone are not the cue; if Roger can respond when B.B.E. is not present ("when I am not in sight"), we may entertain the idea that there are other ways of sensing beside vision.

B.B.E. concludes the analysis with this remark: "The theory of motor suggestion is impossible because he does not know the name of the cards. It seems to me that the visual picture is the only one which will help him." B.B.E. understands the controls to indicate that, if suggestion by means of presumably unconscious motor behavior is excluded—the movement of the eyes, muscle changes in the throat—then the only remaining possibility is that Roger was able to "read" the visual images possessed by human beings. The interpretation is supported by B.B.E.'s own testimony that his or her visual images were unusually powerful, however he or she might know this. So, B.B.E. concludes that it was the vividness of his or her own mental images that alerts and instructs the dog Roger. We need not dwell on the problems with this conclusion, for we have learned from other examples how easily conclusions represent the ideas and thinking characteristic of the times and of our own views of ourself. Some humility is in order as we read these passages if we recognize that our own thinking about B.B.E. and Roger, and Van and Lubbock for that matter, must represent the thinking of our own times and our own intellectual inheritances.

When B.B.E. closes his eyes, he is putting into effect a critical control, but the choice of this control is dictated solely by theory. Theory is handmaiden. It leads us to select the alternatives and then stays around to help us clean up our hypothesis. To apply this notion to B.B.E.'s analysis, with the derivation of the hypothesis that Roger was able to grasp human visual images, B.B.E. moves to examine aspects of the idea:

"In the first place, Is it any peculiarity in myself or in the dog which makes this case unique and unlikely to be repeated? So far as I myself am concerned I think it is due simply to a sympathetic understanding of animals, and to the ability to concentrate my mind clearly on one object. So far as Roger is concerned I think he is remarkable only because he is exceedingly sensitive. He is naturally very intelligent but perhaps no more so than many dogs."

Notice that here two new assumptions or hypotheses are introduced into the analysis, the acceptance of which leads to the sensible conclusion that Roger was sensitive and intelligent. One is that animals have "sympathetic understanding." How can we know whether they do, or do not have this understanding, or, if they do, to what degree, unless we know what the words "sympathetic understanding" mean? We are reminded of Mr. Chattock's dog, Dash, who showed "profound disgust." Edwin Landseer's painting, shown on page 79, assuredly showed something that we human beings might call "sympathetic" and "understanding," but these are human terms used for apparent human emotions. If these motions are characteristic of other human beings as well as dogs and other animals, we need to know enough about the behavioral manifestation of the emotion to note and examine it.

The second hypothesis is that B.B.E. has an unusual ability to concentrate the mind fully on one object. Whether B.B.E. had some unusual talent in regard to mental concentration is a testable matter.

"Another question is," writes B.B.E., "Why have such cases not occurred oftener? So far as I can discover, there has never been a dog which could do the same tricks except when signs were employed. The only similar cases are those of trained horses which go about giving exhibitions. The claim is usually made that the horse reasons out the problem for himself, but the true explanation [i.e., visual images] is probably the same [for the horses as for Roger]. I have examined books on animal training, but nowhere have I found any hint that an animal can be trained in this way. In some cases an elaborate system of signs, difficult for a person to learn, is suggested as a means of guiding the animal to choose the right card. I cannot understand why any animal, after a little training, would not come under the mental control of his trainer just as Roger has done."[12]

B.B.E. concludes that Roger could not perform the mental acts by himself and that some form of human-animal communication was occurring. Lubbock reached this same conclusion regarding himself, and the dog Van. B.B.E. did a simple experiment showing that Roger was responding to some human behavior. Communication requires meaning, and as meaning is not possible here, inasmuch as Roger had not learned the meaningfulness of letters, it follows (writes B.B.E.) that Roger was communicating by attending to the rather special talents of concentration and visual imagery possessed by the human, B.B.E. Because not all the evidence supports this interpretation, B.B.E. concludes: "The whole thing has worked out so simply that it seems to me one of the most natural things in the world, and yet it seems utterly inexplicable. . . . May it not be possible that between our minds and the minds of the lower animals there is a deep and quite subtle connection which may yet be explained in the future, but only by the use of the utmost sympathy and love?"[13]

ENTER YERKES

Robert M. Yerkes (1876–1956), an instructor at Harvard University but later the most distinguished U.S. investigator of animal behavior, was invited by the editor of *Century Magazine* to comment on B.B.E.'s report on Roger. Yerkes opens the commentary by reminding the reader that, when we consider reports by human beings of animal life, we would do well to remember the difference between honesty and reliability. To carry forward this distinction, an honest observer can be unreliable; a reliable observer can be dishonest. We should not assume that honesty produces reliable observations or that unreliable observations emanate from dishonesty. B.B.E. may be the most honest of people but may still attend to unhelpful aspects of Roger's behavior; he or she may be honestly fooled by adhering to ideas of the day ("suggestion," for example or mental imagery) and by already held beliefs about how animals think and act.

Yerkes tells the readers of the example of Clever Hans,[14] which then, of course, was a newsworthy story, and reminds the reader, inaccurately as we now know, that Hans did his clever tricks by responding to the trainer's eye and hand movement. Yerkes thinks this explanation to be the most appropriate way to explain Roger and B.B.E.'s successes. "So it turns out," writes Yerkes, "that, in all probability, Clever Hans possesses no mathematical ability; that he does not understand human language to any considerable extent, and that he has no power of logical thought. In place of these human endowments he has an excellent power of observation by means of sight. Baron [?] von Osten's laborious efforts have resulted in teaching him to watch for extremely minute movements which are made unintentionally by his questioner, and to respond appropriately to such movements. Clever Hans has learned to read the mind of his trainer, and, without understanding what is in that mind, to interpret its contents in terms of horse response. He has been trained to a most unusual degree of exactitude in answering questions, but he exhibits nothing which demands rationality as an explanation.

"Now, what does the dog Roger do which suggests comparison with Clever Hans? My reply is that, in many instances, he answers questions correctly when they are given to him for the first time, and that he does this in spite of the fact that he has not been taught to recognize all of the letters and numerals which he has to use in giving answers. This fact at once suggests to the student of animal behavior and comparative psychology the existence of sense guidance."

Yerkes undertakes a set of observations intended to "control" for the several ways in which Roger might be responding to cues from B.B.E.

"I failed to discover anything which could account for the large proportion of correct answers which were given. That Roger must be able to see what his trainer thought of seemed, as Mr. E———[presumably, the surname of B.B.E. was known to Yerkes] has stated, to be the only convenient explanation; but as this was extremely improbable, to say the least, I decided to investigate the matter more fully at a later date."

Yerkes visited B.B.E. and Roger, and found that Roger was able to add, subtract, and spell. Yerkes confesses that he could see no visible movements on the part of the trainer, B.B.E. Three months later, in September 1907, Yerkes revisited Roger and B.B.E., who had then been separated for six weeks, as B.B.E. had been away. Roger was alert, but "as a result of his vacation, Roger was out of practice. He watched Mr. E———much more intently than he had observed previously, and he made a larger percentage of mistakes."

Since Roger and B.B.E. were out of practice with one another, Yerkes expected to see more errors; the nature of these errors, in turn, might suggest the ways in which communication was occurring between them. "Roger often looked at Mr. E———, but he rarely looked at the cards which were before him on the floor. It soon became clear to me that he did not compare whatever he may have seen when he looked at his trainer with what he saw on the floor. This disposes of the theory that his response is due to the fact that he can see what Mr. E——— is thinking about. *I further demonstrated the fact that he could not answer a question correctly unless he could see the questioner* [Italics added, for emphasis].

". . . By my observation of Mr. E——— during the performance, I discovered that he aided the dog by a variety of movements. (1) He looked directly and steadily at the right card, and when I asked that he [B.B.E.] look at one card and think of another, he said that it was extremely difficult to do so. Now, of course, the dog might perfectly well detect and be guided by the direction of the questioner's gaze. (2) At times, Mr. E———'s hands, arms, and even his entire body moved noticeably in the direction of the card which Roger was expected to touch with his paw. (3) . . . I noticed that Mr. E——— frequently straightened the dog up and placed him in a favorable position for reaching the cards. In this there would seem to be no harm, but I clearly saw that the animal was pressed in the right direction. These . . . movements, and doubtless many more that I failed to see, were made by the trainer without the intention of aiding Roger."

"In justice to the description written by Mr. E———, I must add, however, that these movements are not readily seen by the observer when Roger is in practice and does his best. *It is highly probable that the dog's visual sensitiveness to movements is greater than ours.*" I italicize this comment because it should make us think: Why do we assume that the dog, horse, or other human being's sensory world is no more acute, no more precise, no differently organized from our own? Understandably, we assess their abilities in terms of how we understand ourselves to think and feel. We assume that their brains, their sensory systems, their sensory and motor connections are copies of our own. Yet we know from our investigations of animal life that many, if not most, have sensory capacities beyond our own. The hearing by the guinea pig, the seeing by the hawk, and the smelling by the dog are well-known examples of animals' sensory abilities that far exceed our own. When a horse or dog seems able to communicate to us, why do we not ask what sensory system the animal might use that is unknown, unfamiliar to us, and unexamined by us? It is, after all, one thing to understand that we can only know another being through our own minds, and quite another to suppose that no other mind has any capacities beyond those that compose our way of understanding.

It is a solecism of the behaviorist way of thinking about and examining behavioral phenomena that the animal is the responder and, therefore, by definition, the human being the trainer. We are amused, if we have any sense of irony, when we see that the question of who is the instructor and who the pupil, who the trainer and who the learner, is not merely inexact but circular. Roger's lectures in the School for Canine Psychologists must surely emphasize how human beings can be shaped to supply morsels by the dog performing certain mindless, behavioral acts.

The pioneering experimental behaviorism as practiced by Pfungst, Lubbock, B.B.E., and Yerkes is to be commended for its understanding that causal relationships cannot be gleaned by logic alone and that controlled experimentation locates the appropriate causal connection as the one that is invariant. These experimentalists determined causal relationships by eliminating alternatives. The method of elimination demands, logically, that all the alternatives be known. If not, elimination of alternatives leads only to a false sense of security—indeed, to the acceptance of an untested hypothesis as true because it is the only alternative remaining. In their own way, Pfungst, Lubbock, and Yerkes all fell into the trap of accepting an untested hypothesis. However much and whatever behaviorism teaches us—and it does provide a splendid method for isolating variables and their results—it forces our way of thinking about living beings into a world of straightforward and unbending relationships: a human leaning leads to a horse's tap, a printed card to food. Do we imagine the animal mind to be a directory of such associations, or might we think of the associations of the mind to be formed and expressed less directly, as some sort of library card catalog or map of the mental terrain?

WHAT LAURA DID

Laura Bridgman served as the impetus for a dramatic shift in how human beings go about asking the question, "How do we contact the mind of others?" The way in which the shift occurred is subtle; at least, I cannot identify it. Yet, clearly, the kinds of questions asked of Peter or Kaspar are very different from those asked of the wolf-children and the animal subjects, mostly domesticated pets, described in this chapter. It is as if the nature-nurture question that prompted kings and savants to examine feral children became a frozen subject that refused to soften and liquify no matter how much light was put on it.

Laura gave Lubbock the idea of communicating with silent minds by training the individual to translate the contents of the mind by some behavior. For Laura, this meant using her skin as a perceiver of information that is conveyed to most people by sight and hearing and speech. For Van and Roger, communication meant using the paw or mouth and jaw to fetch or touch some thing. When Lubbock saw that what Dr. Howe had done to Laura could be done by anyone to anyone else, when he saw that the key was to understand training behavior as learning to communicate, rather than training only to see what the animal could learn, when he learned to ask the animal, "Tell me, what do you know?" rather than, "Show me, what can you do?" the door was opened to what would become a major scientific industry of our times—namely, communicating with unspeaking minds, whether people or animals.

But the pathway needed clearing. Along the way, the value of Lubbock's notions was forgotten, as was Darwin's comment on the importance of Laura Bridgman. The study of animal life would therefore swing from the study of performance to the measurement of learning and back again. The reasons for the shifts are related to our changing view of how we think of the minds of animals, and, just perhaps, of how we think of the minds of our fellow human beings. But that argument is best held aside for the next section. First, we want to see how Laura and Lubbock, and Van and Roger, contributed to the continuing development of an idea about the nature of the mind that had been evolving on the Continent but without much notice in the English-speaking world.

PERCEPTION IN TIME AND SPACE

All animals possess the ability to use space, to move, explore, migrate, to home. Because animals are not random users of space, we think of them as directed, as being motivated and intentional. The ability is a remarkable one. For example, when I am hungry, I know or can remember where I am likely to find food; when I wish companionship, I can show a highly developed sense of where to find it. Pigeons, cats, and dogs are renowned for their ability to return to a specific place, a home, a territory; some birds, fish, and mammals migrate over astonishing distances, using cues year after year that people cannot perceive or fully understand.

With sporadic exceptions, behaviorism has had remarkably little to say about how living beings utilize and learn about space. The chief thrust of behaviorism has been to measure responses over time, not within space. Perhaps the three-dimensionality involved in the human view of space is not easily investigated by a two-dimensional model, such as behaviorism, which looks for a connection between stimulus and response. Perhaps it is not too much of an exaggeration to say that behaviorism is more comfortable measuring time than measuring space. The stimulus-response connection, after all, is a temporal one, and the temporal relationship between these two has furnished behaviorism with its most productive and compelling explanations, such as the widespread influence of the schedules of reinforcement. Space, to the contrary, has only a small role in behaviorist interpretations, but the sense and utilization of space have been the province of those who prefer to emphasize how behavior is dependent on perception. Perception seems more akin to space than to time, for what we perceive, we perceive as existing first in a spatial environment.

Consider a road map. It is, of course, a perceptual representation of space. From its markings, I can think about a path between *A* and *B*. The connection between *A* and *B* is not necessarily direct, and there are alternative ways to move between these points. As I examine the map, I note the contour of the road between *A* and *B;* I examine the kind of road; I think about whether I may want to get to *B* from *A* by way of point *C,* this being a place lower on my travel priority, but one that I might visit if the road between *A* and *B* takes me nearby. I study the road map while considering many categories that affect my decision. The direct line is not that which is physically the less distance: the direct line is one, to my mind, that gives weight to factors such as time, comfort, what may be visited or seen along the way, the time of day, how much time I can offer the trip—on and on go the possible categories that include material that is assessed and examined as I plan my route.

The mind, as some analaysts conceptualize it, resembles a roadmap. Both have many connections that may be used to get from *A* to *B*. The choice of which is selected depends on the value placed on aspects. Neither the mind nor the roadmap merely calculates the most efficient routes: it makes use of experience, of memory. The mind may say, I did that route the last four days; today I shall take another one. The mind is not, according to this conceptualization, a series of one-way streets as behaviorism or a simplistic model of the nervous system implies. Rather, it represents all these categories, and many more, in a representation of the phenomena, these being figurative representations of the present, known through sensation, perception, and memory.

Reading a diagram of the relationship between a stimulus and a response, as does the behaviorist, is a simple matter for a human being. Our perceptual apparatus permits us to do so with pencil and paper, for I need only draw a line between the two to show the connection. Here:

Sight of von Osten leaning forward ★★★→ Tap hoof

Sight of Lubbock gaze   ★★★★★★★★★★★★→ Pick up card

Pick up card   ★★★★★★★★★★★★★★★★★★★★→ Take it to Lubbock

B.B.E.'s voice, line of gaze   ★★★★★★★★★→ Set right paw

These are useful shorthand symbols that demonstrate the nature of connections between perceptions and behavior, between events and responses. The experimental technique used by Pfungst, Lubbock, B.B.E., and Yerkes, a technique later to be developed, refined, and much used by behaviorists, was to define and examine alternatives and refinements of the connections with the hope of eventually determining that one stimulus that led to that one response.

Can we imagine the nature of the animal's mind—as to what it sees, hears, smells, tastes, touches—as the total cumulation of all the sensory systems impinging at once on "the mind"? As I communicate with you, my eyes see the writing, my fingers feel the pen or the keys, my ears hear the scratch of the point or the hum of the machinery, I smell the ink, something electrical, dinner, the dog, fresh paint, Eucalyptus; I taste my teeth, something a little metallic; my head is turned, a little upward and to the left. These are the stimuli, the sensations and perceptions, that define "this moment, now" for me. You can teach me to associate these rather precisely: when the smell of dinner reaches a certain threshold, I lay down my pen; when I hear a ring, I look for the telephone. These single connections are certainly teachable and learnable, but their presence shows only my ability to learn or yours to teach. The connection does not show the range of sensations and perceptions to which I am able to learn responses or attach meaning.

If you were watching me, you would say, see, the phone rings, he lifts the receiver. Ring—Lift the Receiver; there is the stimulus-response connection. Almost every time the ring of the phone elicits my action. But other aspects appear in my perceptual world. I am answering the phone; I am a little put out to feel I must leave my writing at this moment; a little worried that I cannot resume this sentence as I had phrased it in my mind, before my hands recorded it fully; the ring of the phone reminds me that I must do something about dinner; that the dog is surely hungry as well; that I need light now to see the page.

The ring of the phone, the ring that you record as an auditory stimulus that leads to a behavioral response from me, is far more to me than the association of auditory stimulus and motor response. For one thing, the ring shifts my attention, and with the ring, each of my senses shifts in some way. By this account, a stimulus is far more than a one-dimensional aspect of one sensory system: it is, if you will, the primary cause of a shift in all my sensory systems, a reorganization of my perceptual world. The response of lifting the telephone receiver is the usual behavior that follows the ring, just as for Clever Hans the tap of the foot was the usual response to human being's leaning forward. When you say that you are recording the frequency with which I lift the receiver, you are selecting for recording one response from all those I make, for I also clear my throat, pitch my voice, raise my eyebrow, shift my eyes.

How can I represent this "sensory consciousness"? I cannot draw it with the same ease that I can draw a two-dimensional stimulus-response connection. I cannot decide how to represent the differing contributions of the differing senses, or how to catch the perceptions in place all at one time for each second brings a shift somewhere. If I cannot conceptualize this mind in some meaningful way, how can I discuss it, or ask questions of it, much less try to understand the ways in which animals and people communicate?

Psychoanalysis is chiefly a two-dimensional conceptualization in that its strength lies in its assertion that identifiable aspects of the past affect the future. In this strength it is similar to evolution. Behaviorism shares this strength, for its goal is to plot the temporal relationship between stimulus and response. Phenomenology offers three dimensions, however, for it emphasizes changes in sensations and perceptions. This distinction opens the discussion by example of the third chief epistemology of our time regarding the mind and behavior.

PERCEPTION AS EXPLANATION

Any entry to the mind is surely by way of the sensory systems. Yet strict behaviorism promotes the view that only that which can be observed, the behavior—not the mind itself—constitutes the useful data from which an understanding of behavior can be built. *What* the mind is, *where* it is—these are indeed questions of importance. To the behaviorist, however, questions are answerable by observing behavior, not by postulating constructs, such as the mind, that are themselves unobservable and thereby untestable and unverifiable. Perhaps because of the emphasis on observable behavior, behaviorism has found that the examination of how and what we perceive is of the greatest importance in understanding behavior.

Certain other ways of thinking about behavior emphasize the importance of the senses, both as sensation, the unmindful form of stimulation, and as perception, whatever the mind, experience, and learning adds or subtracts from the base sensory experience. Because these ways of thinking share an interest in perception, they are sometimes casually and loosely grouped as offshoots of phenomenology. However, they are at once both more and less than this category implies. They are less in that they concentrate on the perceptual experience of a single group of animals; they are more in that they attempt to construct a perceptual road map that may explain behavior. The danger is that by constructing an accurate and beautiful map, one might assume that the job is done, that the map explains it well because it is colorful. To do this is rather like using a roadmap to understand geological evolution: the task of constructing the map, however complex, is but the beginning of the process. Describing is primary, but describing behavior is not explaining it. It remains to be shown how behavior is produced and altered.

Phenomenology takes its name from the emphasis on the nature of the perception—the phenomena. It gives form to a particular way of studying animal behavior, known today as ethology, although the term has undergone serious revision at least twice. We must be careful to stipulate the kind of ethology that we mean when we use the word. In its earliest form, the word "ethology" came from "ethos," a term meaning a "trait." It is my ethos to be mercurial, phlegmatic, or sanguine in temperament, to be clever, intelligent, or slow in intelligence, and so on. These traits were seen as relatively characteristic, unchangeable aspects of a species' or individual's behavior. Just as wings are a physical aspect of birds, but not of mammals, so responsiveness to the sight of an egg might be a trait of fowl but not of apes. A collection of such ethoses can define a particular species, just as we define the species itself by naming its collection of physical attributes.

In the middle part of this century, ethology became a way of defining species by their behavioral traits. Because these traits, by definition, are relatively invariant, it follows that invariancy is solely genetic. In its early form, ethology, by emphasizing invariancy, intentionally or not, adopted a strong stance in favor of determinism of the gene over the utility of learning from experience. Nonetheless, behavior comes about because of a perception: it does not occur without the background or prompting of how the environment is perceived. Thus, while ethology surely observes behavior, it must also have an interest in determining what kinds of perception accompany what kind of behavior.[15] Here is an oft-told example:

The graylag goose, when sitting on eggs, will expend notable effort to retrieve an egg should one become dislodged from the nest. The goose does not seem to learn this complex behavior, for it appears to perform this act on the very first occasion of an egg being dislodged. The action involved in retrieving the egg is remarkably alike in its repetition. Learning seems neither to be responsible for the act nor to have much effect on it as the situation is repeated. The believer in ethology refers to the act, and the repetitions of it, as a fixed action pattern (the FAP, or *Erbkoordination,* as it was called originally). The FAP is a behavioral pattern that is identifiably similar each time it occurs. It is like a reflex perhaps, but a reflex that is set off by a perception, by the innate releasing mechanism, or IRM. The excitement at categorizing FAPs sometimes obscures the fact that no FAP is possible without an accompanying perception. While the midmodern psychologist may search for, find, and categorize FAPs, she or he is also assuming the presence of a perception responsible for the onset of the FAP. If the IRM receives far less attention, it is because perceptions are brain mechanisms and therefore far more difficult to identify than verifiable behavior.

Let us explore the relationship. What triggers or fires or sets off the FAP? In English, one asks what *releases* the FAP, the verb suggesting that something is held in storage, awaiting exposure but firing when a threshold has been reached. While the behaviorist thinks of the *connection* between the stimulus and response, the ethologist thinks of what *releases* the FAP. What releases the FAP is a perception, the IRM—here the perception of something of a certain color and a certain shape, moving in a certain way. The perception is the releaser, not the perception of the "egg" as such, but of its sensory qualities, its shape and color.

In order to define the nature of the IRM, the midcentury ethologist experimented by substituting aspects of the perception: a yellow, white, brown, or purple egg; an "egg-shaped," round, or square object; one that moves in a straight line, one that rotates, one that waddles back and forth. Which of these combinations releases the FAP—and which does not? By ordering the qualities of the perception, the ethologist determines the exact nature of the IRM, and thereby the nature of the perception that serves to release the FAP. By observing and cataloguing FAPs and their accompanying IRM, the ethologist constructs a picture of IRMs and FAPs that defines the animal by defining the traits of its behavior. This method was most clearly explained by Konrad Lorenz, originally in the 1930s,[16] but it is an idea with philosophical antecedents easily found in the ideas on categorical perception by Immanuel Kant. It is also a framework that considers the being as just as mechanical as the view ascribed to the behaviorist. This framework has been refined by other observers and thinkers to produce a way of describing the perceptual aspects of the mind, as we will now read.

PERCEPTION AND THE ANIMAL MIND

This modern form of ethology provides a means for identifying and specifying perception. If it thinks of perception at all, behaviorism prefers to specify perception in terms of thresholds and the like. In contrast, ethology should want to specify the aspects of a perception necessary to release the FAP. During this century, attempts to describe and represent the perceptual world of beings have been a mainstay of European thinking about the mind; little of this work appears to have reached the eyes and ears of English-speaking audiences.[17]

Let us illustrate how phenomenology may be applied to the study of animal life in a way that reflects the assumptions and ways of thinking of the ethologist. Consider the German Jacob von Uexküll's opening sentences in the book *A Stroll Through the Worlds of Animals and Men* (1934):

"This little monograph does not claim to point the way to a new science. Perhaps it should be called a stroll into unfamiliar worlds; worlds strange to us but known to other creatures, manifold and varied as the animals themselves. The best time to set out on such an adventure is on a sunny day. The place, a flower-strewn meadow, humming with insects, fluttering with butterflies. Here we may glimpse the worlds of the lowly dwellers of the meadow. To do so, we must first blow, in fancy, a soap bubble around each creature to represent its own world, filled with the perceptions which it alone knows. When we ourselves then step into one of these bubbles, the familiar meadow is transformed. Many of its colorful features disappear, others no longer belong together but appear in new relationships. A new world comes into being. Through the bubble we see the world of the burrowing worm, of the butterfly, or of the field mouse; the world as it appears to the animals themselves, not as it appears to us. This we may call the phenomenal world or the self-world of the animal."[18]

The analogy of the bubble is brilliant in its aptness: we all live in such a bubble, a bubble whose ends and edges are created by the limits of our perceptions. Just as you and I view our universe as being our bubble, so each animal views its life as contained, but shaped and modified, by its bubble. Each bubble is created by the limits of our sensations, while its shape and contours are created by changes in our perceptual environment accompanying our experiences with our perceptual world. The bubble both contains and *is* the phenomenon of our self-world. The self-world is the "umwelt," the inside or self-world *(welt)*.

"The mechanists," continues von Uexküll (he would include the behaviorists among them, surely), "have pieced together the sensory and motor organs of animals, like so many parts in a machine, ignoring their real function of perceiving and itching. . . . According to the behaviorists, man's own sensations and will are mere appearance, to be considered, if at all, only as disturbing static. We . . . see in animals as well not only the mechanical structure, but also the operator, who is built in to their organs, as we are into our bodies. We no longer regard animals as mere machines, but as subjects whose essential activity consists of perceiving and acting."[19]

Von Uexküll asks us to take a walk with our dog through the meadow. Along the way, on this sunny day, we encounter the tick, surely not one of man's or dog's favorite companion animals. The tick is a living machine, a machine with a sensory system that is receptive to the chemicals of mammal sweat (butyric acid) and to warmth; other perceptions are minimal, less acute, screened out. The tick is blind and deaf to the wavelengths mammals perceive readily. When the tick "smells" butyric acid, its muscles release and it falls, finding itself now on something warm. The warmth sets off feeding, as the tick pumps itself with the blood found by plunging into the warmth. One could build a machine that would react to these stimuli and perform these functions. But the machine would not yet be a tick because it could not perceive, could not change its reactions ever so slightly to accommodate to its goal. It is the process of perception that distinguishes the living tick from the mechanical invention.

Where does this perception occur? The answer is that each and every living nerve cell is a receptor that both perceives and causes action. Cells cluster in a way that makes the cluster sensitive to a certain kind of stimulus alone—a triangle, a color, a movement. When such a cluster is affected by the sensation, when it senses, the cluster reacts. Here, presumably, is the physiological analog of the innate releasing mechanism. In 1962, fifty years after von Uexküll suggested the necessity of such structuring, the research of David Hubel and Torsten Weisel demonstrated that the mammalian brain does in fact react specifically to certain perceptual features, such as shapes, colors, and motion, and their several combinations.[20]

Like the person who eats only the frosting from the cake, or, as von Uexküll puts it, like the gourmet who picks the raisins out of a cake, the tick selects one aspect of its environment. In this example, it selects butyric acid or warmth: "The first task of Umwelt research (of building a model of the Umwelt of any being) is to identify each animal's perceptual cues among all the stimuli in its environment and to build up the animal's specific world with them. The raisin-stimulus leaves the tick quite cold, whereas the indication of butyric acid is of eminent importance to her. . . . As the spider spins its threads, every subject spins his relations to certain characters of the things around him, and weaves them into a firm web which carries his existence."[21]

Each animal, each individual, has conceptions of many aspects of space. I have tactile-space, the space I use when I write with a pen or pick out an object to grasp; taste-space, a different taste in my mouth depending where I put my tongue; a smell-space; an auditory-space, one with softnesses and loudnesses that signal distance and help me locate the presence of other objects within this space; and visual-space, a space that uses clues about the shading and interposition of objects to tell me something of their size and distance. These spaces are united into one grand bubble, the total umwelt, an umwelt characteristic of me, one that changes constantly and yet represents my knowledge of my universe. It becomes nothing less than my private universe of experience, knowledge, associations, and memories, along with the here and now of sensory experience. The tick does not have my visual-space, at least not in the sense that I do when I use my eyes, yet it has a chemical-space that both characterizes and overwhelms the abilities of my umwelt. Consider that insects and mammals who are nocturnal live in an umwelt composed mostly of tactile and auditory space, while primates, including human beings, live in a highly visual umwelt.

When von Uexküll constructs umwelts for different species, we find something akin to those shown in Figure 7.2, and 7.3. a visual umwelt as seen by a human being, a candle-lighted chandelier, is rephotographed to show how it is perceived visually by a fly. We may say, my, how poor the vision of the fly; how indistinct the result! But to say this is to miss the point. If the fly's neural circuitry contains a cluster that evaluates the gradient of light and dark, the difference between the lighted and structural part of the surround is remarkable even to the human eye. Just as the tick picks butyric acid from the available smells, so the fly may respond not to those objects that give structure to a human being's visual experience, but to a composition that gives structure to the eye and mind of the fly. The fly may well respond not to static structure, as represented by the photograph, but to motion, or to some cross between chemical cue and motion. Human fly-swatting is an apt example of the difference between the perceptual umwelts of the two species involved.

My umwelt is of the buzzing, annoying tactile fly flitting through my universe which, at that moment is dedicated to whatever else is for that

Figure 7.2   A phenomenological approach to animal behavior is concerned with the animal's perception of the world: this composite is known as the perceptual umwelt. The figure shows the human beings' perception of a chandelier (under low illumination). Compare it with Figure 7.3.

moment occupying my umwelt. That of the fly is—I do not know. I can, like von Uexküll, try to imagine the umwelt of the fly by photographic means. While this technique may make the point to the human observer, it surely does not match the full umwelt of the fly, for these necessarily involve motion, shape, intensity, and smell: Who knows which and in what degrees?

The umwelt of the individual of any species is changing constantly; such change is what living means. Each change in environment is a recast umwelt. I recognize a previously experienced umwelt and react according to my own motivational state (I am hungry, frightened, eager and antic-

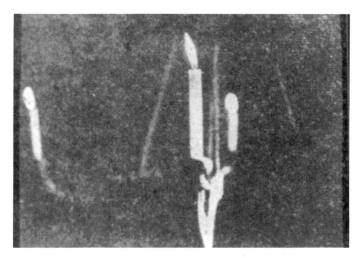

Figure 7.3   The perceptual umwelt of a fly. The image is that of how a fly would perceive the image of the chandelier shown in Figure 7.3.

ipating). To understand the behavior, then, we start with a description of the perception, for it is the perceptual world that determines the umwelt. We then determine with which FAPs the perceptions are associated, and we build a map of these relationships. The map of the umwelt is our definition of the species.

The problem for any approach that tries to create the perception of an animal (or human being, for that matter) is that we cannot agree on how to do it. The behaviorist measures rate of response; the psychoanalyist translates the obvious content into the unconscious wish; the phenomenologist and, I think, the ethologist, must look at how to represent the perception in a way that permits it to be translated into the mind of the observer. Let us examine one such attempt, returning, for this event, to the life of the dog.

BUYTENDIJK'S DOG

Perhaps the greatest mystery of the mind is that of *attention*—namely, how does a living being shift from one perception to the next, making perception of an object once loved now loathsome, ignoring food one hour and smelling it clearly the next? How do we account for changes in perception in terms of attention? Surely, the ring of the telephone, the siren, and the oncoming train are examples of seemingly sudden shifts in the perceptual environment that radically alter the umwelt. Some causes of such shifts appear to be motivated by internal factors, such as hunger. Other times, I move through my life's activities with constantly changing umwelts. Now I feel my fingers writing; I hear music; and, if I attend to it, I can feel the tactile pressure on my behind. What explains my ability to shift attention—to alter the umwelt? Or is it the reverse: Does a change in the umwelt appear to me as a shift in my attention?

When Lubbock trained Van, or B.B.E. and Yerkes asked Roger to perform, what was occurring in the umwelt of the dog? (Or in the umwelt of the human investigator, for that matter?) One guess is that the dog's attention shifted toward the sight of food, along with its smell and potential taste. Surely the training focuses attention on the aspect of the situation that signals a change in perception. Just as the tick senses the acid, so does the dog see a change in the trainer's configuration. To the human being, the change is so slight that it is often unseen, but as one aspect of that configuration comes to have significance—indeed, comes to predict the probability of food or a pet—that aspect of the umwelt fills the bubble. Roger and Van now have an umwelt (restricted, no doubt, to this situation) that sees, hears, and feels little compared to the perception of a twitch of the human mouth, the direction of a gaze, the leaning forward. Like the umwelt of the fly shown in Figure 7.3, the significant aspects of the umwelt are merely whatever becomes significant to the perceiver. It is not my task to tell the fly what it should be seeing. Perhaps the result of training such as that given Roger, and Van, and Hans is to concentrate aspects of the umwelt. Attention is turned to some form now sensed by the animal, but not one perceived in like form by the human being who is occupied with the aspects of her or his umwelt.

F. J. J. Buytendijk (1887–1974), who taught at Groningen just before World War I to just before World War II, attempted to show how one could measure and construct the umwelt of an animal. His choice of the dog is fortuitous for our discussion. His choice of methods showed what can be done, and done well, when the methods of experimental science are applied to the tenets and goals of phenomenology.

As we know, some dog breeds are especially domesticated for their ability to smell. In Buytendijk's study of the dog's sense of smell, chemicals were placed behind doors, with counterbalancing done to eliminate the role of experience, and the dogs were trained to search for a specific odor. Clearly, the designated odor could not be found unless its smell was within the animal's sensory range. Buytendijk found that a police dog was able to discriminate an odor from a nonodor at a threshold approximately one hundred times less than that possible for human beings. This investigator became intrigued with determining how dogs are able to track a smell, but the chief problem in such studies has always been how to separate the smell of the person leaving the odiforous object from the object itself.

Another investigator, Hr. Konrad Most, had arranged a trapeze that ran along a rail by pulleys: the trapeze was fitted with a seat, so that a person could travel along the route, leaving odors wherever, while presumably keeping her or his smell from ground level. Most's results, some of which should be criticized on methodological grounds, were taken to show that the dog trained to follow human odors did so by responding to change—that is, not to the stimulus, but to a change in its intensity, say, from strong to weak. The greater the change, the more likely was the dog to shift course and follow; the smaller the change, whether the stimulus itself was strong or weak, the less able was the dog to trace. These data nicely suit the phenomenologist's idea that it is the *change* in the umwelt which is the stimulus for a change in behavior, not the nature and composition of the umwelt alone, and surely not a simple stimulus-response association.

To study the dog's vision, a set of eleven boxes was constructed, each placed in a fan-shaped kennel run, with the boxes at the more opened end. Buytendijk found that when he placed food behind the zinc (to eliminate odor) door of number 6, it took the dog sixty opportunities to learn to return to box number 6. That is, the dog tried other boxes among these first sixty. It is tempting to think of these attempts not as errors, but as results of shifts in the umwelt that lead to the trying and testing of additional hypotheses regarding the likely location of food. Buytendijk also tested a human child and refers to data on apes in a like situation. He points out that all of these creatures showed delay in learning the simple response of going from the kennel to box number 6. They routinely tried other doors, before settling on going directly to number 6, every time.

Why should animals try other solutions? Why do they not merely return to the place in which food was last found? Is the trying of other doors merely "error," the kind of mistake that occurs whenever we learn by repetition, whether it be how to form letters or how to remember telephone numbers or to spell? Or can it be that the error itself is significant because the trying of alternatives tells us something very important about how the perceptions test the imagination and how the mind works? What were Clever Hans and Van and Roger and, yes, Little Hans really learning?

To some human observers, the beings were learning a task that permitted them to communicate between the umwelts of members of different species. But to the creatures themselves the human cues were but one more aspect of the total umwelt, a slight change in it that came to have a meaning distinct from the other components of the perception. Presumably there were many aspects to the umwelt of Clever Hans. The card itself may have been far less salient than the movement of the men's bodies in Hans's umwelt. Hans may have come to respond to the movement, not to the card. The signal to Hans was a shift in the characteristics of the umwelt; but to the human observer the shift came from the men's movements of their chests and trunks.

RELIABILITY AND VALIDITY

Our usual way of evaluating these assessors does not work when we examine theories based on perception. If we do not think deeply, we may reach the absolutely wrong conclusion that the theories are thereby themselves unreliable and invalid. That our usual way of thinking is inapplicable may be the obvious explanation as to why this form of conceptualization has never been popular in scientific analysis of a certain kind; namely, the deductive method. These are many "sciences"—many ways of knowing—and we would be parochial if we limited our understanding of ourselves to that which is deductive.

We human beings must come to know or picture the umwelt, for any serious development of this line of thinking requires that we have some means to model the characteristics of the umwelt. Otherwise we have theory without any possibility of data to test it. Phenomenology has not been popular in places and times that emphasize the collection of data, because no matter how attractive the idea of an umwelt may be, without a way to test the idea, its success cannot be measured.

Theories based on the nature of perceptual phenomena have more in common with evolution than with behaviorism. Like evolution, perceptual theories work to explain many seemingly unrelated acts and observations. The validity of the model is unlikely to be proved or disproved by a single experiment, no matter how well designed it may be. The validity of such theories is based largely on the logical structure that holds together the framework and that gives it substance on which to hang new discoveries. But the study of the perception of people and animals is worthwhile not merely because it is heuristic. An understanding of perception is logically necessary to any theory of phenomenology and, I argue, ethology. What restrains us is the problem of mental-mapping, for it is as if we are asked to formulate our two-dimensional understanding of the world, that of time and space, into a multidimensional universe whose coordinates we do not see.

PEEKING INTO THE FUTURE

We have now considered three ways to think about the mind: the psychoanalytic, the behavioristic, and the perceptual. We have examined experiments and observations performed by people on animals and children that illustrate the kinds of thinking that provide examples of these ways of thinking and investigating. When compared, they are not contradictory, for upon examination they are seen to share commonalities of both spirit and method.

Meanwhile, in the early years of this century, when these methods were being developed in Europe, a number of chimpanzees, Old World monkeys, New World monkeys, dogs, cats, chickens, fish, and people, especially those in the United States, had been involved in showing experimenters what the animals, could do and learn. Interest in human-animal interaction had reached a peak among researchers. The Mental Ladders (First mentioned in Chapter 3) were being constructed as models of the different amounts and degrees of intelligence enjoyed by different species and genera. The relationship between people and primates, the location of the "missing link," the wish to communicate with animals—these goals came to the fore. On the one hand, people began to see themselves as members of the world of living animals; on the other, they worked to show how, intellectually, they were separate. The nature-nurture questions asked of the feral; the questions asked about animals' abilities to read and do math, to do activities that we people think unique to ourselves, were giving way to a different set of questions.

We begin the analysis in the next chapter with Peter, a chimpanzee who finds himself reaching, grasping, and performing to furnish people with an example of "the missing link." But, of course, the link is only missing if one thinks there ought to be one.

The Mental Ladder   PART III

Charles Darwin's collecting and accounting of compelling
evidence for the effect of natural selection on the evolution of
species necessarily emphasized the structure of animal species.
But species may be characterized as well by their behavior and,
perhaps, by their thoughts and feelings. Darwin knew the
potential power of this idea, for his last book was concerned
with the evolution of human and animal emotions.

Some readers believed that the nature of animals' structures
could be used to erect a metaphorical ladder, this being a
hierarchy of animals organized, for example, by order, genus, and
species. If such a ladder of physical structure could be organized,
could we also organize a *mental* ladder, one that arranged animal
life by its kind or quality of thinking or feeling?

The idea of a Mental Ladder is similar to that of the Chain of
Being, but the apparent similarities may deceive us. To illustrate,
this part considers the importance attached to monkeys, our
fellow primates, in the building of the Mental Ladder.

Peter and Moses: Chimpanzees Who Write

8

T HE WISH TO design and erect a Mental Ladder, one that arranged animal species by their intelligence, followed by half a century the understanding of how to build such a ladder based on structure. Darwin's initial understanding of the evolution of the physical aspects of species set in motion the building of a metaphorical ladder of structure, a hierarchy of species based on physical traits.

Although Darwin clearly believed that a like-constructed Mental Ladder was feasible—his 1872 book on emotions makes this evident—the attempt to build one was surprisingly slow in coming. While naturalists rushed to find the similarities and differences among animals in order to identify genera and ancestral relationships, attempts to make such comparisons for mental activity were scarce. The lack of a fossil record for mentality might be a chief reason, but so might the fact that human beings regard mental activity as more mystical, and less measurable, than bones and their structure.

The Ladder of Mental Life comprises different degrees of development of animal life represented by the rungs. The upright poles form the scaffold that holds the rungs in place from top and bottom, from, say, amoebae to ourselves. The idea originated with Aristotle and was rejuvenated and diagrammed by the Victorians, most likely because Darwinism gave the idea of the Mental Ladder a certain prominence and respectability. Romanes, writing in 1882, tried to meld the newly formed principles of evolution into a social philosophy, and offered a ladder of mental abilities of species and genera and the emotional potentials of humankind. His ladder is re-created in Figures 8.1, 8.2, and 8.3. These figures display, in a way that mere words cannot, the goal and attributes of the late Victorian Mental Ladder.

The proposed ladder of intellectual development starts with the lower rungs representing protoplasmic movements (performed by protoplasmic organisms, such as paramecia and amoebae), "rises" through various levels of neural activity, and reaches memory with the echinoderms. Higher organisms are capable of instinct, association, recognition, rea-

	Products of Intellectual Development	The Phycological Scale
50		
49		
48		
47		
46		
45		
44		
43		
42		
41		
40		
39		
38		
37		
36		
35		
34		
33		
32		
31		
30		
29		
28	Indefinite morality	Anthropoid Apes and Dog
27	Use of tools	Monkeys and Elephant
26	Understanding of mechanisms	Carnivora, Rodents, and Ruminants
25	Recognition of pictures, understanding of words, dreaming	Birds
24	Communication of ideas	Hymenoptera
23	Recognition of persons	Reptiles and Cephalopods
22	Reason	Higher Crustacia
21	Association by similarity	Fish and Barrachia
20	Recognition of offspring, secondary instincts	Insects and Spiders
19	Association by contiguity	Mollusca
18	Primary instincts	Larvae of Insects, Annelida
17	Memory	Echinodermata
16	Pleasures and pains	
15		Coelenterata
14	Nervous adjustments	
13		
12		
11	Partly nervous adjustments	Unknown animals, probably
10		Coelenterata, perhaps extinct
9		
8		
7	Non-nervous adjustments	Unicellular organisms
6		
5		
4		
3	Protoplasmic movements	Protoplasmic organisms
2		
1		

Consciousness (spanning rows 14–18)

son, and, eventually, the capacity of "imagining indefinite mortality." Such imagining is characteristic of ape and dog but not of monkey, elephant, or those lower on the scale. Consciousness begins with the coelenterates. So says the phylogenetic ladder that places different phyla on the rungs.

Emotional development in humankind also progresses stepwise and rungwise until, by fifteen months of age, the emotional feelings and reactions known to human beings are present. Here is the ontogenetic ladder that traces the development of the individual of the species. Romanes does not say so directly, but he apparently assumes that emotions arise from one another: that, for example, cruelty arises by differentiation from emulation. The result of such scheduling is seen in Figure 8.2., this being the tree that shows the organization of mental and emotional life.

The separation into emotion, will (or motivation, as we would call it today), and the intellect (or reason) is a longstanding categorization of human ability. Curiously, theories of the mind have often been divided into three parts, the number "three" appearing to hold a magical import for intellectual philosophers. The three-category notion arose during the Middle Ages and reappeared in nineteenth-century thought, when it became known as faculty psychology. Each of the three aspects, or faculties, of the mind—reason, emotion, and will—was now thought of as a separate faculty. Cognition (reason), emotion, and motivation (the will) remain central in our times, as examination of any university curriculum in the study of psychology will show. Freud propounded one of the most influential theories of our times—psychoanalysis—when he used the concepts of the id, ego, and superego to represent separate but interactive mental faculties.

The Great Chain of Being, and its step-sibling (but not genetic twin), the concept of the Ladder of Mental Life, is central to the development of humankind's thinking about animals. A link can exist, however, only if some sort of separation is to be found among individuals, species, and genera. Without a chain or ladder, there is no need to look for a link, but as differences in form and mentality are measurable, there must be a chain, and a chain requires links. Because there seem to be no links, and if we continue to believe in the concept, it must be that it is the links themselves that are missing. Therefore, goes the logic, we should search

Figure 8.1   A late-nineteenth-century view of a mental ladder, this one demonstrating a comparison of emotion, the "intellect" (cognition, in today's language) and the will (motivation). 8.1 arranges animal forms in terms of their intellectual development. Note the position given the (domestic) dog, who is thought to be aware of the concept of indefinite morality. "Reason" begins with Crustacea, and "recognition" of others with reptiles and octopus. The drawing is by Georges Romanes who set out the categories and the "animal ladder" shown.

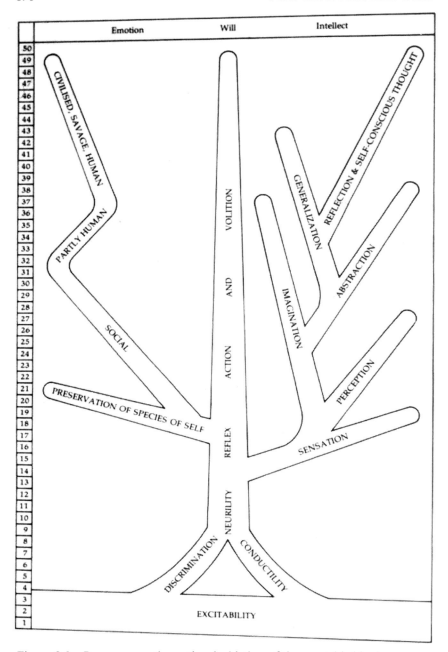

Figure 8.2   Romanes continues the elucidation of the mental ladder by
showing the relationships among emotion, the will (motivation), and the
intellect (cognition).

	Products of Emotional Development	Psychogenesis of Man
50		
49		
48		
47		
46		
45		
44		
43		
42		
41		
40		
39		
38		
37		
36		
35		
34		
33		
32		
31		
30		
29		
28	Shame, remorse, deceitfulness, ludicrous	
27	Revenge, rage	15 months
26	Grief, hate, cruelty, benevolence	12 months
25	Emulation, pride, resentment, aesthetic love of ornament, terror	10 months
24	Sympathy	8 months
23		5 months
22	Affection	4 months
21	Jealousy, anger, play	14 weeks
20	Parental affection, social feelings, sexual selection, pugnacity, industry, curiosity	12 weeks
19	Sexual emotions without sexual selection	10 weeks
18	Surprise, fear	7 weeks
17		3 weeks
16		1 week
15		Birth
14		
13		
12		
11		
10		Embryo
9		
8		
7		
6		
5		
4		
3		Ovum and
2		
1		Spermatozoa

Figure 8.3   The emotions, as we know them, begin with the larvae of insects (capable of surprise and fear) and make their way "up" the ladder to human beings and dogs, who show shame and deceitfulness, among other emotions.

for what is missing—namely, the missing link. From the assumption flows the ensuing logic; from the logic flows the search for the links between the minds of animals.

To illustrate the idea of building a Mental Ladder, the next three chapters describe how monkeys came to be understood as the chief missing link in the chain or ladder. The wish to design and erect such a ladder appeared simultaneously with people's discovery of the mental abilities of monkeys, apes, and baboons. Significantly, our closest genetic relatives, the apes, monkeys, and baboons, were the last group of animals to attract our attention. That attention began clearly enough near the turn of this century, and it has passed, to use an industrial metaphor, from being a cottage industry to an international industry that may now be regarded as "big business." Only within the last century have primates become the product and subject of this multinational corporation. In this period, they have become subjects for medical research of interest and value to people, subjects in experiments and demonstrations whose intention is to show the possibility of human-ape communication. They have also become an endangered and threatened form of life. The number of primates used in research each year is staggering. Each newly found human disease offers yet another threat to the animal primate population as we discover that their genetic similarity to us means that they, rather than we, serve as research subjects for our benefit.

AT THE KEITH THEATRE

On the evening of Monday, September 14, 1908, Lightner Witmer (1867–1956), a resident of Philadelphia, Pennsylvania, was visiting Boston, and decided to go to the Keith Theatre. His chief purpose, however, was investigation, not entertainment. Witmer was on the faculty of the University of Pennsylvania in Philadelphia and had become interested in the assessment of intellectual abilities. He had founded and now directed a clinic, said to be the first psychological clinic in the United States, which measured the intellectual ability and potential of persons of uncertain ability. After testing, advice was given as to appropriate placements in school and institutions. Today, such testing is commonplace, and through mental tests decisions are routinely made as to whether children profit most from "special education," whether they are capable of "advanced placement," and, in other cases, whether they should be placed in residential institutions, given foster care, or remain at home.[1]

Witmer's Psychological Clinic, seen from our contemporary viewpoint, was the first representative of what would become another major industry—testing. This industry was not limited to the placement of children into educational facilities, but also included the assessment of those who would enter college, graduate schools, medical schools, law schools, and the like. Millions of tests are administered annually in order to provide useful information as to achievement and ability; each of us carries with us a lifetime of scores that have shaped our future. Our intelligence has been measured by a test at various times in schooling. College Board scores have much to say about which college accepts us for further education, and our success at training for a profession is determined in part by the score we achieve on standardized tests. When Witmer was introducing the idea of intelligence testing as a means of aiding society and the people who comprise it, he could not have foreseen how tests would be used in our time. His interest, like that of Alfred Binet (1857–1911) in France, was in the intellectual differences between and among people and in the practical issues of how to best utilize, not exploit, the talents of individual people. He saw his approach as progressive for both individuals and society.

On the night when Lightner Witmer was preparing himself for an evening at the theatre, where, by the way, a chimpanzee was to appear, Little Hans, the child, had just reached the age of six, his fear of horses having only just been solved, and Clever Hans's career with von Osten had only just come to an end with von Osten's illness. The European and North American mind was alert to the implications of Darwinism, for the generation that followed Darwin had keenly understood the implications of natural selection for the organization of society, the importance of genetics, survival of the fittest, and extinction of the weak. The notion of animal intelligence was very much in the background of established thought. The importance of knowing how to measure the traits of individuals was now evident in the United States and a matter of societal and national concern. How else could the measurement of the traits of individuals be translated into social good for human beings unless the individuals' traits could be measured?

The evening at the theater was not Witmer's first experience at observing primates. Earlier in his professional life, Witmer had observed an infant orangutan as part of his clinical studies. It was his belief, expressed to the investigator William Furness[2] who acquired the orang for him, that "there was no reason why an ape who could be trained to ride a bicycle might not be taught to articulate at least a few of the elements of language, and I [Witmer] expressed the desire to undertake this experiment [teaching an ape to communicate] the outcome of which, whether successful or not, I thought would be an important contribution to animal psychology."[3]

With this remark, the modern study of ape language was under way. Within the century, these studies would become an expensive and productive industry. Having begun experiments with the orang in Philadelphia, Witmer was intrigued to see the show advertised for the Keith Theatre in Boston which featured a chimpanzee, Peter by name. Peter, said the advertisement, was *"born a monkey and made himself a man,"*[4] thereby reversing the more common occurrence.

Although there is no specific evidence that Witmer knew about Clever Hans at this time, he was aware, he writes, "of the difficulty of judging the intelligence of an animal from a stage performance. So-called educated horses and even educated seals and fleas have made their appeal in large numbers to the credulity of the public. Can any animal below man be educated in the proper sense of the word? Or is the animal mind susceptible of nothing more than a mechanical training, and only given the specious counterfeit of an educated intelligence when under the direct control of the trainer?"[5]

There were good questions, indeed. Pfungst and Krall and Witmer would have had much to say to one another had they had the opportunity to meet. Clearly, Witmer, knew how to ask the right question, and his motives were known: His reasons for studying children and animals were related to his wish to assure each child the education and training of which she or he was capable. He appeared to have faith that a democratic United States would lead to a new world order, one based on government providing for the intellectual needs of its individual citizens. He was an intellectual soulmate of his fellow citizen and (suburban) Philadelphian, Walt Whitman. He believed a Psychological Clinic was needed, not to be able to show that numbers could be placed squarely and accurately on the abilities and potentials of people and animals, but to demonstrate that an understanding of differences between people and between species would help us understand our natural world and enhance individual potential.

Let us proceed now to the Keith Theatre, and to Witmer's account:

"I knew before I entered the theatre that I should be unable to judge from a mere stage performance whether Peter had been in any real sense educated or simply trained in the blind performance of a few tricks. Nevertheless the performance deeply impressed me, as it must anyone, with the expertness of the animal in skating, riding a bicycle, drinking from a tumbler and eating with a fork, threading a needle, lighting and smoking a cigarette."[6] Figure 8.4 shows Peter, in evening dress, at dinner, while Figure 8.5 shows Peter sharing the lighting of a cigarette.[7]

Witmer was a skeptical and conservative viewer. Not only was he mindful that stage performances say nothing of importance about whether an animal (or human, for that matter) is educated or trained, but he was also aware that even idiots are capable of astonishing feats. That is, training and education are not necessarily related. "Idiots capable of unusual dexterity and possessed of musical or even mathematical ability, yet in other respects markedly subnormal in intelligence, are not unknown," he writes.[8] Further documentation of such idiot-savants reinforces our appreciation of the awesome abilities shown by some of these people who are incapable of normal living in other respects.

Peter's performance at the theatre suggested to Witmer, however, that intelligence, not mere performance, was to be seen.

". . . in riding up an inclined plain and down a small flight of steps, he [Peter the chimpanzee] allowed himself to go very close to the flies, and to save himself put out one hand and cleverly pushed himself away, while still retaining his balance. I also saw him ride as close as possible

Figure 8.4   Peter at dinner. The original photograph is labeled "An anxious moment: Don't I get my drink?"

to the side wall of the stage setting, and take a very short turn with the evident purpose of seeing whether or not he could do it. From time to time I observed that he made the work more difficult for himself than was needful, seemingly out of a mere bravado and in pure enjoyment of the task."[9]

This sort of intelligence—the ability to execute a task, that was complex even for human beings, the seeming alteration for the "pure enjoyment of the task"—interested Witmer most. To ascribe an IQ number to a person may well determine the kind of training or education by which the person can profit, but it never quite tells us what we want to know about the person's motivation, ability to do something more, or interest in doing something imaginative.

Test scores have not earned our full trust because we know that intelligence has another component, one that tells us something essential about the person or animal. Witmer understood this. He saw Peter doing five performances. Then he made arrangements to take Peter to the Psychological Clinic in Philadelphia where Peter was "tested" in the same way that Witmer had measured children's abilities and traits. By so doing,

Figure 8.5   Peter shares a cigarette. The human being is unidentified, but he is not Witmer. There is no evidence that Peter was a habitual smoker.

Witmer had begun the construction of the Ladder of Animal Mentation or, at least that part of it concerned with the placement of the primates, human and nonhuman.

PETER AT THE PSYCHOLOGICAL CLINIC

Mr. and Mrs. McArdle were Peter's trainers and owners, but they knew very little about Peter's background, having acquired him only a few months previously.

Peter arrived in the United States by ship from England. On shipboard, in order to keep Peter from climbing the rigging where he did damage and was impossible to capture, roller skates were put on his feet. Undaunted but skilled and imaginative, Peter learned to skate on deck within two days. By the time he arrived in the United States, he was an accomplished skater, so much so that it was decided he should go on the stage. The McArdles claimed that Peter threaded a needle the first time he tried to do so and that he acquired the skills in using human tools, such as hammer and screwdriver, seemingly without training. He was also able to select from a number of keys the proper key for a lock. Peter astonished Witmer while he was talking to the McArdles at the Clinic by dashing away to use a faucet to wash his hands and take a drink of water. Witmer gave Peter some tests, the same tests, as Witmer noted, that he used to study the mind of the child.

Witmer explored the notion that Peter, and perhaps chimpanzees as a species, was the "missing link" on the ladder, that rung between human beings and the higher apes. Witmer was aware of the then available fossil evidence supporting the descriptions of the Neanderthal man and Java man as prehuman. While the years since would produce the discovery of additional fossil evidence and witness the postulation of a number of extinct primate forms linking humankind with the other primates, the gap remains large, if not so large as it was at the turn of the century when the fossil evidence supporting the idea of a Great Chain of Being was first being gathered. The search for the missing link was among the most engaging of scientific topics of the day. Witmer was neither wrong nor silly to think that a fossil or living being would be found who showed the fine gradation in form and intelligence to be found between human-kind and the fellow primates. Witmer and others of his time assumed that the finding of the link was almost in their grasp. They were wrong.

To return to the Clinic, "On the morning of October ninth [1908] Peter skated into the clinic with a breezy rush. He was clad in black cloth trousers, waistcoat, and Tuxedo coat, and wore a starched shirt and collar with a red neckerchief instead of a cravat, which he later pulled off. He had on striped socks and patent leather oxfords, as well as skates, and on his head he wore a small silk hat kept on by elastic. He dashed straight for an open window, which had to be closed immediately. During his stay at the clinic he looked often at the window, apparently interested in passing cars which he could plainly hear and just catch a glimpse of. His excursion to the window over, he skated about the room, apparently ready to shake hands with the company present. My secretary, who took stenographic notes of all that occurred, bent down and offered her right hand; he took it, and after giving it a shake, put the back of her hand to his lips in the most courtly and gallant manner. Then he skated off, round and round a platform, pursued by Mr. McArdle, turning expertly and dodging with remarkable celerity, from time to time stopping to thump the platform in apparent fun and bravado. He then climbed upon a chair and began to examine a camera with great interest, tried to turn the screws, squeezed the bulb, manipulated the shutter, and felt the bellows. These movements were executed with precision and dispatch, and with no attempt at destructiveness, but rather in a spirit of pure investigation. During his stay at the clinic he skated at intervals about the room, ap-parently for the sheer love of it. After one test which involved a consid-erable strain upon his attention his trainer said, 'You may not get down and run around and play.' He instantly darted off, skating round and round the room, from time to time inciting Mr. McArdle to pursuit by thumping upon the platform.

"During the tests he sat upon a small three-legged stool, eight or nine inches high, which was placed upon a low kindergarten table. Being on skates he was thus confined to a small area. He stood and moved about on the table from time to time, never once slipping or losing his bal-ance."[10]

Here are the tests administered and descriptions of Peter's behavior:

The cigarette test. Peter was offered a cork-tipped cigarette. Peter reversed it so the tip was in his mouth. The cigarette was taken from him and offered to him, again, in various ways. Peter always manipulated the cigarette so the tip went in the mouth. When given matches, Peter tried to strike them. Told to spit, he spat on his shoe; told to spit elsewhere, he spat on the table. The trainer wiped Peter's shoe; when given the cloth for wiping, Peter wiped the table clean.

Stringing beads. Witmer used a task that he employed with the children to assess their intelligence. Witmer took a string and showed him three times how to place the string through the bead. Peter promptly put the bead in his mouth, whereupon Mr. MacArdle said, "No, no, it's not a cherry" at which point Peter took it from his mouth and put the string through it. Mr. MacArdle criticized Witmer, saying that the task was made too easy, that it was enough to show Peter the box of beads and string.

Peg board. Mr. Witmer put three pegs in a row. When Peter had the opportunity, he placed the pegs in the holes, but the locations were irregular and unformed to the human eye.

Lock opening. During his stage performances, Peter used a key to open a lock. In the Psychological Clinic he was offered a smaller padlock of a different type, one that required a different motion to open. The padlock had a bar (unlike spring locks, now common) that required removal and replacement. Mr. Witmer showed Peter how to do this, and Peter did so promptly. Mr. Witmer then showed Peter how to reinsert the bar. Peter reinserted the bar properly and returned the key to Mr. Witmer upon request. Peter showed like ability in using a screwdriver and opening a box.

Form board. The formboard was, and is, much used to distinguish degrees of intelligence among human beings. The task requires the placement of the forms into the corresponding holes. The kind of perception and movement involved remains a standard measure of intelligence in intelligence tests. Peter did not succeed in placing the blocks in their places. When he was shown how to do so, he merely moved the block around the board. When a piece was placed in the appropriate square and pounded twice to help it fit, Peter repeated the pounding, but placed the block incorrectly. He seemed pleased that he had pounded the block twice, and he showed no further interest in the task. Mr. Witmer suggested that Peter might be nearsighted, or that he took the task to be one of imitation, rather than correct placement. (Witmer's grasp of the importance of psychophysics was excellent: he had written a laboratory manual on the subject.) The lack of success led to the next task, the writing test.

The writing test. Figure 8.6 shows the blackboard, as rearranged and drawn by Professor Twitmyer, which depicts the results of this important task. Mr. Witmer's description best conveys the task and the results:

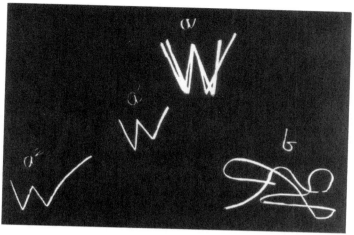

Figure 8.6   Peter's blackboard writing. A, the letter **W** twice drawn, tracing one **W** over the original; B, the scrawl produced on the first attempt; A1, Peter's copy after the second trace; A2, the second effort when Peter is asked to again make the **W**.

"I drew forward a blackboard, the writing surface of which he could easily reach when standing upon the table. He took a piece of chalk eagerly, and before I had made any mark upon the board, began to scrawl in a corner of it. I took the chalk from him and said, 'Peter, I want you to do this,' and rapidly made the letter **W** in four strokes. Peter's attention had not been fully given while I made the letter. He took the chalk and scrawled beneath in much the same manner as he had done before. I picked upon another piece of chalk and said, now look, this is what I want you to do,' and traced another **W** over the one which I had just drawn. Peter watched the operation intently, then with the chalk in his hand, he quickly made the four movements and drew a fairly perfect letter beneath the **W** which I had traced. After a brief interruption due to the excitement of the spectators at this performance, Peter's interest in the board still remaining as appeared from his continued scrawling, I asked him to try again, and he made at some distance from the first letter another **W**, somewhat less perfectly formed."[11]

Dr. Witmer concluded his description by commenting that Peter seemed to him to have been "motor-minded." By this he meant that Peter imitated motor actions. When seeming to write the letter, he imitated, Witmer thought, the movement of the hand, not the perceptual outcome. "I doubt," writes Witmer, "whether he could copy a **W** on the board if he had not first seen me make it."[12] The distinction between the ability to copy a motor action and the ability to watch a result and copy the result is, indeed, an important one. The latter requires some form of memory storage in which the brain stores the image until the hand can perform the actions. As Witmer remarks, the child first learns to copy by imitating the motor movements; only later, if at all, does the child come to reproduce an image.

Peter, the potential missing link between humankind and animalkind, had now demonstrated what he could not do, and Witmer found a clue as to the way in which Peter was not human, or at least not a "normal" human. The normal human mind, when coupled with functioning motor behavior, is able to copy images by both storing them and retrieving them. In this limited test, Peter was not able to do so. So Witmer was led to ask about Peter's ability to speak, speech being, as some still believe, the ability that sharply separates humankind from animal. But Witmer's conclusion about Peter was a provocative one. His comment has echoed down the ages to us, for it is possibly of more importance to today's evaluations of animal behavior than it was in Witmer's day.

"*I now believe,*" Witmer writes a year later, "*that in a very real sense the animal is himself giving the stage performance.* He knows what he is doing, he delights in it, he varies it from time to time, he understands the succession of tricks which are being called for, he is guided by word of mouth without any signal open or concealed, and the function of his trainer is exercised mainly to steady and control."[13] But let us make our own judgments: let us review what occurs to Peter, his trainers (Mr. and Mrs. McArdle), and the observers.

Apparently, it was Mrs. McArdle who was involved in trying to teach Peter to speak. Her technique was to have ready a glass of water, water being something that Peter wanted often. She said "Mama." When Peter was judged to say the word more and more clearly, he was given a drink of water. Witmer judged that Peter was able to articulate the word "water," but that he did so with "great effort" and "unwillingness." We might note, from the advantage of a century, that at the time it was believed that apes could never speak, because they did not have the requisite equipment either in the brain or in the vocal apparatus. Witmer tells us that Peter's articulation of "Mama" was weak indeed. Only the "M" sound was clear, and the ah sound was of doubtful accuracy; the voice was a "loud whisper," not a "voiced articulation."

"Peter's chief fault," remarks Witmer, "is one that I have found occasionally in young children showing an arrest of speech development. He tries to speak with the inspired and not with the expired breath."[14] Witmer, in private, taught Peter to shape the mouth appropriately for expired speech and reports that he was able to do so: Peter was able to articulate in humanlike fashion, but he was not trained to do so. He could articulate, with appropriate training, this presumably being of the kind, that the child receives from being around speaking persons.

Could Peter understand language as well as articulate? "When Peter is asked, 'Where is Mama?' he points to Mrs. McArdle. When asked, 'Where's Dada?' he points to Mr. McArdle. When asked, 'Where's Peter?' he taps his shirt front."[15]

THE MISSING LINK: TOOLS AND LANGUAGE

Witmer's view of the missing link was shaped, of course, by the beliefs and intellectual myths of the times. If one believes that there is gradation among and between animal species, especially between animals and humankind, then a seemingly large gap can be made to appear smaller if some intermediate form of life can be found. Because a large gap exists between the great apes and human beings, as we human beings understand it, a link may be found between the two: it is not absent, we guess, but merely missing. The search for a missing link represents a shift in our human understanding of the order of nature. Surely Itard never thought of himself as working with a missing link between animalkind and humankind; rather, he believed he was working with a miniature, socially deformed, but educable, human being.

The link between the great apes and humankind was expected because the gap itself was created by the idea: human beings are articulate and communicate by articulation of the mouth. Here are the data as they were understood at the time. Great apes give calls but they do not learn language, presumably because they lack the requisite centers of the brain and the required tools to alter air into sounds; human beings invent, manufacture, and use tools, ranging from screwdrivers to automobiles. Some animals species may be opportunistic users of tools, but none creates, and surely not ever to the degree common among human people.

Language and tool use were aspects of the tests administered by Witmer; these aspects remain in use today. They are the chief characteristics that we human beings ascribe to ourselves in order to distinguish ourselves from the animals. If Peter could be shown to use some language and some tools, then it could be claimed that his abilities were neither solely apelike or solely humanlike, but somewhere between them. Peter, and perhaps all members of his species then, were links between existing animal forms and, at the same time, an explanation of the history of when and how human beings came to be what they are.

Witmer presented this case: the ability of human beings to invent and use tools "made intellectual achievement a controlling factor in natural selection and survival. The discovery and use of fire carried him far beyond the mere animal intellect and made possible human civilization and culture. Steam and electricity have initiated a new era of intellectual development."[16] Note the implication that intellectual development evolved from invention and, presumably, not that invention arose from intellectual development. Here Witmer anticipated an issue that will concern us in Chapter 9—the cause-and-effect relationship between culture and intelligence. And, Witmer continued, in regard to language, "Reason may appear as an attribute of the animal mind, but in the absence of language its manifestations must remain so insignificant as to be practically negligible in comparison with its varied employment by man."[17]

As Witmer summarized: Peter, the chimpanzee, had a tool-use ability well within the range of human beings. He used a hammer; and he used fire (to light the cigarette). He was probably right-handed, a fact that Witmer assumed to be more related to higher intelligence than left-handedness was: Peter passed the tool-use test. But could he learn to use language? Witmer observed that Peter already understood spoken language (but, then, so did Clever Hans, or so it seemed to some, including the September Commission) and, as we will learn, so it is to be claimed, does a gorilla (Chapter 11). That Peter was shown to be able to engage in what Witmer considered to be the three modes of language—articulation, hearing, and writing—suggests that Peter could learn to use language. His ability to read; however, was untested. Witmer concluded his analysis of Peter as tool-user and language-user:

"If Peter had a human form and were brought to me as a backward child and this child responded to my tests as creditably as Peter did, I should unhesitatingly say that I could teach him to speak, to write, and to read, within a year's time. But Peter has not a human form, and what limitations his ape's brain may disclose after a persistent effort to educate him, it is impossible to foretell. His behavior, however, is sufficiently intelligent to make this educational experiment well worth the expenditure of time and effort."[18] During the century following Dr. Witmer's conclusion, much time, effort, money, and argumentation would be spent doing just as Witmer anticipated, if not for the reasons that Witmer would have found sensible.

Witmer concluded his article in *The Psychological Clinic* of 1909, with awesome foresight: *"I venture to predict that within a few years chimpanzees will be taken early in life and subjected for purposes of scientific investigation to a course or procedure more closely resembling that which is accorded the human child"* (italics added).[19] Indeed, attempts to understand the mind and to communicate with the great apes, especially the chimpanzee and gorilla, would continue. By the last decade of the century, some ninety years after Witmer's investigation, chimpanzees in laboratories and zoos were undergoing language training with the expectation that human beings could learn to communicate with them.

Peter's act and abilities are all but forgotten in the present canons of animal lore. But the approach, that of presenting Peter tasks and determining how well or how poorly he accomplished or solved them, was to become the kind of experiment that would headline the decades immediately ahead. Peter was the pioneering primate in the United States, a country that fifty years later would see the government overseeing the construction and support of six research facilities devoted especially to the Order Primates. He was the first primate to be tested in a rigorous manner in order to measure his mental abilities, the first to have his abilities compared to those of human beings, and, some would claim, the first to write a letter of one of the human alphabets.

Witmer's interest in the missing link may just as well be understood as a search for the "missing rung," that piece of the metaphor of the ladder that associated humankind with the next animal group. To the human mind at the turn of the century, there appeared to be a missing rung between monkeys and people, between baboons and people. An understanding of the great apes offered that generation, as it offers ours, the promise of filling that vacancy. We moderns have not lost our belief in the ladder. While we appear to have accepted the notion of evolution as the frame on which we base our understanding of the structure of ourselves and other beings, we are sometimes ill at ease with using evolution as an explanation of mental abilities.

The effort to fill the missing link of mentality between ape and human was a slow one, chiefly because people seemed satisfied with the explanation that the gap was a large one, as could be seen clearly by the humans' ability to use speech and construct tools. But Witmer was not alone in his recognition that nonhuman primates, a group much ignored by human investigators, were worthy of experimental attention. The other person who recognized their importance was Richard Lynch Garner who not only undertook the study of ape and monkey speech in U.S. zoos, but also lived in a cage in West Africa, the better to communicate with the animals.

THE MAN IN THE CAGE

Peter was not quite the Adam of primate mentality. The story now shifts backward in time but forward in thinking. Previously, Richard Lynch Garner (1848–1920) had observed chimpanzees, not in the comparative comfort of the Psychological Clinic, but in French Gabon and French Congo (now Gabon and Congo), in the natural state, with Garner confining himself to living in a cage.

Karl Krall knew of Garner and included in his book such information as he knew about Garner's work, including several pictures of Garner and chimpanzees to illustrate Garner's attempts to help chimpanzees to speak. Garner was well known in the United States and Britain for his books about his experiences with chimpanzees and was in demand as a public speaker. The attention was well founded, for Garner had lived in West Africa in a cage from which he studied the behavior of chimpanzees and gorillas. He had taught one ape to speak words in several languages and another to write, and prompted serious study of the mind of the ape. But Garner's name is unknown today, and his pioneer work rarely mentioned and even more rarely read. There is a reason.

Garner was born in 1848 and participated in the U.S. Civil War in the Confederate Army. In the 1880s and 1890s he demonstrated an interest in monkeys and apes, but few specimens were available for study anywhere. He studied the few monkeys he could find at the zoos of Cincinnati, Chicago, and later, New York. He became interested in primate speech patterns and appears to have invented the technique of playing to primates recorded sounds of other primates. He recorded their sounds and played the sounds back to the monkeys to see if the animals recognized their sounds or those of species-mates. His idea was reinvented a century later, when "playback" experiments would be considered newly devised and clever ways to study animal cognition. Garner summarized these pioneering studies in a book published in both the United States and England in 1892.[20]

Garner sailed from New York City, with his photographic and phonographic equipment, for Africa via England on July 9, 1892. He arrived at French Gabon on October 18 of the same year and searched for a suitable camp site. Locating one, he set up the cage that would become his "abode," as he called it, in April 1893, and, along with the ape Moses and a native boy, who remains otherwise nameless, he took up residence in the cage. He was forty-five years old. He and the native boy (and, perhaps, others) remained at this site, using the cage as home for 112 days, or until late August 1893. Their experiences, including the return trip to England, are told in the book *Gorillas and Chimpanzees* (1896).[21]

Garner wrote within the sensibility of his time, and only his words can convey that tone. Yet his voice is more in tune with our own time. He writes of the primates with a clear and evident appreciation for their being, for their independence, for their deserving of human care, for their intelligence and their place in nature. Never is there a hint of condescension or, for that matter, of romanticism. The romanticism would come later and it is probably for that reason that his work became obscure and literally banned by a later generation of monkey-watchers.

The preface to *Gorillas and Chimpanzees* tells the prospective reader that Garner studied "these animals in the freedom of their native jungle" and that "this type of study had not hitherto been enjoyed by any student of Nature." Later, Garner made the point about himself that "no white man has ever seen so many gorillas and chimpanzees." This, too, is almost certainly so.

The introduction assures us that "the author has refrained from rash deductions and abtruse theories, but has sought to place the animals here treated in their true light, believing that to dignify the apes is not to degrade man, but to exalt him even more. It is hoped that a more perfect knowledge of these animals may bring man into closer fellowship and deeper sympathy with Nature, and cause him to realise that all creatures think and feel in some degree, however small."[22] Garner's chief interest was to study animal speech. "It was logical to infer that the anthropoid apes, being next to man in the scale of nature, must have the faculty of speech developed to a corresponding degree. . . . As the chief object of my studies was to learn the language of the monkeys, the great apes appeared to be the best subjects for that purpose, so I turned my attention to them.

"The gorilla was said to be the most like man, and the chimpanzee next. There were none of the former in captivity, and but few of the latter, and they were kept under conditions that forbade all efforts to do anything in that line.[23] [He means that none or few were available in zoos.] As the gorillas and chimpanzees could both be found in the same section of tropical Africa, I selected that as the field of operation. . . . The part selected was along the equator, and south of it by about two degrees. The locality is infested with fevers, insects, serpents and wild beasts of diverse kinds. To ignore such dangers would be folly, but there was no way to see these apes in their freedom, except to go and live among them.

"To lessen, in a degree, the dangers incurred by such an adventure, I devised a cage of steel wire, woven into a lattice with a mesh one inch and a half wide. This was made in twenty-four panels, three feet three inches square, set in a frame of narrow iron strips. Each side of the panel was provided with half-hinges, so arranged as to fit any side of every other panel. These could be quickly bolted together with small iron rods, and, when so bolted, formed a cage of cubical shape, six feet six inches square. Any one or more of the panels could be swung open as a door, and the whole structure was painted a dingy green, so that when erected by the forest it was almost invisible among the foliage.[24]

"While it was not strong enough to withstand a prolonged siege, it afforded a certain immunity from being surprised by the fierce and stealthy beasts of the jungle, and would allow the occupant time to kill an assailant before the wires would yield to anything except an elephant. . . . Over this frail fortress was a roof of bamboo leaves, and it was provided with curtains of canvas to be hung up in case of rain. The floor was of thin boards, steeped in tar, and the structure was set up about two feet from the ground, on nine small posts.

"It was furnished with a bed, made of heavy canvas supported by two poles of bamboo, attached to the edge of it. One of these poles was lashed fast to the side of the cage, and the other was suspended at night by strong wire hooks, hung on the top of it. During the day, the bed was rolled up on one of the poles, so that it was out of the way. I had a light camp chair, which folded up, and a table was improvised by a broad, short board hung on wires. This could be set up by the wall of the cage at night, out of the way. To this meagre outfit was added a small kerosene stove and a swinging shelf. [See Figure 8.7.]

"A few tin cases contained my wearing apparel, blanket, pillow, photograph camera and supplies, medicines, and an ample store of canned meats, crackers, &c. A magazine rifle, revolver, ammunition and a few useful tools, such as a hammer, saw, pliers, files, and a heavy bush knife, completed my stock, except some tin platters, cups and spoons."[25]

"I went up the Ogowe River about two hundred miles, and through the lake region on the south side of it. After some weeks of travel and inquiry, I arrived at the lake of Ferran Vaz, in the territory of the Nkami tribe.[26] The lake is about thirty miles long, by eight or ten wide, and interspersed with a few islands of large size, covered with a dense growth of tropical vegetation. The country around the lake is mostly low and marshy, traversed by creeks, lagoons, and rivers. Most of the land is covered by a deep and dreary jungle, with a few sandy plains at intervals.

Figure 8.7   Richard Lynch Garner and Native Boy "Starting for a Stroll,"
evidently having abandoned the notion of living continuously in the cage. The
drawing is labeled in the original, as "from a photograph," leaving the question
as to who took the picture, for Garner writes as if he and Native Boy were
alone. Note that Garner carries a gun, Native Boy a spear.

. . . In the depths of this gloomy forest, reeking with the effluvia of
decaying plants, and teeming with insect life, the gorilla dwells in safety
and seclusion. In the same forest the chimpanzee makes his abode, but is
less timid and retiring. . . . On the south side of this lake, not quite
two degrees below the equator, and within some twenty miles of the
ocean, I selected a place in the heart of the primeval forest, erected my
little fortress, and gave it the name *Fort Gorilla*. . . . My sole companion
was a young chimpanzee, that I named Moses, and, from time to time,
a native boy as a servant.[27]

AT FORT GORILLA

"Seated in this cage, in the silence of the great forest, I have seen the
gorilla in all his majesty, strolling at leisure through his sultry domain,
in quest of food. I have seen the chimpanzee under like conditions, and
the happy, chattering monkey in the freedom of his jungle home."[28]

Moses, who was bought from a trader met along the way, was to
become a denizen of this household. Information on his cleverness and
life-style occupies much of the information Garner is able to provide

about the chimpanzee. Because modern techniques of statistical analysis, systematic collection of data, and sampling of the population being studied were not invented, Garner merely walked and watched. All of his simple experiments with the animals were with Moses and with three nonhuman primates who joined the party later, Aaron, Elishiba, and Consul.

The experiments were intended to determine whether the animal could be domesticated to learn human skills and human ways. Garner may have been the first North American to learn that observing primates requires not merely the availability of the animals, but also huge amounts of time and patience on the part of the human observer. He was to establish the importance of studying animals in their natural state and did so with minimal assistance, financial and otherwise. He actually welcomed the lack of the luxuries of communication that we think necessary for modern survival.

But something happened to Garner in Africa that changed his character. Just as Dr. Itard came to regard the child Victor differently and to rethink the nature of the questions he wanted to ask of the boy, so Garner's contact with his subjects changed his attitude toward primates and people. When doing playback experiments at U.S. zoos, Garner was the very model of a careful thinker, experimenter, and writer. After the Africa experience, his claims became excessive, his accounts questionable, his findings seemingly contrived. Whether it was the fame from the playback experiments that changed him, or whether it was his inability to repeat the circumstances that led to his first fame, we cannot know. Whatever the cause, Garner's observations began to change in quality and sensitivity.

Much of Garner's description of the time spent in the cage is not of the animals, but of the typical day in which the care of oneself—feeding, bathing, sleeping—seems to the reader to occupy inordinate attention, but to any experienced field worker, constitutes a natural and fair description of the routine and boredom of field work, where so many small acts so easily and efficiently accomplished at home come to occupy the better part of the day. In the early morning, all three take a stroll. We see a picture of Garner, the native boy, and Moses doing so in Figure 8.7. From mid-morning until mid-afternoon, the jungle becomes still; not truly inactive, surely, but quiet. It is wisest to make use of this quiet time for lunch and a nap. The afternoon stroll will produce different sightings, different combinations of animals, but the animals are doing mostly the same things, searching for food. Some animals are engaging in social acts. It is wise to be back to camp early, for the sun sets quickly. It is wise to prepare dinner and be ready for bed before the sun goes down, and there is little twilight. The night is made longer for human beings perhaps by the noises of the animals, animals who now cannot be seen but can only be imagined. Twelve hours of bed, with a kerosene light for some difficult reading, and then the sounds, perhaps sleep, perchance dreams. Morning comes suddenly, and the three are off again for the morning stroll. And so it goes for 112 days, except for a trip that, as we shall see, led to very sad consequences for all and to the death of one of the trio.

Garner tells us much about chimpanzees—for example, about the source of their name (from the Fiot language, meaning, loosely, "small bushman") and that there are two types, *ntyigo* and *kulu,* a notion recognized today. We should be careful about separating animals into species based on very small differences, he reminds us, such as in baldness, or pelage, or the range of vocalizations. While these two kinds of chimpanzee are different to the human eye, this alone is not a reason for thinking of them as separate species. These are not, writes Garner, different species, as some would have it, but different races, like "white and negro" of a common stock. Those who specialize in the identification of species are sometimes thought to be either "lumpers" or "splitters." The former recognize fewer species, seeing pelage color or the nature of the eye orbit, for example, as but variations; the latter use these marks to identify new species or subspecies. Garner comes very, very close to regarding chimpanzees as a species of human being. This view is championed today as there is so little difference between chimpanzees and people in DNA.

In Garner's time, it was common to think of human beings as being of three types, each based on skin color and some less evident characteristics. It is too strong to say that Negroes and Caucasians were seen as different species, but many equated race with species. Garner was a lumper: in this he stands apart from his contemporaries. Indeed, later, he appears to be placing the great apes in the same category as human beings. In short, he went from wanting to communicate with the apes by language in order to understand them to ascribing them qualities not far removed from those frequently given to people alone.

Garner summarizes his knowledge of the mentality of the chimpanzee with a comment pertinent to our attempts to analyze the metaphor of the Mental Ladder: "It is difficult to compare the mental status of the ape to that of man, because there is no common basis upon which the two rest. Their modes of life are so unlike, as to afford no common unit of measure. Their faculties are developed along different lines. The two have but few problems in common to solve. . . . There are, perhaps, instances in which the mind of the ape excels that of man, by reason of its adaptation to certain conditions. It is not a safe and infallible guide to measure all things by the standard of man's opinion of himself. It is quite true that, by such a unit of measure, the comparison is much in favor of the human, but the conclusion is neither just nor adequate.

"It is a problem of great interest, however, to compare them in this manner, and the result would indicate that a fair specimen of the ape is in about the same mental horizon as a child of one year old. But if the operation were reversed, and man were placed under the natural conditions of the ape, the comparison would be much less in favour. There is no common mental unit between them."[29]

GARNER'S PROBLEM

The idea of a human being living in comparative captivity while the animals lived in their natural state was a compelling one that would be reinvented to great acclaim in our times. But unlike modern observers, Garner discovered that the animals were uncooperative. They were hardly to be seen, not because he had chosen the wrong locale, but because they avoided the cage and its captives. The animals had no compelling reason to behave naturally, or any other way, for the benefit of a caged human. Garner appears to have assumed that the apes would continue to live on without attention to his presence, or to the comings and goings of the native boy, himself, and whatever supply lines were organized. A decade later, Yerkes, investigator of Roger the dog, would remark that one chimp is no chimp, meaning that in the absence of social interaction social animals cease to be of their species. Garner discovered that a chimp under observation by a human is also no chimp at all, but in his disappointment at having no data after so much promise and effort, the message did not occur to him. But news of his effort was spreading. His letters to zoo directors and newspaper editors were creating more and more attention, as Garner observed, since somehow newspapers arrived via the post at the cage.

In order to obtain information about the chimpanzees and gorillas, Garner abandoned his cage and turned to the study of the individual animal, one held captive either by force or by being tamed. He also began to rely on the observations of others whom he had met on his way. Garner had now reversed his purpose: The charm of his idea had been that *he* was the captive and the animals free; now, in order to make observations, he found it necessary to make the animal captive. When observations were scarce, and his need and wish to publish powerful, he turned to what he had heard, not what he had seen. He was not necessarily wrong to make these adjustments, but in his writings he failed to make evident the distinction between observation and hearsay. The newspapers did not care; the books continued to sell; and fellow investigators, whatever their personal motives, were watching Garner.

Garner's record is mixed. Hear, for example, his description of one social behavior known as the *kanjo:*

"One of the most remarkable of all the social habits of the chimpanzee, is the kanjo, as it is called in the native tongue. The word does not mean 'dance' in the sense of saltatory gyrations, but implies more the idea of 'carnival.' It is believed that more than one family takes part in these festivities.

"Here and there in the jungle is found a small spot of sonorous earth. It is irregular in shape, but is about two feet across. The surface is of clay, and is artificial. It is superimposed upon a kind of peat bed, which, being very porous, acts as a resonance cavity and intensifies the sound. This constitutes a kind of drum. It yields rather a dead sound, but of considerable volume.

"This queer drum is made by chimpanzees, who secure the clay along the bank of some stream in the vicinity. They carry it by hand, and deposit it while in a plastic state, spread it over the place selected, and let it dry. I have, in my possession, a part of one that I brought home with me from the Nkami forest. It shows the finger-prints of the apes, which were impressed in it while the mud was yet soft.

"After the drum is quite dry, the chimpanzees assemble by night in great numbers, and the carnival begins. One or two will beat violently on this dry clay; while others jump up and down in a wild and grotesque manner. Some of them utter long, rolling sounds, as if trying to sing. When one tires of beating the drum, another relieves him, and the festivities continue in this fashion for hours.

"I know nothing like this in the social economy of any other animal, but what it signifies, or what its origin was, is quite beyond my knowledge."[30]

The issue of the "carnival" is to arise sixty years later when V. and F. Reynolds, a young and just married team, reinvented for modern times the notion of observing apes in the wild by staying in the Budongo forest of Uganda.[31] In 1965, they reported evidence of the "carnival," a set of acts very much like those first reported by Garner but one not seen, or at least not reported, in the years between their observations. Garner was criticized for romanticism when he published news of the kanjo, and the derision is not entirely undeserved as the description is, I deduce, not something he saw, and certainly not something he saw from his cage. Rather, it appears to be a vivid description of something he was told; but, by whom, and when, we do not know. In Garner's books, conspicuously more in the later than in the earlier, stories told to him about ape behavior become part of the narrative to the point that it is impossible to separate observations for which he can take responsibility from those that are, to say the best for them, and him, embellished hearsay. That Garner failed to distinguish what he saw from what he heard from others is the core reason why his work was scorned and eventually neglected, for this lack of distinction came to be accompanied by a lack of trust on the part of those also familiar with animals and their social ways. Yet, the kanjo, or something like it, is seen again by the Reynoldses in our age, and these are people who have shown themselves to be dedicated and considerate observers of chimpanzee and human behavior.

In our times, observers would pride themselves on joining the chimpanzee or gorilla troop, at least in some psychological way. To my knowledge, J. Goodall[32] never claimed such a relationship for herself, but others have done so for her. The eventual partial taming of the chimpanzees she observed by providing food for them surely altered the relationship, just as Itard and Victor, among others, came to blur the relationship between observer and observed. D. Fossey,[33] more than any other observer, appears to me to have followed Garner's path, although this career was marked by the observer's coming to believe, I think, that she had more in common with the gorillas she studied than with the human beings with whom she worked. Both the lives of Garner and Fossey have the characteristics of Greek or, better, Shakespearean tragedy, in which, as Bradley[34] explained it in his commentaries on the tragedies, the "tragic flaw" deepens and overwhelms the character. For Garner, as it was, is, and will be for others, the flaw was a wish for continuing fame and attention.

MOSES, THE CAPTIVE

"He was found all alone in a wild papyrus swamp of the Ogowe River. No one knew who his parents were, or how he ever came to be left in that dismal place. The low bush in which he was crouched when discovered was surrounded by water, and the poor little waif was cut off from the adjacent dry land.

"As the native who captured him approached, the timid little ape tried to climb up among the vines above him and escape, but the agile hunter seized him before he could do so. At first the chimpanzee screamed, and struggled to get away, because he had perhaps never before seen a man, but when he found that he was not going to be hurt, he put his frail arms around his captor, and clung to him as a friend. . . . The next day he was sold to a trader. About this time I passed upon the river on my way to the jungle in search of the gorilla and other apes. Stopping at the station of the trader, I bought him, and took, him along with me. We soon became the best of friends and constant companions.

"I designed to bring Moses up in the way that good chimpanzees ought to be brought up, so I began to teach him good manners in the hope that some day he would be a shining light to his race, and to aid me in my work among them. To that end, I took great care of him, and devoted much time to the study of his natural manners, and to improving them as much as his nature would allow.

"I built him a neat little house within a few feet of my cage. It was enclosed with a thin cloth, and had a curtain hung at the door, to keep out mosquitoes and other insects. It was supplied with plenty of soft, clean leaves, and some canvas bed-clothing. It was covered over with a bamboo roof, and suspended a few feet from the ground, so as to keep out the ants.

"Moses soon learned to adjust the curtain, and go to bed without my aid. He would lie in bed in the morning until he heard me or the boy stirring about the cage, when he would poke his little black head out, and begin to jabber for his breakfast. Then he would climb out, and come to the cages to see what was going on.

"He was jealous of the boy, and the boy was jealous of him, especially when it came to the question of eating. Neither of them seemed to want the other to eat anything that they mutually liked, and I had to act as umpire in many of their disputes on that grave subject, which seemed to be the central thought of both of them.

"I frequently allowed Moses to dine with me, and I never knew him to refuse, or to be late in coming on such occasions, but his table etiquette was not of the best order. I gave him a tin plate and a wooden spoon, but he did not like to use the latter, and seemed to think that it was pure affectation for anyone to eat with such an awkward thing. He always held it with one hand, while he ate with the other, or he drank the soup out of his plate.

"When he would first take his place at the table, he behaved in a nice and becoming manner; but having eaten till he was quite satisfied, he usually became rude and saucy. He would slyly put his foot up over the edge of the table, and catch hold of the paper (used as a tablecloth), meanwhile watching me closely, to see if I was going to scold him. If I remained quiet he would tear it just a little more, but keep watching my face to see when I discovered it. . . . When he carried his fun too far, I made him get down from the table and sit on the floor. This humiliation he did not like at best, but when the boy would grin at him for it, he would resent it with as much temper as if he had been poked with a stick. He certainly was sensitive on this point, and evidenced an undoubted dislike to being laughed at. . . .

"The only thing that he cared much to play with was a tin can that I kept some nails in. For this he had a kind of mania, and never tired of trying to remove the lid. When given the hammer and a nail, he knew what they were for, and would set to work to drive the nail into the floor of the cage or the table; but he hurt his fingers a few times, and after that he stood the nail on its flat head, removed his fingers and struck it with the hammer, but, of course, never succeeded in driving it into anything. . . .

"From time to time I received newspapers sent me from home. Moses could not understand what induced me to sit holding the thing before me, but he wished to try it, and see. He would take a leaf of it, and hold it up before him with both hands, just as he saw me do; but instead of looking at the paper, he kept his eyes, most of the time, on me. When I would turn mine over, he did the same thing, but half the time had it upside down. He did not appear to care for the pictures, or notice them, except a few times he tried to pick them off the paper; and one large cut of a dog's head, when held at a short distance from him, he appeared to regard with a little interest, as if he recognized it as that of an animal of some kind, but I cannot say just what his ideas concerning it really were. . . . One thing that Moses liked was to play peek-a-boo with me or the boy. He did not try to conceal his body from view, but would hide his eyes, and then peep. . . .[35]

"It was never any part of my purpose to teach a monkey to talk [it should be remembered that Garner's purpose was to study natural monkey-speech in the natural environment; the acquisition of Moses was by happenstance, and we are not told the origins of the native boy], but after I became familiar with the qualities and range of the voice of Moses, I determined to see if he might not be taught to speak a few simple words of human speech. To affect this in the easiest way and shortest time, I carefully observed the movements of his lips and vocal organs in order to select such words for him to try as were best adapted to his ability. . . .

"I selected the word mamma, which may almost be considered a universal word of human speech; the French word, feu, fire; the German word wie, howl; and the native Nkami word nkgwe, mother. Every day I took him on my lap and tried to induce him to say one or more of these words. For a long time he made no effort to learn them, but after some weeks of persistent labour and a bribe of corned beef, he began to see dimly what I wanted him to do. . . .

"In his attempt to say mamma he only worked his lips without making any sound, although he really tried to do so, and I believe that in the course of time he would have succeeded. He observed the movements of my lips, and tried to imitate them, but seemed to think that the lips alone produced the sound.

"With *feu* he succeeded fairly well, except that the consonant element as he uttered it resembled 'v' more than 'f,'[36] so that the sound was more like vu making the u short as in 'nut.' It was quite as perfect as most people of other tongues ever learn to speak the same word in French, and if it had been uttered in a sentence, any one knowing that language would recognize it as meaning fire. . . . In his efforts to pronounce wie he always gave the vowel element the German 'u' with the umlaut, but the 'w' element was more like the English than the German sound of the letter."[37]

Garner was probably not the first human being who tried to teach a chimpanzee to talk, although he may have been the first Europeanized Westerner to do so. We have no known record of the African people's interaction with the great apes, although there must have been a good deal of it. Nonetheless, I think we have no earlier written record in Western languages of a stipulated procedure and results. Communication with the animal mind is, in our times, a scientific industry whose origins appear to be Garner's procedure.

If Moses was not the first chimpanzee to communicate, or at least to copy selected human speech, he was more likely the first to have served as a legal witness to a written document. "While living in the jungle, I received a letter enclosing a contract to be signed by myself and a witness. Having no means of finding a witness to sign the paper, I called Moses from the bushes, placed him at the table, gave him a pen and had him sign the document as witness. He did not write his name himself, as he had not yet mastered the art of writing, but he made his cross mark between the names, as many a good man had done before him. I wrote in the blank [space] the name . . . and had him with his own hand make the cross as it is legally done to all people who cannot write. With this signature the contract was returned in good faith to stand the test of the law courts of civilization, and thus for the first time in the history of the race a chimpanzee signed his name."[38]

ANOTHER CAPTIVE

"Having arranged my affairs so as to make a journey across the great forest that lies to the south of Nkami country and separates it from that of the Esyira tribe," writes Garner, "I set out by canoe to a point on the Rembo about three days from the place where I had so long lived in my cage. I disembarked, and after a journey of five days and a delay of three more days caused by an attack of the fever, I arrived at a trading station near the head of a small river called Ndogo. . . . About the time I reached here, two Esyria hunters came from a distant village, and brought with them a smart young chimpanzee of the kind known in that country as the kulu-kamba. He was quite the finest specimen of that race I have ever seen. . . . As soon as I saw this little ape I expressed a desire to own him, so the trader in charge brought him and presented him to me. As it was intended that he should be the friend and ally of Moses, although not his brother, we conferred upon him the name of Aaron. . . . At the time of his capture his mother was killed in the act of defending him from the cruel hunters, and when she fell to the earth, mortally wounded, this brave little fellow stood by her trembling body, defending it against her slayers, until he was overcome by superior force, seized by his captors, bound with strips of bark, and carried away into captivity. . . . It is true that it is often difficult, and sometimes impossible, to secure the young by other means; but the manner of getting them often mars the pleasure of having them, and while Aaron was, to me, a charming pet and a valuable subject for study, I confess the story of his capture always touched me in a tender spot.

"Before leaving the village where I secured him, I made a kind of sling for him to be carried in. . . . From there to the Rembo was a journey of five days on foot. . . . The only means of passing these dismal swamps is to wade through the thin slimy mud, often more than knee-deep, and sometimes extending many hundred feet in width. . . . Aaron did not realize how severe the task of the carrier was in trudging his way through such places, and the little rogue often added to the labour by seizing hold of limbs or vines that hung within his reach in passing, and thus retarded the progress of the boy, who strongly protested against the ape amusing himself in this manner. . . . The quarrel went on until we reached the river, but by that time each of them had imbibed a hatred for the other that nothing in the future ever allayed. . . . The boy gave vent to his dislike by making ugly faces at the ape, which the latter resented by screaming and trying to bite him. Aaron refused to eat any food given him by the boy, and the boy would not give him a morsel except when required to do so. At times the feud became ridiculous, and it only ended in their final separation. The last time I ever saw the boy I asked him if he wanted to go with me to my country to take care of Aaron, but he shook his head, and said, 'He's a bad man'."[39]

No further mention is made of the boy shown in Figure 8.7, the young man who accompanied Garner to the area, stayed with him, and served him. This parting ends our knowledge of the relationship. I honor his contribution and, by representation, the other Africans who served European and Western explorers, by making him part of the dedication of this book.

MOSES AND AARON

When Garner left for the trip on which Aaron was collected, he left Moses in the charge of a local missionary. (The native boy had now separated from service.) During Garner's absence from the cage, the missionary who was living with Moses became ill with fever. Moses was therefore left with a native boy who lived at the mission. The boy kept him tied by a short rope attached to a cage. Soon, Moses developed "a severe cold, which soon developed into acute pulmonary troubles of a complex type, and he began to decline."

"After an absence of three weeks and three days [during which time Aaron was acquired], I [Garner] returned to find him [Moses] in a condition beyond the reach of treatment. He was emaciated into a living skeleton: his eyes were sunken deep into their orbits, and his steps were feeble and tottering; his voice was hoarse and piping; his appetite was gone, and he was utterly indifferent to anything around him.

"When he discovered me approaching, he rose up and began to call me as he had been wont to do before I left him, but his weak voice was like a death-knell to my ears. My heart sunk within me as I saw him trying to reach out his long, bony arms to welcome my return. Poor, faithful Moses! I could not repress the tears of pity and regret at this sudden change, for to me it was the work of a moment. I had last seen him in the vigour of a strong and robust youth, but now I beheld him in the decrepitude of a feeble senility. When Aaron was set down before him, he merely gave the stranger a casual glance, but held out his long lean arms to me to take him into mine. His wish was granted, and I indulged him in a long stroll. He [Aaron] was like a small boy when there is a new baby in the house. He cuddled up close to Moses and made many overtures to become friends. . . . Aaron tried in many ways to attract his attention, or elicit some sign of approval, but it was in vain. . . . At length he [Aaron] lifted the fruit to the lips of the invalid and uttered a low sound, but the kindness was not accepted. . . . Failing to get any sign of attention from Moses, he moved up closer to his side and put his arms around him. . . . During the days that followed, he sat hour after hour in this same attitude, and refused to allow any one except myself to touch his patient. . . . I gave Moses a tabloid of quinine and iron twice a day. These were dissolved in a little water and given to him in a small tin cup. Aaron soon learned the use of it, and whenever I would go to Moses, he would climb up the post and bring me the cup to administer the medicine. At night when they were put to rest, they lay cuddled up in each other's arms, and in the morning they were always found in the same close embrace.

"My conscience smote me for having left him, yet I felt that I had not done wrong. It was not neglect or cruelty for me to leave him while I went in pursuit of the chief object of my research . . . . Hour after hour during that time he lay silent and content upon my lap. . . . With his long fingers he stroked my face, as if to say that he was happy again.

"His suffering was not intense, but he bore it like a philosopher. The last spark of life passed away in the night. It was not attended by acute pain or struggling, but, falling into a deep and quiet sleep, he woke no more.

"Moses was dead. His cold body lay in its usual place [in the hut used by Moses and Aaron], but [the body] was covered over with the piece of canvas kept in the cage for bed-clothing. I do not know whether Aaron had covered him up or not. . . . I had the body removed [Garner did not do it himself] and placed on a bench about thirty feet away, in order to dissect and prepare the skin and skeleton to preserve them. When I proceeded to do this, I had Aaron confined to his cage, lest he should annoy and hinder me at the work; but he cried and fretted until he was released. . . . When released, he came and took his seat near the dead body, where he sat the whole day long and watched the operation.

"Moses will live in history. He deserves to do so, because he was the first of his race that ever spoke a word of human speech; because he was the first that ever conversed in his own language with a human being; and because he was the first that ever signed his name to any document; and Fame will not deny him a niche in her temple among the heroes who have led the races of the world."[40]

Was Moses the first to speak human speech, to converse with a human being, to sign a legal document? The "witnessing" of the document, or at least the human's acquiring the written X, may well be a first, although what the occasion says about primate talents is difficult to decide. Many chimpanzees could do this task, but few are asked. As for speech: Were these human words, or were they sounds heard by a human as words? The difference is critical. Garner had made something of a career of listening to primate vocalizations and categorizing them, but speech, after all, is in the ears of the hearer as well as the vocal mechanism of the speaker. In our times, honest observers have heard animals to give sounds recognizably human, but we cannot say whether this ability demonstrates the intelligence of the animal or the adaptability of the hearer.

Did Moses and Garner converse? It takes two to converse: Garner thought they did so, but what did Moses think? All of this is said not to be merely skeptical, but in Chapters 10 to 12 we will see that the issues pointed to require constant thought. The techniques and ways of interpreting have a way of reappearing in new disguises for new generations.

Our times are marked by seemingly reasonable investigations of talking and conversing apes, dogs, and parrots. If Garner was gullible and uncritical, what can we say about the meaning and veracity of the modern findings regarding human-ape communication?

Moses' epitaph is written with grace, clarity, and simplicity, yet clearly with emotion. Let it stand at least as a testament to the chimpanzee's inclusion on the Mental Ladder not as a separate rung, but as a "race" of humankind, an assignment that Garner, no doubt, meant as a compliment.

HUSBAND AND WIFE

"Four days after the death of Moses I secured a passage on a trading-boat that came into the lake. . . . I found room in one of the canoes to set the cage I had provided for Aaron, stowed the rest of my effects wherever space permitted, and embarked for the coast." And so ends Garner's caged residence at Fort Gorilla; so begins the return to Liverpool and the United States and the beginning of a second story, one that touches only lightly on the issues of animal mentality. Yet it has much to say about the relationship between chimpanzee and man, and, to be sure, of chimpanzee and chimpanzee. Let us push forward:

"After a delay of eight days at Cape Lopez, we secured passage on a small French gunboat, called the Komo, by which we came to Gabon, where I found another *kulu-kamba* in the hands of a generous friend, Mr. Adolph Strohm, who presented her to me; and I gave her to Aaron as a wife, and called her Elisheba, after the name of the wife of the great high-priest. [They are shown in Figure 8.8.]

"It would be difficult to find any two human beings more unlike in taste and temperament than these two apes were. Aaron was one of the most amiable creatures: he was affectionate and faithful to those who treated him kindly; he was merry and playful by nature, and often evinced a marked sense of humour; he was fond of human society, and strongly averse to solitude and confinement.

"Elisheba was a perfect shrew, and often reminded me of certain women that I have seen who had soured on the world. She was treacherous, ungrateful, and cruel in every thought and act; she was utterly devoid of affection; she was selfish, sullen, and morose at all times; she was often vicious and always obstinate; she was indifferent to caresses, and quite as well content when alone as in the best of company.

"It is true that she was in poor health, and had been badly treated before she fell into my hands, but she was by nature endowed with a bad temper and depraved instincts."[41]

There was a wait for the boat to England, and the time allowed Garner to construct caging for Aaron and Elisheba, as he wished to take them along. During the wait, the chimps came into contact with animals

Figure 8.8   Aaron and Elishiba, whose relationship is a love story along the nature of Tristan and Isolde or Aida and Radames. The end is the same for all these lovers. What is Garner's role in the tragic story of Aaron and Elishiba? Is he savior? Or assassin?

they had not seen before. Here is what happened while the threesome waited for the boat and embarked:

"Mr. Strohm, the trader with whom I found hospitality at this place, kept a cow in the lot where the cage was. She was a small black animal, and the first that Aaron had ever seen. He never ceased to contemplate her with wonder and with fear. If she came near the cage when no one was about he hurried into his box, and from there peeped out in silence until she went away. The cow was equally amazed at the cage and its strange occupants, though less afraid, and free until it came near to inspect them. . . . When taken out of the cage, Aaron had special delight in driving the cow away, and if she was around he would grasp me by the hand and start towards her. He would stamp the ground with his foot, strike with all force with his long arm, slap the ground with his hand, and scream at her at the top of his voice. . . . Elisheba never seemed to take any special notice of the cow except when she approached too near the cage, and then it was due to the conduct of Aaron that she made any fuss about it."[42]

We can but imagine the forty-two day trip to Liverpool on this ark. Some passengers, and certainly some crew, passed the time by teasing the animals, much to Garner's distress. The teasing may have been less fear-provoking than the sight that awaited the two apes on disembarking, for here human beings quite distinct even from those they had seen before waited. "On reaching the landing-stage in Liverpool, some friends who met us there expressed a desire to see them, and I opened their cage in the waiting-room for that purpose. [This was foolish and not to Garner's credit, to be sure.] When they [the apes] beheld the throng of huge figures with white faces, long skirts, and big coats, they were almost frantic with fear. . . . In their own country they had never seen anything like this, for the natives to whom they were accustomed wear no clothing as a rule, except a small piece of cloth tied around the waist, and the few white men they had seen were mostly dressed in white; but there was a great crowd of skirts and overcoats, and I had no doubt that to them it was a startling sight for the first time."

The tolerating of teasing, the sight of people, a new climate, all took its toll. Garner had now removed himself as observer of primate social life where it occurs naturally to become an animal keeper—a displayer of previously unseen apes. He was known as the man who had learned to communicate with apes, a man whose adventures had been reported by his letters to the newspapers of London and New York, a man who, as Singh expected to do, introduced his charges to the civilization of Europe and North America. On the one hand, his writing suggests love, more than respect, I think, for his animals. On the other, he had become very close to becoming a pitchman for these same animals. The theme of the curious love of the captor for the captive is not an unknown one in either literature or life. Garner's account continues;

"During the first two weeks, [Elisheba] developed a cold. A deep, dry cough, attended by pains in the chest and sides, together with a piping hoarseness, betrayed the nature of her disease, and gave just cause for apprehension. During frequent paroxysms of coughing she pressed her hands upon her breast or side to arrest the shock, and thus lessen the pain it caused. . . . The sympathy and forbearance of Aaron were again called into action, and the demand was not in vain. Hour after hour he sat with her locked in his arms, as he is seen in the portrait [Figure 8.8]. Even the brawny men who work about the place paused to watch him in his tender offices to her. . . .

"On the morning of her decease I found him sitting by her as usual. At my approach he quietly rose to his feet, and advanced to the front of the cage. Opening the door, I put my arm in and caressed him. He looked into my face and then at the prostrate form of his mate. The last dim sparks of life were not yet gone out, as the slight motion of the breast betrayed, but the limbs were cold and limp. When I leaned over to examine more closely, he crouched down by her side and watched with deep concern to see the result. I laid my hand upon her heart to ascertain if the last hope was gone; he looked at me, and then placed his own hand by the side of mine, and held it there as if he knew the purpose of the act.

"At length the breast grew still and the feeble beating of the heart ceased. The sturdy keeper came to remove the body from the cage; but Aaron clung to it, and refused to allow him to touch it. I took the little mourner in my arms, but he watched the keeper jealously, and did not want him to remove or disturb the body. . . . How I pitied him! How I wished that he was again in his native land, where he might find friends of his own race!"[43]

Care was given; the concern was real. ". . . the keeper put a young monkey in the cage with him for company. Thus he passed his time for a few weeks, when he was seized by a sudden cold, which in a few days developed into an acute type of pneumonia. I was in London at the time and not aware of this, but feeling anxious about him, I wrote to Dr. Cross, in whose care he was left, and received a note in reply, stating that Aaron was very ill, and not expected to live. I prepared to go visit him the next day, but just before I left the hotel I received a telegram stating that he was dead.

"Poor little Aaron! In the brief span of half a year he had seen his own mother die at the hands of the cruel hunters; he had been seized and sold into captivity; he had seen the lingering torch of life go out of the frail body of Moses; he had watched the demon of death bind his cold shackles on Elisheba; and now he had, himself, passed through the deep shadows of this ordeal.

"I have all of them preserved, and when I look at them the past comes back to me, and I recall so vividly the scenes in which they played the leading roles—it is like a panorama of their lives."[44]

AND GARNER

Garner died in January 1920, alone and unrecognized in a hotel in Chattanooga, Tennessee. *The New York Times* apologized for the late obituary of January 24, pointing out that Garner's fame was not appreciated at the hotel, and the importance of his death was therefore unrecognized. There is irony here, for Garner had spent his last twenty years working only to reify his fame won earlier.

Garner was an inventive and intelligent researcher when he undertook to use that new invention, the phonograph, to record monkey sounds; clever to see that it could be used to test the possibility of primate speech and communication; and courageous to face the west African jungle by living in a cage, albeit evidently with more support than he mentions. Many of his views and accomplishments were at least a century ahead of their time.

In middle age, he began to reinterpret his work by making arrogant claims for his successes. He annoyed academics both with his transparently false claims and his lack of academic credentials other than teaching high school in Tennessee. The writing of books and visiting of Africa gave way to lectures, at first to professional groups and then, eventually, to anyone who would listen. Other researchers spoke against his work, and the attitude changed from merely being challenging to being derisive. It is a pity, for Garner's primary work deserves credit and appreciation. He clearly regarded apes as a race of human beings, and he treated them with the concern that is not always evident among researchers today. Dr. Witmer's visit to the Keith Theatre was not merely a matter of his going to the theater for entertainment and, by chance, seeing Peter's performance and seizing the opportunity to investigate the missing mental link. Such is the account Witmer provides us in his complete description of his work with Peter, but the account must be shaped by his memory to suggest a lucky meeting coupled with opportunistic thinking. We know that Witmer had already procured an orang for similar study and that he was responsible for showing at least one other chimp, Mimi, in local playhouses. While Garner found himself discredited in part for showmanship, no such accusation was flung at Witmer: certainly none stuck. And yet Witmer is grossly undervalued today both for his pioneering contribution to clinical and educational psychology and for his work with the apes, and Garner is ignored.

Witmer and Garner pioneered competing strategies that would remain with us to this day, strategies that are philosophically incompatible. Is the Psychological Clinic the place to search for the mental abilities of the missing link, or should we look to the animals in their natural state? Do the tests in the clinic measure, and thereby limit, and mask the animals' abilities; or might the animal in the natural state never have occasion to show the human observer what it knows?

In some ways the contrast between Witmer's and Garner's approaches serves as a model of alternatives of how we decode the mind of our animal relatives. Witmer uses the laboratory, for it provides control; Garner goes to the apes' home and confines himself. Witmer compares the mental abilities of ape and human being directly; Garner eschews such comparison. Witmer has a social agenda, namely, that of giving every child an educational opportunity equal to his or her abilities; Garner appears to be interested in basic, unapplied science. Witmer understands Peter's abilities to be motor, and he stops short of postulating human mental abilities for Peter, even when Peter "wrote" a letter of a human alphabet. Garner claims the writing of an X by a chimpanzee to be a sign of humanlike intelligence. Garner's ways of studying the chimpanzees, especially those he had captured, did not have happy results for the animals involved. Witmer wants to "test" because he has a purpose, a way to use the information he finds for practical affairs. Garner shows no interest other than in knowledge for its own sake. Witmer's future is with the Psychological Clinic, with teaching teachers and classifying those ab-

normal aspects that interfere with human education. Garner came to regard his achievements as evidence of his own distinction.

The comparison of Witmer and Garner suggests two themes that are
to occupy us in the chapters that follow. First, What is the difference, if
any, between human and ape? Second, What measures do we use, and
where do we use them, laboratory or nature, to search for the similarities
or differences?

Exploiting the Missing Link 9

I F T H E R E exists a Mental Ladder that separates animal forms in terms of their abilities to think, remember, feel, solve problems, and the like, might there not be like separations to be found *within* species? Indeed, might not the individual members of a species be separable in terms of their mental capacities? If such distinctions can be made reliably, might they not be correlated with some other factors, especially genetic ones, for, after all, are not such capacities determined or at least limited by genes?

Just as Itard set out to test theories of the intellect current during his times with Victor and Singh to worry about the education of the soul during his times, so Witmer's Psychological Clinic was testing a nascent theory to be found at the turn of this century in the United States— namely, that U.S.-style democracy, when appropriately arranged and applied, was capable of giving each human being the degree and kind of education suitable to that child's talent and ability. It was an audacious and a daring idea, for it argued tacitly that the effect of genetic variation was unrelated to class. Or, to press the point, that class status was unrelated to ability. Witmer did not develop the intellectual and political ramifications of this view, at least not in the journal he founded and edited, *The Psychological Clinic.* By his actions, however, he did demonstrate his view that mental ability could be measured in such a way that each human being could be offered an educational process that utilized her or his abilities, regardless of the wealth and social background of the parents. Hear Witmer make the point:

"One does not expect figs to grow from thistles, and the slum child seems naturally destined by the force of heredity to grow into an inefficient adult. There are many reasons, however, for repudiating this belief in the potency of heredity. The different races of men are not separated from one another as are the fig tree and the thistle. The different social classes of the white races constitute more nearly a single human family. Modern research, such as the parliamentary investigation into the physical deterioration of the English people, indicates that the degeneracy which is systematically associated with slum life . . . is the result of the treatment received during infancy and childhood. Children of the rich, of the moderately well-to-do, and of the poor are, as it were, representatives of the same species of plant growing under diverse influences of soil, sunshine, air, and moisture. You would doubtless consider it dangerous for your own child to spend a single night in a typical home of the slum. Children born into slum life, of slum parents, apparently do not differ very greatly from your own. A few of them, but only a few, are strong enough to fight their way out of their environment into better conditions, and even these bear permanently the scars of the battle. Many succumb quickly to the unfavorable conditions. The majority are irretrievably damaged in early life and their physical, mental and moral development is more or less seriously retarded. Shall we shut our eyes in order not to see that grinding poverty is slowly executing a death sentence upon many of these children, a sentence which is only the more cruel because it takes so many years to be finally carried out?

"It is only through the persistent effort to restore defective children of the slums to normal physical and mental condition that we may expect to throw light upon the causes which are producing degeneracy. . . . The Psychological Clinic has undertaken this work of restoration."[1]

In short, the purpose of the Clinic was to learn to measure human ability, to separate innate ability from the deprivations of the environment, to provide both the slum-child and the well-to-do child with opportunity suitable to their native ability. Witmer's arranging for Peter to visit the Psychological Clinic was a sign of the change in human opinion about the relationship of humankind and animals. A new opinion, given credence by Darwin's thinking of a half-century before, was that animals and human beings should be perceived as existing on a continuum of life, a continuum of structure and of mental ability. Whatever mental processes could be seen in human behavior were therefore to be found, to however slight a degree, in animal life, although a wide gulf might separate human judgment from animal judgment. Yet, besides the metaphor of the Mental Ladder, what precisely was the relationship between animal and human being?

ONTOGENY AND PHYLOGENY

One view had achieved some scientific prominence. This was the theory promoted by Ernst Heinrich Philipp August Haeckel (1834–1919) which set forth a detailed argument that "ontogeny recapitulates phylogeny," as it is summarized. The idea is that within the development of the individual human is to be found the development of the "lower" species, "lower" genera, and "lower" orders.[2] Within its own development, the human embryo, it is suggested, shows the development of other orders: in its initial stages, the one-celled becomes two, the two becomes four, the four eight; and thus the embryo is at first fishlike, with the prominent tail, then mammal-like, and, eventually, ape- and humanlike. That which is true of the embryo is true of the remainder of development from birth onward, for the genes of other animals are represented, however slightly, in the development of the full human being.

There is nothing irretrievably awkward about this view of human and animal development. Indeed, it catches and extends the imagination. It is believable. It explains how things came to be the way they are. Like the concept of evolution itself, the idea that ontogeny recapitulates phylogeny takes the known observations and explains their probable history, while leaving the meaning of the explanation to individual taste. Does the fishlike period of my embryonic development account for my ability to swim, as the shape of my middle ear can be seen to have developed from a fishlike structure? Is my childlike play a vestige of the adult behavior of apes? Darwin's understanding of how natural selection affects populations does not require or demand such an interpretation, yet the notion that ontogeny recapitulates phylogeny has a seeming correspondence to the Aristotelian and Darwinian description of the Great Chain of Being that relates all animals to one another within one kingdom. Whether or not the concept of the Great Chain is true depends less on the rigors and excellence of experimentation than on the perception and values of the reader. It is not difficult for human beings to be able to see in the development of the human embryo simpler life forms. The recapitulation notion is neither true nor false for the reason that it cannot be verified or disproved. It is, rather, a human myth that may help us to organize our knowledge.

The observant human mind, working in the first years of the twentieth century, had intellectual reason to trust the view that development, like the invention and production of species, was a slow and steady process, one that repeated the paths chosen because of the direction of evolution itself. That Peter should be tested in a Psychological Clinic speaks, first, of the acceptance of the view that there was some humanlike quality in him to be found and tested. The pre-Darwin mind might not have seen the utility of testing and measuring this mind, for it would not have assumed the connection between ape qualities and human qualities. The notion that it is not only the phylogeny and ontogeny of animal structure that can be seen to be represented in all creatures, but *behavior* as well— that human behavior, for example, has rudiments of the behavior of "simpler" creatures—forever changes our concept of what it means to be a human being. Now a human being becomes a beast that contains within it all other forms of animal life. It is a tempting image, especially when we people act in ways that shame us.

What is added to Darwinism is the notion that mental aspects can be measured and quantified. To the modern mind, which has been treated to a century of such measurement, to the point that it has become an expected part of our social order, the goal of the Psychological Clinic produces skepticism. For we all have experienced the inequities of testing human beings. We are skeptics, even if we accept mental measurement as a means of discriminating ability and education. Our private view of testing, of applying numbers to mental performance, is that it is necessary, sometimes unreliable, beside the point, and rarely descriptive of *us* as individuals.

MEASURING MENTATION

The science of mental measurement itself passes the tests of reliability and validity: it is neither wrong nor right. It may be used, of course, to promote or demote, as our human society wishes. Mental measurement developed coincidentally with the notion of the Mental Ladder: whether these concepts fed on one another is impossible to unravel, but that they mix together easily is obvious. There is no ladder if species or individual differences cannot be measured, and there is no science of measurement if there are no differences to measure. If the idea of a Mental Ladder disturbs us, we are apt to be alarmed by the notion of testing for differences in ability. If we find the idea of a Mental Ladder comforting, we are apt to search for ways to find and measure the differences it demands. We may find these concepts to be liberating, or confining.

To the human mind of the first decade of this century, and so to Witmer, the act of measuring capacities was seen as liberating, not as confining. Testing provided an apparatus and method to provide each child with her or his intellectual and vocational needs, to show the democratic nature of mental ability. To a culture accustomed to using children for the repetitive work of factory or farm, the notion that one's abilities and potential for learning could be measured and predicted provided an opportunity to liberate mind and body. What better application of the liberating fruits of democracy?

Witmer's clinic took children from the streets and factories and accumulated information and measurements of their mental and physical abilities. To measure, one must have some capacity to measure; some such capacity is named as a category (e.g., arithmetic ability, memory for digits); and these categories come to be named by the task of measurement itself. Some of these categories (digit span, spatial arrangements) found their way to formulating averages and norms that could be used everywhere for assessing a particular child's standing in relation to its age group. The idea of measuring categories to establish potential for training and education would reach fuller force two decades later with the development of intelligence tests. That Peter's abilities were measured attests to the new acceptance both of the reality of a ladder of animal mental life and of the use of mental testing to reach the goal of equality of opportunity and destroy the notion of ability as a genetic trait. Dr. Itard would have understood the distinction.

PETER'S AND MOSES'S COLLEAGUES

Peter was not the first animal to be asked to demonstrate his ability in the laboratory environment. Despite diverse beginnings, the notions of mental testing and the testing of animal species came together by 1900. The search for quantification of species' abilities and the abilities of individual animals was under way. The one, it seemed, would provide a quantification of phylogenetic differences and the other, in some ways, of ontogenetic differences.

The honor of being the first primates to be tested in a U.S. laboratory for purposes of measuring mental ability appears to belong to three *Cebus apella*. *Cebus* are best known as the organ-grinder's monkey, although hardly anyone alive has ever seen an organ-grinder at work. *Cebus* are clever primates, and the ease with which they learn new tasks is one reason why they are used as surrogate human-tools for persons without control of their arms and legs. These South American monkeys, small in stature compared to the terrestrial apes, baboons, and monkeys of the Eastern Hemisphere, are trained to assist the paraplegic by delivering food, water, and other objects on command.

The *Cebus* were residents in a laboratory at Columbia University in New York City and were tested by Edward L. Thorndike; the results of these tests were first published in 1901. These *Cebus* were housed by Thorndike in his home near the University. Thorndike had long maintained animals so that he could conduct research on their abilities. While spending some time at Harvard University between 1895 and 1897, studying with William James, he had kept chickens in his lodgings, and he had maintained and "tested" dogs and cats sometimes at his lodgings and sometimes, seemingly, at the school.[3] Thorndike wanted to test *Cebus* because little was known about monkeys—so little, in fact that the standard nomenclature of the genera and species followed Linnaeus in classifying some baboons as dogs. They had been removed to Boston from the rain-forests of Peru and Bolivia to the laboratory in which they now lived in a room measuring 8 by 9 feet, in cages 2 by 3 feet, as described by Thorndike.[4]

Thorndike began to study animals because, as he stated: "psychology has never decided whether it wants to talk of consciousness or of behavior." The nineteenth century, he said, emphasized consciousness at the expense of behavior. Now it was time to study behavior, for, unlike the mind, it was measurable. Both Thorndike and Freud, writing at the same time, were suspicious and skeptical of the psychologies they had studied. They found ways to correct the inadequacies they saw, ways that would be of longstanding significance. Just as Freud's ways of thinking founded psychoanalysis, so Thorndike's, along with others in the United States and Russia, founded behaviorism.

Thorndike began his report of his work with chickens, cats, and dogs by noting that our knowledge of animal life was chiefly limited to their

sensory systems, through the comparative study of anatomy and physiology, and we could only guess as to the nature of their instincts. We
needed knowledge that was both more and different: namely, how they
learn and how they acquire and build new behavior through experience.
The suggestion that the important aspect of animals was that they modified their behavior, learned, and created new kinds of behavior by reacting to happenings in their environment, gave impetus to the movement
toward behaviorism.

Thorndike did not appear to have understood the importance of his
own proposal. He did not immediately grasp that he was arguing for a
complete shift in the way we think of animal and human capabilities.
The human concept of animal life was one in which instincts predomi-
nated. To be sure, some were awed by the ability of animals to perform
clever and seemingly thoughtful tasks, but the awe derived from seeing
such feats against the backdrop of the instincts, which were understood
as fixed and limiting. The issue of how much of behavior occurs by nature, by instinct, and by nurture is an issue seemingly as longstanding as
the invention of the written word. Yet it is not the complete issue: There
remain the questions of how much is instinctual and unchangeable, how
much is alterable by the organism, and how much instinct and learned
behavior may interact.

The reliance on instinct for explanations was a European notion that
had been expounded by ethology: It fit well with the notion of distinguishing animals by class (by naming genera) and with assuming that
once assigned to a rung on the ladder, there was little chance to move
up or downward, for ability was determined by one's birth. At least in
his youth and early days as a student, Thorndike was a believer in a
progressive democracy, the same values that appear to underlie Witmer's
researches. In this earlier view, Thorndike saw learning as the single capacity that permitted change. Learning was a potential ability that tran-
scended genetic limitations, not because what was learned could be transmitted to the offspring genetically, but because what we learned influenced
our reproductive success, thereby determining which genes were passed
on. The animal who learns most rapidly, the group that outwits its rival
groups, the species that is more successful than rival species, the genus
that succeeds where another fails—all are examples of ways in which the
ladder of mental ability can be reconstructed. If the way in which natural
selection works on populations that Darwin suggested is true, then change,
not stasis, needs to be understood, for change is the rhythm of the natural universe.

Thorndike's analysis and proposition—that there could and should be
a psychology of behavior as well as, or alongside of a psychology of
consciousness—is a splendid example of the new view that stated that a
ladder of mental ability could be created by using the devices of mental
measurement. Thorndike argued that this could be done not merely by
speculating about consciousness or watching animals perform unusual

feats, but by inventing tests, the results of which could be measured and used to assign the animal places on the ladder.

In the introduction to the 1898 paper, he explained the purposes of his animal studies: "The main purpose of the study of the animal mind is to learn the development of mental life down through the phylum, to trace in particular the origin of human faculty."[5] Thorndike was scathing in his criticism of those who search for consciousness in animal life. He writes, "[They do not] . . . give us a psychology, but rather a *eulogy* of animals. They have all been about animal intelligence, never about animal *stupidity*."[6] Thorndike believes that animals learn by forming associations: "Does the kitten feel *"sound of call, memory-image of milk in a saucer in the kitchen, thought of running to the house, a feeling, finally, of 'I will run in'?"* Does he perhaps feel only the sound of the bell and an impulse to run in, similar in quality to the impulses which make a tennis player run to and fro when playing?[7] With the youthful sarcasm that sometimes passes for wit, he continues:

"They [animals] used to be wonderful because of the mysterious, God-given faculty of instinct, which could almost remove mountains. More lately they have been wondered at because of their marvelous mental powers in profiting by experience."[8] It is to the experience, not to the marvelous mental powers, that Thorndike directed his and our attention. He did so by setting a task for the animal, which he believed could demonstrate the presence of mental associations. The task would require no unusual motion from the animal, save a new physical sequence in response to the environment, and one that Thorndike believed, lent itself to the study of imitation.

ANIMAL IMITATION

Thorndike probably did not see himself as inventing techniques to measure the mental ability of animals for the purpose of filling in a Mental Ladder. Yet if he never precisely expressed the goal, the reader understands that this was what Thorndike was about. His interest in whether animals imitate was vital to the erecting of the Mental Ladder, not so much because of what he discovered about imitation itself, but because his method of investigating created a reaction among scholars elsewhere who then built laboratories and recruited animals to test Thorndike's findings.

Human beings seem charmed when an animal imitates human activity: how else to explain our fascination and perhaps appreciation of the chimp dressed as a person riding a tricycle, walking a tightwire, or, like Peter, using tools in a human way. Thorndike thought imitation to be of the greatest importance to our understanding of animals because "a good test of the intelligence of any animal," he wrote in 1901, "is its ability to learn to do a thing by being shown it or being put through the requisite movements. Human adults would learn readily in either of these ways, because we thus get ideas of what to do and how to do it and modify our actions in accordance with these ideas. If the reader had never seen a glass or a faucet, he would nevertheless learn how to get a drink by turning the faucet and holding the glass beneath it, if he saw some one else do it, or if some one took his hands and put them through the movements. The intelligence required in such cases is not of a very advanced sort; it is not the power of abstract reasoning or of seeing the relationship of facts, but is simply the capacity to have ideas and to progress from the idea of doing such a thing to the act itself."[9]

Romanes's ladder of animal abilities (Figures 8.1 to 8.3) projected tool use to be an ability chiefly of the monkey and elephant, and learning by "association in time" to be a capability of the mollusks. But Romanes never mentioned imitation as a distinct mental ability. Thorndike's interest in imitation as a critical measure of intelligence, or "stupidity" to use the term he preferred, led him to ask: "Which animal forms are capable of learning by association? (Hence his study involving chickens, cats, and dogs.) And, in a second set of studies: Do monkeys learn by imitation?

To anticipate his results by just a little, we note that Thorndike's monkeys, along with the chicks, dogs, and cats that served in his experiments, failed to show the ability to imitate. The report of this failure set off investigations of monkeys' abilities by other experimenters, and the major purpose of these methods came to be to determine whether monkeys are capable of imitation. Through simple observation, however, everyone knows perfectly well that monkeys imitate. We think of this ability, I suspect, as an evident characteristic of the primates we have observed. Why, then, should Thorndike have become convinced that monkeys had no imitative ability? And why should various experiments and well-controlled observations have tended to agree with this conclusion? That question is the moral of the tale.

When experiments fail to match observation, we must pause to consider. What is to be instructive in this story is not so much whether or not monkeys can imitate, but why our common sense and observation should appear to be so wrong—why it should be so different from what careful experimentation and credible control and analysis tell us.

THE PUZZLE BOX

Thorndike's observations on chickens, cats, and dogs were based on his belief that descriptions of wonderful feats by animals (the "eulogizing of animals," as he called it) tell us little about animal capacities. Such stories may be awesome at the hearing, and they may inspire in us an appreciation of animal life, but a unique achievement itself tells us little about animal life in general and very little about the life of the individual animal. In like fashion, I am in awe of human beings who walk on hot coals or who scale the walls of buildings, or who run or swim with great speed, who may be eight feet tall, or two feet tall, but these variations, awesome as they may be, tell me far more about the ranges of human characteristics than about how to define and understand humanness. The unusual achievements of animals draw our attention as well, but alone they do not tell us much about animal life unless we investigate them systematically. Clever Hans was indeed an interesting horse, but he became a horse whose capacities were of value to our understanding of animal life only when systematic and intentional observation was undertaken. Similarly, Little Hans may or may not have been a child representative of universal behavior or feelings, but the systematic explanation of the causes of his behavior is of universal value.

In order to measure animals' abilities to learn by the association of ideas and motions, Thorndike designed what was later called a puzzle box. The name is an unfortunate one, for Thorndike clearly believed that animals did not "solve puzzles." Rather, he was persuaded, they learned by mental association and, perhaps, by imitation. To the human observer, the suddenness with which an animal tries and tries, and then appears to find the "correct" behavior is evidence for the "insightful" solving of a problem. Thorndike, however, thought that this kind of thinking, this eulogizing of animals, merely represented a sloppiness of human thought, an unwillingness to determine precisely *how* the animal was learning.

The puzzle box or, as it should have been called, the "association box," is shown by Thorndike's original drawing in Figure 9.1. There were, in fact, several such boxes, each differing from the other in size and in the ways the door could be opened. The box shown required the animal to perform separate acts in order for the door to open: a platform had to be depressed in order to open a bolt; the bolt had to be raised; a second bolt could be raised by pulling a string attached to a pulley attached to

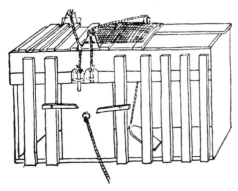

Figure 9.1   Thorndike's first puzzle box, or "association box," as it might have been called with greater accuracy. Note the pedal and its connections. The puzzle boxes and their releasing-apparatus differed in size to accommodate different species.

the second bolt; then both bolts had to be raised in order for the animal to open the door, leave the box, and attain the food.

Variations in the size of the boxes were mandated by the nature of the animal tested, notably, chickens, cats, or dogs. Why should the animal want to get out of the box? Thorndike tells us that he set food outside the door, and for the cats and dogs (but not for the chickens) saw to it that the animals were "kept in a uniform state of hunger, which was practically utter hunger.' In a later edition of the text, Thorndike explains in a long and defensive footnote that the term "utter hunger" meant only that the "animal would still eat a hearty meal at the end of the day."[10]

Thorndike presents first his findings regarding cats. By plotting the duration of time taken to leave the box as a function of the passage of time, Thorndike draws our attention to the behavior described by Figure 9.2 here redrawn to show the units measured. The figure is one of six displayed, and it is the one I have chosen for discussion, for it best describes what Thorndike thought he had discovered—and what we contemporaries think the data imply. We may want to revise this view, for although Thorndike's discoveries are said to be that animals learn by trial and error, not by insightful learning or imitation, it is important to read how Thorndike describes the animals' behavior:

"Starting, then, with its store of instinctive impulses, the cat hits upon the successful movement, and gradually associates it with the sense impression of the interior of the box until the connection is perfect, so that it performs the act as soon as confronted with the sense-impression. The formation of each association may be represented graphically by a "time curve."[11] The time curves are shown, and the one reprinted here as Figure 9.2 was chosen by Thorndike for detailed comment. "The time-

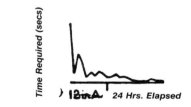

Figure 9.2   The cat's "time-curve," as Thorndike called it. The function would come to be called a "learning curve." The learning curve is characteristic of all species, and its discovery prompts meaningful questions. For example, why do living beings learn more about an association when first presented with it than they do later in the series of presentations? Why does the 24-hour lapse have so little effect on memory?

curve," he writes, "is obviously a fair representation of the progress of the formation of the association." [12]

The interpretation, according to Thorndike, is that the cat learns the association gradually while coming to associate the sense-impression with the act. Thorndike sees the learning as an association being acquired and built upon. What the cat demonstrates to Thorndike, and by way of him to us, put in modern terms, is that the learning is gradual, not abrupt, and that *the animal (thereby) learns more in the early trials and less in later trials* (hence the slope of the curve). Notice that the curve is not linear: learning does not occur, say these data, in equal units. Rather, more learning occurs early on. Why should this be? Is there something about the workings of the nervous system, the mind itself, that learns most at first association? Why should there be a learning *curve?* And, while Thorndike was yet to grasp this fact, why should all animals of every species display the same learning curve? The question of why the course of learning should be universally alike is an intriguing one.

Whether the cat has come to associate mental images with acts is not possible to say. In Thorndike's time, at the turn of the century, the presence of mental images was taken for granted. We moderns are less sure of the ubiquity of mental images. The cat and Thorndike have shown the course of learning to be nonlinear, gradual, and negatively decelerating, with "more" occurring at first and "less" as learning continues. The next century would show this basic pattern to be the course of learning for all living forms. The discovery of the generality of the learning curve— this shape is a characteristic behavior of *all* animals—would be understood by the later twentieth century as among the most important discoveries regarding behavior.

Now look at the patterns displayed by the other cats. Figure 9.3 shows the original page from which the cat's learning curve was selected. Does each cat represented here show the same course of learning? Can the performance of the single cat described (Figure 9.2) be said to be typical, to represent the performance of other cats? Are all curves gradual, not abrupt, and characterized by a slope that is negatively decelerated? Thorndike was working, interpreting, and writing thirty years before statistical techniques would become available to help answer these questions. Therefore, he could not answer these questions by any technique more substantial than his opinion. Because mathematics was not yet being applied successfully to the data of behavior, Thorndike did with his data what you and I just did. We looked at them and formed our own opinion, as did he.

The seeming generality of gradualness that Thorndike saw and we confirm is worthy of our attention in Figure 9.3. The performances of cats B, C, D, F, and maybe E might be said to show an abrupt, rather than a gradual, pattern of learning when we examine the drop in performance during the early trials. Later investigators would take such abrupt drops as evidence for *insight,* as evidence that the animal solved

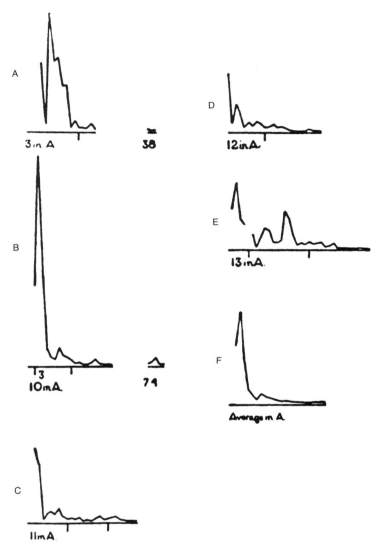

Figure 9.3   Figure 9.2 showed the learning curve Thorndike discovered from his researches with cats. This figure, 9.3, now shows the six time-curves shown by Thorndike. The one used for Figure 9.2 is in the upper right. Are the remaining five curves representative?

a problem by perceiving the situation and then suddenly seeing a solution. But Thorndike's hypothesis was that learning is gradual, although it may appear to the human observer to be sudden and insightful. From these results, he saw support for his hypothesis that learning is gradual but also abrupt. Over time, only this one graph (D) came to be reproduced in books and texts as evidence for the gradualness of the course of learning.

When dogs were placed in a like puzzle box, they behaved differently from the cats. Instead of patrolling the box, as the cat was seen to do, the dogs moved as near as they could to the food dish and spent their time in this location. Their unwillingness to move about meant that they had less of an opportunity to do one of the acts that opened the door. The cats' moving about, to the contrary, provided greater opportunity for them to perform, even if randomly, one of the acts necessary to open the door and, to be seen as solving the problem. Nor, Thorndike reports, do dogs try as hard as cats to get out.

Thorndike seems to have been a little annoyed by these differences. He seems not to have expected them, or he seems to have thought that his puzzle box was a universal test that equated animals' instincts and thereby measured their mental ability. Yet the differences in the conduct of the animals, even in the same puzzle box, make it difficult for the test to work equally well for the two kinds of animals. In our times, investigators would see these species-specific reactions as central to each species' unique way of responding—the cats' moving about, the dog's staying near the reward. But Thorndike wanted to reduce the learning experience to its core, to its simplest element, much as our knowledge of chemistry reduced the compound to the atom. The perfect apparatus, the perfect puzzle, should be so arranged as to yield identical elements of behavior from all animals. Later, B. F. Skinner (1904–1990) would see the significance of successfully reducing the unit of behavior to one in which all animals could be seen to engage. The training of such behavior, the single response, would become the scientific unit of his behaviorism.

As Thorndike notes, the dogs were not as well motivated as the cats. For example, they could not be put in a state of "utter hunger" because the barking that resulted upset the neighbors.[13] Even with the behavioral differences and the need for different kinds and degrees of hunger, the dogs showed learning curves like those of the cats, as shown in Figure 9.4. Some might suspect that the initial points of the resulting data, the ones that Thorndike explained as being due to the less motivated state of dogs as compared to cats, might just as well show a greater decrease in time required between the first few trials. That is, they show that the dogs were more insightful, but without the availability of the original data, we cannot test the issue of degree of motivation.

The chicks were tested somewhat differently. They were placed in "a small pen arranged with two exits, one leading to the enclosure where the other chicks and food were, one leading to another pen with no exit."[14] The results complete the trilogy of animal mental testing: chicks, too, display the learning curve. To my eye, the curves look more like those of the dogs, but without the availability of statistical inference, it is merely our eyes that are the judges, as they were in Thorndike's time. What can be said with certainty is that Thorndike's animals displayed curves that were alike, although the precise differences and similarities among the curves might suggest interpretations of the animals' mental

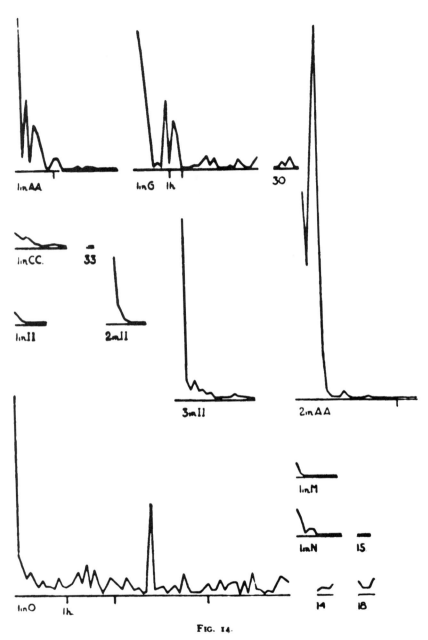

Fig. 14.

Figure 9.4 Thorndike's evidence regarding the "time-curve" in dogs. Thorndike is less happy with these variable results than he was with results from cats, but accounts for their variability by pointing out that few dogs were tested, and that they were not equally hungry. The problem of arranging for equal motivation from hunger would present Thorndike with an ethical dilemma.

states different from those implied by an explanation based on imitation and association.[15]

What Thorndike achieved with the invention of the puzzle box was a demonstration of the idea that mental tests could be used with animals, as Witmer was to show with striking results with Peter a decade later. Thorndike used the puzzle box to augment his position that animals' capacities should be measured, not guessed at. He had also discovered a universal law: that all living beings show the same courses of learning by association—the same learning curve results whatever the species. The discovery of a universal law of animal learning gave credence to the view that the mental ability of animals could be and that profound discoveries would result regarding the kinds and qualities of animal thinking and learning.

Thorndike now turned his attention directly to imitation. He was aware of some evident examples of imitation, such as the singing of birds, but by imitation he meant something quite different: he meant the learning by one animal by observing the actions of another. First, Thorndike built a pen for his chicks and looked for cases of imitation. For example, a hole in the screen was found, and one chick used it: Did the others, having perhaps seen it used, use it as well? That is, could they learn to use it by imitating another chick? Despite watching many such examples, only once did he see imitation, and he was convinced that the response required here (stepping on a platform when done) was truly accidental and not "true" imitation. Second, to further study imitation, this time with cats, two cats were placed in the puzzle box: one opened the door, then left; then the second also left. But had the second cat learned to open the door by observing the first? Did he merely follow the first? Or did he make use of the sight of the open door? When the cat who did not open the door originally, but who may have imitated, was placed in the box alone, this cat did *not* open the door. This was seeming evidence that the cat had not *learned* by imitation. Dogs, also failed to show evidence of learning by means of imitating one another when this method and technique were employed.

Thorndike's conclusion was that none of these animals learned by imitation. Data do not always tell us what we want to hear, but if we ask the question properly, well-collected data always speak to some point, as we will see. Thorndike's question was answered by nature with a seeming "No," a seeming experimental deadend. But remember that Thorndike's question was whether animals learned from observing one another. The question will be raised again by other investigators for the reason that the seemingly negative answer—that animals do not learn by imitation—was a function of the way in which imitation was here defined and denies common sense—as do many observations placed under the scrutiny of scientific methods.

Thorndike began the studies with chickens, dogs, and cats assuming that animals possessed mental properties. Such was the accepted notion of the day, and Thorndike's experiments were meant not to deny animal mentality, but to show the conditions under which they could be observed and measured. At first, in the 1897 paper, Thorndike was satisfied to report that these animals did not show imitation. Thorndike's report of his findings has the tone of surprise. The tone is, at once, arrogant ("I am the first to have thought of this elegant way of measuring mentality") and a little uncertain and apologetic ("I can't believe my own results") and a little fearful ("Just maybe there is something wrong with this technique, for common sense says that animals imitate; at least some of them, at least some of the time. Maybe someone with more experience or more clever will see where I have gone wrong and I will be made foolish.")

Maybe.

Animals do not learn by imitation; they do not learn by being made to imitate the acts of others. A rat does not learn to press a bar by our taking his paw in hand and demonstrating what is wanted. Instead, animals learn to repeat *whatever* behavior leads to satisfactory outcomes, such as release from confinement, access to food, and the like. With this statement, Thorndike has postulated the general rule of animal learning. The only mental capacity needed to explain all of the learned behavior of animals we see is this: Animals will repeat whatever motor pattern results in pleasant, nonpunishing outcomes. The most complex form of animal behavior, even the seemingly unique feat that excites our awe and admiration, is part of a chain of learned motor responses. To say this is not to deprecate the act, but to show the marvel of its components being chained together. So-called unique examples of animal intelligence are either such chains (Clever Hans's abilities, for example) or they are random and unrepeatable (the story of Dash's seemingly emotional response to a picture). Thorndike's analysis has grown from the particular explanation (imitation) to the general (behavior is stamped in by the outcomes of behavior). His effort gave forthcoming generations a tool to be used by the developing notions that would emerge as behaviorism.

Thorndike's new method for analyzing the animal mind launched a century of productive investigation on the issue of precisely how animals learn. It also acted as a counterweight against the sort of unspecific eulogy to animal mentalism that we find in writers such as Romanes writing in the immediate post-Darwin era and, among our contemporaries, in those observers who believe that animal minds are denigrated when we strive to understand the components of their learning. Thorndike's insight as to how to measure animal behavior led to the formulation of general laws of learning that describe the behavior of all living beings. It also established the experimental method as the best means to compare the abilities of species; it provided a convincing test of ability by which the Mental Ladder could be built; and, not unimportantly, it encouraged for a century the study of animal life under controllable laboratory conditions.

The laboratory, the puzzle box, the equipment, the measuring, the artificial alteration of the animal's environment, its emotional and mental state—all may be seen either as among the necessities of scientific methodology or as cruelty to our fellow creatures on this planet. Is what we have now come to understand about the learning curve worth the imposition on animal life? And, even if it is, is the laboratory method the best way to assess animal mentation, or, by using the laboratory, do we not merely demean the animal and, thereby, the experimenter, while making certain that we will fail to appreciate the intelligence of the animal, as it is not to be seen under such constrictions? The question has been with us all along, of course, for the unspeaking mind has no way to tell us directly what it wants, feels, and thinks.

ANIMAL MENTATION OR EXPLOITATION?

Thorndike's approach attracted disapproval in his own time. The most thoughtful critic was T. Wesley Mills, then at McGill University in Montreal, Canada. Mills expressed his contempt for Thorndike's view of animal mental life by writing a criticism of Thorndike's dog, cat, and chicken experiments. Mills, like Thorndike, had written a book[16] on animal intelligence that both criticized the eulogizing of animals characteristic of British writers of the post-Darwin generation and searched for scientifically acceptable concepts that would describe and explain the animal mind. Thorndike did not care for Mills's book, and he wrote a book review of it so saying; Mills, in like fashion, did not care for Thorndike's comments on his (Mills's) book.

In those days, arguing ad hominem was common. "Dr. Thorndike," writes Mills, "has not been hampered in his researches by any of that respect for workers of the past of any complexion which usually causes men to pause before differing radically from them, not to say gleefully consigning them to the psychological flames."[17] The image of Thorndike as Savonarola, at first consigning humankind's finer achievements such as art and music to the flames in Florence and, later in the same spot, being himself so consigned, does warm the reader.

Mills's arguments are important, especially today, because they set out what has been a centuries-old battle between what appear to be two poles of thought. If these poles gave us absolute alternatives, a choice between the poles could be made. Hypotheses should be falsifiable. But often what appear to be contrasting theories are not, for they are not true alternatives. Whether to study the mentation of animals in the laboratory or in the field condition of the natural state ostensibly presents an alternative, but in fact it is a mixture of choices. The reliability and validity of such studies is one issue, but the question of whether doing so exploits the animals is another. Most arguments on the issue confuse aspects of both, thereby making it impossible logically to distinguish or evaluate true alternatives. Validity is one concern; exploitation another.

Mills and Thorndike had difficulty remembering the distinctions, and we do little better in our time.

Mills begins his arguments against Thorndike's methods by classifying people who are interested in animal life. There are those who (1) "see in the animal mind only a sort of weaker human intellect; who look chiefly for evidences of intelligence and take no account of the failures and stupidity of animals"; Romanes was a founding member of this group, and both Thorndike and Mills found this approach silly. (2) "Those who recognize that the animal mind is not the equivalent of the human mind . . . but who nevertheless recognize the resemblance." The membership of this fellowship is now large. It includes most of us who see in animal conduct something of the emotions and categories of thought that shape our own behavior. (3) "Those who approach more or less closely to the view that animals are automata . . . [or who] consider animal consciousness as utterly different from human consciousness."[18] This position, thinks Mills, is that which Thorndike evinces by the conclusions to his experiments. Notice that so far none of these "types" speaks directly to the question of the validity of natural study or laboratory study.

Mills then presents an argument that has become increasingly powerful and familiar to the contemporary reader—namely, that the *artificiality* of Thorndike's work, what with chickens in pens and cats and dogs in puzzle boxes, disqualifies the value of the results. These confinements are not natural environments for these animals. Because of the unnaturalness of the environment and the test, the animals reflect nothing of importance about their abilities. This argument does *not* say that such matters cannot be measured in the laboratory, or that the field condition is the only reasonable place to measure animal behavior. Today, some argue that all laboratory work is invalid, and, presumably, all work in the natural state, work that is "ecologically valid" to use a popular term, is to be lauded. But this is not Mills's argument here. Mills now loses sight of his own logic and turns to vividness to enhance his point:

"As well enclose a living man in a coffin, lower him, against his will, into the earth, and attempt to deduce normal psychology from his conduct,"[19] writes Mills regarding laboratory work that tests animal mentation. While Thorndike honors the scientificness of the controlled environment and its equipment, the puzzle box, for example, Mills honors the behavior of animals left to create and solve their own problems, as they would in the natural state. Learning a puzzle box is not analogous to finding a path over a flooding stream. It can say nothing reliable about how problems are solved, because the problem is not naturally encountered. Thorndike looks at charts and says: Here! See the common element? The learning curve, similar for all animals, will look like this! Mills, for his part, continues by writing that it is individual differences that are of most significance—that, when recognized in nature, the individual differences are what enable one to see the general principle. For Mills, the *differences* in the curves are of interest, not the similarities, whereas for

Thorndike, it is his perception of the similarities among them that leads to generality and understanding. Where Thorndike sees a gradual curve, a curve representative of learning, Mills asks, "So what does your curve tell me about *this* animal or *that* one?"

Significantly, Mills argues that the experiment with chickens was at once the most misleading and the most valuable. Birds, being lower on the phyletic scale, Mills argues, have greater need of association and imitation, for the pen, however confining, is not as unlike the natural environment for chickens as is the puzzle box for dogs and cats. But for these chicks, writes Mills, Thorndike gives no data, only casual observation, whereas the drawing of the learning curve is the product only of the cats and, less demonstrably, the dogs. Mills is accurate in his estimate of the ways in which Thorndike's original paper reports the results, for this is exactly what Thorndike does. Thorndike shows learning curves for dogs and cats, but is unable to do so for chicks, evidently because their high degree of sociality means that they do not solve the puzzle-box problem as individuals. The animals, continues Mills, now sending a stake to the heart of Thorndike's conclusions, did not show imitation for the simple reason that the situation inhibited them from doing so. The idea of measuring imitation in an enclosure is silly, says Mills. Does Thorndike really doubt the existence of animal imitation? Of course animals imitate, but not under the stress of confinement and hunger.

It is easy to agree with Mills, for our natural kindness compels us to confess that, however universal, especially when seen against the backdrop of nineteenth-century animal eulogizing, the issue of real importance is whether an animal learns by imitation and, if so, the nature of the requisite conditions. Whether it does so in the lab or the natural state is really not an issue regarding whether there be imitation, although it may be an issue of exploitation. When Thorndike reports that these animals do not show the capacity for imitation, he is going far beyond what his results tell him, for he is generalizing from the limited situation to all situations. The results tell us merely that these animals do not imitate one another under these particular conditions of confinement and hunger. Whether they do under some other conditions is undetermined. The issue that Mills and Thorndike are discussing, an issue hidden in their eventual feelings for one another and in the hyperbole of description, is whether laboratory conditions necessarily limit our ability to generalize or whether they enhance it. Whether imitation is studied in the laboratory or in the open, whether it is studied in captive animals or in free ones, does not affect generality. What determines the success of generality is the design of the experiment or the observations, whether made in the lab or nature.

The issue of the laboratory and nature remains, long after Mills's and Thorndike's time, an issue that haunts and hampers the study of the animal mind. Study of the Mills-Thorndike unpleasantness is still valuable to us living today because it remains unresolved. The issue reappears and is restated and reargued in tiresome fashion and, one supposes, will continue to be so if commentators do not take apart the nature of the argument. Experiments can be done in the natural state just as they can be done in the laboratory, but the design of the experiments is necessarily different and, therefore, the kind and degree of generality is not identical.

The issue is seemingly resolved from time to time by the argument that both laboratory and natural work are needed, but this solution overlooks the need to think very clearly about the kind of generality that may be produced by either. Why not both? Why not take to the laboratory that which, seen in natural conditions, is most promising, most apt to increase or revise our understanding of animal thinking and animal feeling? The sight of chimpanzees fashioning twigs that they then pry into termite mounds is exciting: the observation shows tool use, prescience, the ability to fashion an object to fit a shape. Why can't these elements be investigated separately in the helpful, if confining, environment of the laboratory? Because the difficulties of rearranging the situation are too powerful, because the animal does not respond to, or in, the laboratory environment? How can we balance our need to know about other beings with their need to avoid confinement and hunger? We are reminded of Karl Krall's explanation that his horses became unable to answer when "controlled" by the laboratory—that is, when unsympathetic visitors were present. The same explanation is now given for chimpanzees and gorillas whose sign language and linguistic skills go askew.

A correlate question remains: Which approach measures behavior and mental ability in a reliable and valid way? One pole, one cluster of notions, states that behavior of interest can be seen only under natural conditions; that experiments are indeed useful, but they are most useful only when the animals are responding in natural conditions to natural forms of stimulation. The intriguing aspect of animal mentality and intelligence is the way the animal sees problems and solves them within the natural environment. The other pole, not a strict alternative, points out that the behavioral events need to be isolated so that the components can be traced. By controlling the environment, one can assess how a particular aspect of behavior comes to be, to understand what produces it, what refines it, what makes it whole.

The controversy initiated by Thorndike and Mills will not leave us. It remains at the heart of how we understand the intelligence, mentality, and emotion of animal life. Thorndike himself does not appear to have thought of the Mental Ladder or to have been guided by the metaphor, although it is difficult to believe that in his time his thinking was not shaped by the then-current concept of one. Later in his life, however, his interests turned to educational psychology, a field that is known, in part, for using tests of mental measurement to distinguish people and their abilities. In short, Thorndike turned to what Witmer had begun.

ERECTING THE PRIMATE LADDER

Thorndike's surprising and unexpected results—that chicks, dogs, and cats do not learn to imitate—helped bring monkeys into the laboratories and into countries to which they were not accustomed. The negative results regarding imitation did not forestall, but rather prompted, interest in the mental abilities of primates as measurable in the experimental settings. We are quite accustomed to primates today, but to people living a century ago, monkeys were true exotics, only recently seen and described by European visitors to Africa and India. Some monkeys had been displayed in Europe, but for most Europeans, these animals were unknown. Newspapers vied to reproduce woodcuts of the new species, and the hunt was on. This climate explains, in part, Garner's popularity and success.

The Christian Bible, for all its descriptions of animals interacting with human beings, makes only two references to the nonhuman primates, and this to "apes" in the most general sense.[20] Little Hans may have seen a primate at Schönbrun Zoo, but if he did, he did not mention it. When Witmer saw Peter on the Boston stage, he was among the first to see a primate doing anything but residing in a cage. When Garner played phonographic recordings of monkey sounds to the inhabitants of American zoos, he found subjects in only a few zoos. The time was right to extend the Mental Ladder to include the primates. Once Thorndike had shown a way to bring mental capacities under observation, attention would now be on the mental capacity of primates. Not only were monkeys and apes unique and unknown, but they also seemed likely to be the missing link, to furnish information on the crucial rung that separated animals from people.

Were monkeys and apes "little people" mentally? Were they conscious? Both Thorndike and Witmer had contributed to the notion that animal mentality could be measured: Mental tests could be set; success and failure recorded. Mills argued that however inventive such tests might be, the behavior of animals under unnatural conditions of confinement and hunger rendered any such data uninterpretable and degrading to the animal's abilities. For the time being at least, Mills's argument went unheeded: the race was on to erect the better Mental Ladder, and the first goal was to examine the probable missing link—namely, the primates.

By 1900, three years after the publication of his study on imitation in dog, cat, and chicken, Thorndike, as noted earlier in the chapter, acquired three *Cebus* monkeys. *Cebus* is a genus of the New World, that is to say, Western Hemisphere, monkeys. Like all New World monkeys they have tails that may be used for support or, in the case of monkeys with prehensile tails, as rudimentary hands. By common agreement, *Cebus* are intelligent by human standards. Thorndike reportedly used these monkeys to test their intelligence; he defined intelligence as their ability to reach through the bars of an especially built box, the puzzle box, to unlatch a hook, turn a bar or bolt, or unlatch a wire. Evidently, Thorndike, too, had learned something from his previous work on imitation-learning. He was now interested in how the monkey learned to use the mechanism rather than in imitation alone. His attention had shifted from assuming what the puzzle box tested to investigating what is measured. He put the question of imitation aside in favor of the more general question: How and what do animals learn? Thorndike was wise to make this shift, for surely the discovery of the learning curve represents the lasting value of Thorndike's work.

Thorndike had also learned something from the criticism that his methods were unnatural. He now justified his work to the reader by writing, "In order that such experiments shall be valid tests of the workings of an animal's mind it is necessary that he [the animal] surely desire to get into the box, that he not be disturbed by the surroundings in any way that will alter his mental efficiency, and that the experimenter be able to handle him easily without frightening him or taking his attention away from the box. . . . These desiderata were obtained by testing the monkeys when hungry and using bits of food of which they were fond by experimenting with them after they were used to their habitat and to my presence by getting them into the habit of coming to me and enjoying being handled, having their paws taken, etc.; by showing them the exit or putting them through it only when they were attending to the box."[21] A happy, calm animal is one willing to learn, Thorndike now believed, thereby, in some ways, coming to terms with Mills's position. Thorndike would spend a large part of the remainder of his productive career studying methods of teaching human children.

Thorndike's emerging interest in the techniques of teaching human beings is evident in retrospect, for the task Thorndike set for the monkeys was not whether learning occurred by imitation, but whether the animals might imitate *him*. (The view is not rare among teachers and parents that the student or child learns chiefly by imitating the teacher or parent.) Placing the monkey in the box—the monkey being now, perhaps, unafraid, unhungry, calmed, petted, and placated by Thorndike—Thorndike himself now demonstrated the solution to the puzzle. By manipulating one of the mechanisms, the door opened and (presumably) some food or food reward was then obtainable. In short, Thorndike offered himself as the one to be imitated.

None of the three *Cebus* monkeys showed any sign of copying Thorndike's behavior. His ability to manipulate the latch, if of any interest to the monkeys, remained unimitated by them. Rather than learn by copying the behavior of the human Thorndike, they pleased themselves. In an illuminating passage, Thorndike notes, "Monkey No. 1. apparently enjoyed scratching himself. Among the stimuli which served to set off this act of scratching was the irritation from tobacco smoke. If anyone blew smoke in No. 1's face he would blink his eyes and scratch himself, principally in the back. . . . He was often given a lighted cigar or cigarette to test him for imitation. He formed the habit of rubbing it on his back. After doing so he would scratch himself with great vigor and zest."[22]

Thorndike here adds an observation that qualifies as a pure example of operant conditioning, although Thorndike calls it an example of "associate learning": "For by taking a well-defined position in front of his cage and feeding him whenever he did scratch himself I got him to scratch always within a few seconds after I took that position."[23]

Thorndike had not yet explored his inclination to determine the rate at which an animal learns a new task, but his ideas were preparing him for the undertaking, which would prove to be his lasting contribution to our understanding of animal intelligence. His own mind was stuck on the notion that imitation is the filter that determines which species belongs on which rung of the Mental Ladder. He summarizes the evidence by saying that the monkey mind has much in common with that of the human, while fish learn slowly and are capable of learning only a few habits. These are Thorndike's words: "Dogs and cats learn more than the fish, while monkeys learn more than they. In the number of things he learns, the complex habits he can form, the variety of lines along which he can learn them, and in their permanence when once formed, the monkey justifies his inclusion with man in a separate mental genus."[24] By way of these ideas, Thorndike had come to offer his plan of a Mental Ladder. The construction seems to have had no reliance on learning by imitation, for Thorndike was not able to show that the animals he examined and observed learned in this way. Of what material, then, is the Mental Ladder constructed?

Two persons, skeptical of Thorndike's conclusions, decided to develop more refined ways to examine animal abilities, in the way of Thorndike, to be sure, but with significant and imaginative changes. They were Melvin E. Haggerty (1875–1937) and Gilbert Van Tassel Hamilton (1877–1943).

Haggerty worked in the laboratories at Harvard and at the New York Zoological Gardens (the Bronx Zoo). Like Thorndike, he would spend the longest part of his productive career in education, as dean of a school of education. Hamilton, a physician by education, was later to become interested in Freud, psychoanalysis, and human sexuality. His tabulations of frequencies of various kinds of sexual behaviors predate the more famous Kinsey data. Haggerty had undertaken the psychiatric responsibility for a young man living on a Southern California ranch, a ranch on which apes and monkeys were kept as a collection of exotic species, along with dogs, and, as is to be seen, farm laborers. Although Haggerty and Hamilton undertook pioneering work on the nature of monkey mentality, both appear to have been more interested in primates as an order than in the theoretical issue of the Mental Ladder.

To Haggerty, Thorndike's finding that monkeys do not imitate seemed wrong, for in popular opinion monkeys were copious and inveterate imitators. He suspected that monkeys had no particular reason to imitate human beings (especially those who, like Thorndike, blew cigarette smoke in their faces); that they had no occasion in the wild to imitate people and no special reason to do so in captivity. That they imitated one another Haggerty did not doubt. He said straightforwardly, but without direct criticism of Thorndike, that a monkey watching another monkey lifting a latch is a very different thing from a monkey watching a human lifting a latch. These views led him to two questions: Do monkeys imitate people? Do monkeys imitate one another?

In late 1907, working in the Harvard Laboratories with three *Cebus,* and in the summer of 1908 working in the New York Zoological Park (in this case, the Central Park Zoo which is now part of the "Bronx" Zoo), Haggerty set out to investigate the two questions about monkey imitation. Haggerty was aware that in the ten years since the publication of Thorndike's work other people had examined various species of monkeys in search of the monkeys' ability to imitate human beings. Among these persons were A. J. Kinnaman in 1902, who had found that a female rhesus *(Macaca mulatta)* had copied the performance of a male rhesus in a puzzle box, but that the male did not himself show learning by imitation. The teasing finding that female primates learned by imitation while males did not reappeared nearly a century later in work on how such animals come to manufacture and use tools. Known results that show differences between the sexes in ability to learn are sporadic and based on few animals, but the suspicion is evident and recurring among those who study apes and monkeys. I shall tell you my chance observation on this matter:

While doing an experiment in which baboons *(Papio hamadryas)* were pressing buttons to select visual slides they wished to inspect, I noted that while males controlled the apparatus, females sat a respectful distance and watched, sometimes humanlike, with their arms folded behind their backs. When the males, having completed their task, were removed from their enclosure, the females did not need to be taught how to use the buttons. They went right to them and began to push without the need for training. It would seem that the females learned by imitation of their fellow baboons alone, but that the *possibility* of their demonstrating this behavior was swamped by the males' keeping control of the apparatus and perhaps by maintaining some unrecordable control over their attempt to demonstrate a behavior already well learned. Such is the importance of the difference between learning and performance, between what an animal is seen to be doing and what it may do under appropriate conditions—and such is the silliness of experimenters trapped by the views of their times.[25]

J. B. Watson, later in life to be the founder of behaviorism, used rhesus monkeys *(Macaca mulatta)* in his experimentation. In support of Thorndike's results, Watson found that monkeys showed no signs of learning from one another by imitation. The rhesus, unlike the New World primates that worked for Thorndike, are Old World primates, and differences between the two groups in structure and behavior are obvious. In addition, Thorndike's result with monkeys had been supported by other researchers working in other laboratories with different genera of primates. Not surprisingly, Haggerty remained unconvinced.

Haggerty was able to do a true comparative study—that is, to compare the abilities of more than one species and genera of primates. The work at Harvard employed the perhaps oft-used *Cebus,* but by achieving permission from the director of the Bronx Zoo, Dr. William Hornaday, Haggerty was able to compare the abilities of representatives of both the New and Old World monkey genera. Haggerty knew his animals. He took care to describe, if perhaps unscientifically, his impressions of the general behavior of the animals. "The Cebus are cowards," he writes, "except toward those that they can easily vanquish. One fight is usually enough to settle the supremacy of the cage. The whipped animal seldom makes another effort to rule. The victor, however, often delights in continuing punishment which the vanquished receives with howls and shrieks of fear." [26] The rise of behaviorism a decade later would make this kind of comment sinful. Although the description conveys only the author's perception, those who have watched *Cebus* would have no trouble believing it. At what point do we admit that what is seen is the perception of the human observer—useful for communication to other human beings, but never true in any ultimate or unchallengeable sense?

In the description of the monkeys' cages, Haggerty leaves some reassuring, if inadvertent, traces of his beliefs. He is sympathetic to the "natural" point of view argued by Mills, or, at the very least, he is clearly concerned that the animals be comfortable. A tree and branches were placed in the cages in order to provide some aspects of the environment to which the monkeys were accustomed ("the animals could stretch their tails by wrapping the tip end around a branch and suspending their whole weight from the limbs, a performance apparently as enjoyable to the monkeys as swimming is for the average boy" [27]).

The monkeys did not seem to enjoy being caged alone, and so efforts were made to give them companions. Haggerty, in a section called *Characteristics of Individual Animals,* describes the personalities of each animal in detail in the hope that we can understand the monkeys' reactions if we know something about their history and personalities. The modern reader senses that the descriptions are not solely for the purposes of experiment: Haggerty appears to find the individual differences among the animals delightful as well as of scientific interest. His article lists the food given on each day of the week (Sunday: bananas, yellow corn, sunflower seed). "No effort was made to handle the monkeys with the hands in transferring them from one cage to another. They were allowed to go down a runway or to enter a small box which was then transferred to the larger one." [28] Haggerty takes these niceties for granted, rather as something that anyone would do for the animals or as something necessary to achieve experimental isolation and therefore scientific validity. Figure 9.5, taken from Haggerty's paper, shows one of the *Cebus* caged with the tree and branches.

In Haggerty's experiment, the monkeys were placed in a large version of Thorndike's puzzle box. The same test and the same general problem were used, but Haggerty's understanding of the animals' behavior led

Figure 9.5   A cebus monkey, *Cebus capucinus,* often called the Capuchin monkey or organ-grinder's monkey. This New World primate was the focus of investigations of primate behavior evidently stimulated by Thorndike's experiments.

him to design experiments far different in concept from those of Thorndike's. The view of what constitutes evidence of "imitation" is well expressed: "They [the animals] should be met as nearly as possible on their own ground and presented with problems in which they may have the advantage of their fund of inherited and acquired modes of behavior. At first the elements entirely new should be as few as possible. . . . If . . . they do manifest imitative behavior (or some aspect of it; even humans don't imitate perfectly correctly the first time) the complexity of the problems can be increased and thus by successive steps the range of imitative behavior can be determined."[29] The nature of the task was varied in accordance with this belief, and each monkey had the opportunity to imitate in a variety of ways, for example, with "partial imitation" being observed. The results were recorded as failure, partial success, or success.

A good scientific fable would result from finding that these animals showed learning by imitation, thereby identifying the way in which the animals were treated by the experimenter as the reason why Thorndike could obtain no evidence for imitation. I think that Haggerty would have liked to have shown evidence for animal imitation. But is it not so.

Consider, for example, the results recorded from one test of imitation in which the experimenter went behind a screen where food was kept and examined whether the animal imitated the behavior as shown in Table 9.1.

Based on these results we could conclude that there is little that is new: Thorndike was right; monkeys do not learn by imitating. Or, one might conclude that Thorndike was wrong—that some monkeys, under some conditions, show some learning by imitation. Or that, perhaps worse,

Table 9.1 Results of the Screen Experiment

I	
Number of animals used in the imitation tests	5
Cases of successful imitation	1
Cases of partially successful imitation	2
Cases of failure to imitate	2
II	
Cases of imitation when the initiator was confined during the activity of the initiate	0
Cases of imitation when the two animals were in the cage together	3
III	
Cases of immediate imitation	0
Cases of gradual imitation	3
IV	
Cases of imitation in which the imitating animal did not himself experience the result of the act before performing it	0
Cases of imitation in which the imitating animal did experience the result of the act before performing it	3

Source: Adapted from Haggerty, 1909.

neither is true and both are true, for some animals show some imitation some of the time. The data appear to support the last interpretation. We are tempted to leave, shaking our heads, seeing once again that nothing is stable in the world of investigating the conduct of living beings.

But let us not be so hasty. Nature has a habit of answering only the *precise* questions it is asked. Getting it right the first time is the questioner's problem. So it goes with the experiment that follows the rules of deductive science.

Haggerty, writing in 1909, knew this. So he asked.

Haggerty: "Do those monkeys who imitate do so in the presence of monkeys with whom they are acquainted (while, it is implied, not doing so with unfamiliar monkeys"?)

Nature: "No"

Haggerty: "Is it the reverse? Do they imitate unknown, rather than known and congenial animals?"

Nature: "Yes"

Haggerty: "Why?"

Silence.

Haggerty: "Is it because with unknown animals there is a level of arousal that leads to more activity and thereby more behavior to be imitated?"

Silence.

> *Haggerty:* "Do the monkeys only imitate when very hungry, or the food much prized?"
> *Nature:* "Yes"
> *Haggerty:* "Always?"
> *Nature* (sensing the trap): "No"
> *Haggerty:* "Is too much hunger destructive to imitation?"
> *Nature:* "Yes"
> *Haggerty:* "Is it true that imitation itself *must be learned?*
> *Nature:* "Yes"

Haggerty only hints at the last suggestion, but hint he does. His published report ran to over one hundred pages, much of it describing every monkey in every situation, but to the modern reader, it falls short of its original promise. After pages and pages of detailed description, Haggerty the author offers only weak and seemingly tired and spent interpretations of the results, the "congeniality" of animals being the chief among them. We guess that Haggerty is tired of exploring ambiguity: the results are not as clear as he would like, as the table of results shows. Haggerty senses that the results do not provide a decisive answer from nature as to the presence or importance of imitation. Thorndike is not shown to be wrong; monkeys are not shown unequivocally to be imitators, and the best laid experiments of monkeys and men sometimes fail to tell us what we thought they might.

Our sympathy is with Haggerty, for he did the job properly. He understood the monkeys and their needs. His method of working was correct: start with the simple and work to the complex to discover where imitation begins and ends. Treat the animals with respect. Ask what they can do, not what they can't do. But Haggerty himself could not explain his results, because he did not know what to do with the answers. We sense that he did not see the significance of his own study, lacking, as he must, of course, the historical perspective available to us. We are sorry, because his paper shows that he studied animals because he cared for them, not because of the reputation they might give him. Perhaps his reticence to make claims explains why his work is rarely mentioned or explicated by our contemporaries. This is unfortunate, for unlike Mills, who engaged Thorndike in a lengthy and essentially misguided and thereby nonproductive argument about the validity of the natural experiment and the laboratory experiment, Haggerty set a standard for the students and scientists of the future about how to ask questions of animal life in a way that is both humane and knowledgeable of the abilities of the animals.

Across the American continent near Santa Barbara, California, G. V. Hamilton had a laboratory of living creatures. On a property surrounded by oak trees, a solid fence served to contain a collection of animals, among them the then rarely seen macaque monkeys. The fine weather permitted them to live outdoors year round, although, as Hamilton notes, there was an occasional frost. Nonetheless, it was not sufficiently cold to require the artificial heat used by growers in the South living and cultivating nearer the mountains. Although the animals were not in their natural conditions, they were not housed alone or in small cages. They roamed together in a fenced area, and we may expect, if not the full display of social behavior to be seen in the natural state, some evidence of the complex social behavior that we know to bind primates, such as ourselves, into groups.

Hamilton thinks that the trial and error or, as he calls it, the "try and try again" method by which Thorndike and others measured animal intelligence missed the point of what animals do naturally.[30] Had he ever heard the term *ecological validity* he would have approved of it and used it to describe his efforts, for he was interested in setting for the animal subjects a problem similar to that which they might encounter in nature. Although Hamilton's views of the importance of the natural experiment put him in the same camp with Mills, Hamilton had his own ideas.

Hamilton, like John Lubbock before him, wondered how to develop a task that would measure what an animal could do, rather than one that would measure what it could not do. He wanted to develop a method that was "fair" to species of different sizes and abilities, for, after all, human beings might appear stupid if asked to find their way over or in oceans without navigational equipment, a chore that many birds, fish, and amphibians do routinely and with apparently little learning. Rats learn mazes (and also live in mazelike burrows), whereas people have been known to become confused when navigating one-way streets. What kind of task, asks Hamilton, is one which every animal uses, and what measurement can be made that does not unfairly favor those with wings, hooves, color vision, or ears pitched especially to high frequencies?

The answer relates to an animal's own hypotheses used to solve a problem, the problem being "what logic shall I use to be efficient?" Figure 9.6 shows the "equipment" Hamilton designed. Once in an entrance hall (EH), the door behind is closed: the participant cannot retrace her or his steps. The participant now faces four doors, numbered, from left to right, as ExD1, ExD2, ExD3, and ExD4. Any of the doors can be locked from behind. Assume that I have learned that food or some favored entity is placed behind one of the doors. How will I decide which door to try?

If I assume that the food is behind one and only one door, regardless of how many doors I try, I might try the doors in order, say, from left to right. My chance at success is 25 percent on the first try. If I find nothing, my probability of success increases until the attempt at ExD4, assuming all other attempts failed, when my chance for success on the last door is 100 percent. I shall acquire the reward in at least four attempts if I follow the pattern of trying the doors in order, or if I try each door once.

But I have no way of being certain that the reward remains behind one particular door. That is, I may try ExD1 through ExD4, find nothing, and return to ExD1 where I find the reward. Under this condition,

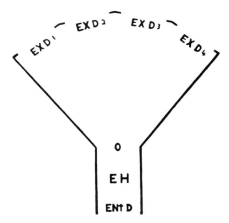

Figure 9.6   Hamilton's apparatus for measuring animal and human abilities.
The animal or person is placed in the entrance hall (EH) by way of the
entrance door (ent D). At point O, the exit doors (EXD1 to EXD4) are
equidistant. In a typical experiment, food, or the ability to exit, is the reward
gained by selecting the proper doorway. After a trial, that doorway is locked.
Can the animal or person learn that the same doorway is never right twice in
succession? If so, do people and animals differ in the ways in which they learn
the principle?

I might decide that the rule of placement is not as I thought. What
pattern of choice is wisest? This device, though simple, is able to tell us
something, albeit indirectly, about *how* animals solve problems. True, the
"apparatus" is simple, but the problem presented is complex. The problem measures in terms of success, but its inventiveness is that it encourages the animal to form and test hypotheses while permitting the human
investigator to *infer* a great deal about the kind and nature of such hypotheses.

Hamilton is sensitive to his animals, their capabilities, their needs, and
their welfare. He goes to some trouble to build different chambers to
suit the different sizes and shapes of the animals that assist him. He spends
time introducing the animals individually to the apparatus, at first with
all the doors open and food scattered about, continuing until, as he says,
the animals accommodate themselves to the surrounding and show no
sign of fear.

Now we are ready for the "formal" trials: On the first, one door could
be opened and the remaining three could not. When the openable door
was opened (by the snout, by the paw), the animal left the chamber
through the doorway and food was given. On the second—and it is here
that Hamilton's cleverness and imagination may be appreciated—the unlocked door on the first trial was now locked, but another, previously
locked, was unlocked. Continuing this pattern for one hundred trials, the
rule of the experiment was that the unlocked door was always one that

had been locked on the previous trial. The rule: the door that could be opened was never the same door twice in succession.

The successful animal, Hamilton thinks, will form an hypothesis regarding the rule. Whether this hypothesis is in some way represented visually or auditorially, or indeed at all, we do not know. We do know that the animal changes its course of action in ways that are reliable and, as we will discover, predictable for the species. The experimental question is, What is the course of learning while the animal comes to find the rule efficiently? Note that the brightest of animals, including the most intelligent human being, will choose the openable door 33 percent of the time if they use the rule. Note also that Hamilton is making assumptions about the animal mind—namely, that one characteristic of the mind is that it forms speculations and predictions about the environment, and tests them, afterward altering the prediction for the next opportunity. What he is asking the animals to do by means of his experiment is to demonstrate *how* they think.

The choice of "subjects" is equally enlightening. The company included human beings, monkeys, dogs, horses, and cats. Hamilton notes carefully the subspecies, species, and genera of the nonhuman animals, and when he describes the human beings, he does so in a way that should be offensive to modern sensibility. Shortly, we will read these descriptions and, in our times, most of us will be offended by the classifications of people according to race, religion, and genetic origin. There is no need to apologize for Hamilton, however, for he wrote in his time, and by contemporary standards he was enlightened. What we learn is the way that an educated, thoughtful, and sensitive person working almost a century ago classified animals and fellow beings. He classifies individuals by their ability, labor, sex, and presumed social status. Here they are:

"Man 1. Age, 34 years. Native (Spanish-Indian) Californian. Ranch laborer in the experimenter's employ. A man of limited education, but of average intelligence for his class. He went through the trials (of the experiment) in the stolid, unemotional manner that characterized his work in the fields. The 'boss' wanted him to walk into and out of an enclosure 100 times, and he did so without asking questions or shirking his task.

"Boy 7. Age, 15 years. American, of original English descent. Grocer's boy. Country school education. He was shy and nervous at the beginning of the experiment, and always seemed to be more or less affected by a fear of appearing stupid.

"Boy 6F1. Age, 13 years. Father Italian, mother Swedish. Country schoolboy. He was less alert mentally than were his brothers, who are described below.

"Boy 5F1 Age, 11 years. Brother of Boy 6F1. Country schoolboy. Volatile, alert, and rather distractible.

"Boy 4F1. Brother. Age 7 years. Country schoolboy. Relatively precocious, and an excellent subject.

"Girl 1, Age 10 years. American of Scotch descent. Until recently a student in the public schools of Cambridge, Mass. Her superiority of school training, and her very considerable degree of mental precocity gave her a decided advantage over all the other human subjects, including, even, the adult Man 1.

"Boy 7. Age 26 months. American of mixed descent (son of the experimenter). At the time of the experiment he could walk; had a limited vocabulary of words, which he did not put into sentences: was able to find his way about the house; and understood many simple commands. Very quick to form new associations.

"Defective Man A. Age 45 years. Native (Spanish-Indian) Californian. Limited school education, but had read history and uncritical works on socialism. He was a nervous, suspicious, 'muddled' person, with a grievance against society in general, and a surprising fund of self-acquired misinterpretation relating to his social environment. His curiosity and his desire to argue matters rendered him available, but he seemed to be in constant dread of the apparatus, and always labored under a suspicion that it was not the simple structure that it pretended to be.

"Defective Boy A. Age 11 years. English. With the exception of occasional perfunctory lessons from a governess, his education was practically nil. He was barely able to read simple words, was unable to respect the conventions of conversation usually recognized by a child of six years and manifested an inordinate fondness for asking questions."[31]

The monkeys were macaques, mostly rhesus *(Macaca mulatta)*, but some of unstated species; the dogs were Boston terrier-mongrel, Boston terrier-English setter; the cats, Manx and house; the horse, western, a gelding.

The first set of results is shown in Figure 9.7 In which the people and animals are arranged into groups. Note the differences both between and within species. Evidently, the experiment distinguishes a Mental Ladder of phylogeny as well as one of ontogeny—that is, of genera differences as well as differences that occur during the development of a member of a single species.

Now consider what else we might learn about how our fellow beings go about solving a problem. Remember that, while the rule remained constant regarding where food and release were to be found in the enclosure, the subject's approach to the rule could vary. Hamilton distinguishes between these types of approaches:

"Type A. All three possible doors tried, *once each;* no effort made to open the impossible door.

"This is the most adequate possible type of classified reactions [writes Hamilton].

"Type B. All four exit doors are tried, once each, and in an irregular order.

"Type C. This reaction can occur only when the door to the extreme right (Door 4) or the one to the extreme left (Door 1) is the unlocked door. It involves trying each of the four exit doors once, and in order

DISTRIBUTION OF CLASSIFIED REACTIONS ACCORDING TO AGE AND
PHYLETIC AVERAGES

Subjects.	Average age.	Total classified reactions.	TypeA %	TypeB %	TypeC %	TypeD %	TypeE %
Older human (8).........	14 years	226	76.11	19.47	2.21	1.77	0.44
Man A (1)..............	45 years	29	48.28	51.72	0.00	0.00	0.00
Boy A (1)..............	11 years	40	62.50	7.50	30.00	0.00	0.00
Infant (1)...............	26 months	38	15.79	5.26	18.42	26.32	34.21
Mature monkeys (2)......	10 years	75	21.33	22.67	20.00	29.33	6.67
Immature monkeys (2)....	1.5 years	71	15.49	22.53	30.99	18.31	12.68
Mature dogs (2)..........	2 years	82	12.20	21.95	0.00	34.15	31.71
Older puppies (6)........	98.50 days	357	13.61	19.84	14.79	17.51	34.24
Younger puppies (8)......	52.75 days	364	14.26	10.16	13.74	23.08	38.76
Mature cats (3)..........	1 year	127	8.66	19.68	3.94	27.56	40.16
Kittens (2)..............	63 days	104	6.73	10.58	0.00	14.42	68.27
Horse (1)..............	8 years	50	8.00	4.00	2.00	24.00	62.00

Figure 9.7   The results of Hamilton's survey of how different species, and human beings of different cages, learn the principles designed for the "puzzle box." The strategies are A: Doesn't go back to door that was successful last, but tries the remaining doors once each. B: Tries all four doors once each, in an irregular order. C: Tries all four in order (left to right or right to left). D: Tries a given door more than once during the trial. E: Tries to escape apparatus.

from left to right or from right to left. Thus if Door 1 be unlocked, the subject tries the doors in the following order: 4, 3, 2, 1.

"Type D. More than a single continuous effort to open a given door during the trial; but between separate efforts to open the same door there must be an effort to open some other door. Thus, the subject tries doors in the following order: 4, 1, 4, 3; or 4, 1, 2, 4, 3 etc.

"Type E. This type includes various highly inappropriate modes of seeking escape from the apparatus. . . . during a given trial the subject tries a door, leaves it, then returns to it and returns to it a second time *without having tried any other door.* During a given trial the subject attacks a group of two or three locked doors two or more times in a regular order."[32]

By examining Figure 9.7, we can see the proportions of the trials during which each of these types was demonstrated. When we look under Column A, we find the proportion to be that used to organize the table by animal type on the extreme left: The older humans use the Type A method 76 percent of the time, the highest percentage for this method. When Hamilton refers to this method as the most logical, he is, as are we, presenting the human view of logic and thereby judging the adult human solution to be the most efficient and intelligent.

Boy A picks this solution somewhat more than does Man A, and puppies select it only slightly more often than do mature dogs. Whether these differences are statistically reliable, we cannot know. The statistical ways of determining this were not available to Hamilton in 1911, and without the original data, we cannot now calculate the answer. As was true of Thorndike's results, we are left to use our judgment and unable to use the inferences that statistical devices now provide.

The study is remarkable, for it is truly comparative in nature, studying phylogeny (the species and genera) alongside ontogeny (the age and mental ability of various members of a species). It stands alone both then and probably now by providing information on the location and nature of the rungs of the Mental Ladder. Hamilton, like Haggerty and Thorndike, however, makes no direct mention of the concept of a Mental Ladder. Hamilton appears to have seen his work as demonstrating how animals solved problems rather than as comparing the species' abilities in ways of forming hypotheses to do so. The references to be found in the work are indicative of Hamilton's thinking; they are to Darwin, Romanes, and Spencer, all nineteenth-century British naturalists of various persuasions, and Thorndike in passing, but, suggestively of Hamilton's own future work; to Sigmund Freud, Carl Jung, Eugen Bleuler, who wrote on suggestion; and to Adolph Meyer, who wrote on "mental disease." Hamilton's future work, conducted at the same site, would lead him to examine monkey sexuality and its relation to psychoanalytic concepts[33] and, by 1927, to publish his views on undergoing his own psychoanalysis. In the late 1920s he became interested in marriage, and in the relationships between married men and women.[34] Haggerty turned to child psychology, specifically to the effect on children of the then worldwide economic depression.[35]

To find another attempt to erect a Mental Ladder, we move forward in time nearly a half century. By this time, the urgency of finding missing links, of determining levels of mental capacity, and of measuring individual differences had become unfashionable. What remained of the age in which the beginning investments were being made regarding how animals think, reason, and feel was an interest in a ladder based not on finding capacities of the mind, but on how animals of different species responded to training and conditioning. "Mental life" had become a bankrupt term, one that was thought to be merely hiding behind the undefinable and largely undemonstrable postulation of mental conceptualizations of both human beings and animals.

Animals did not think, so went the modern version: they behaved. Even if they did think, such information was unknowable by human beings and therefore unmeasurable and inadmissible. Here is one of the achievements of a rigid application of behaviorism. To measure the unmeasurable leads only to false postulations, false definitions building one upon the other until, like the enumeration of possible instincts, the house of cards falls. This, too, is the procedural view of behaviorism, and it is a view that has guided the bulk of twentieth-century thought in the West about animal thinking and animal feeling. It does not say that there is no mentation: only that, like Thorndike's understanding of consciousness, mentation could not be defined or measured and thus had no independent reality, or only a false reality, for those who would care to build a science of animal behavior.

Experimental research on the process of conditioning, to the contrary, provided the experimenter with control over the animals' environment. The resulting curves of learning, unlike those of Thorndike, could be fine tuned, demonstrated without quarrel, and shown to be characteristic of all species. The discovery of the learning curve, along with discovering the generality of schedules of reinforcement, are the two great, lawful findings regarding the behavior of living animals.

SCHEDULING REINFORCEMENT

Schedules of reinforcement are temporal patterns by which an animal is rewarded for a specific act. They provide so much control over behavior that they too produce curves of learning and forgetting that are the same for all animals, whether rat, bear, or fish. The effects of these schedules represent the most powerful control over human and animal behavior yet known. The fact that they produce like results in all animals capable of responding mark them as the ultimate explanation of the way in which living beings modify their behavior to suit the environment. Like the discovery of the learning curve, the discovery of the power and generality of schedules of reinforcement was a major discovery about behavior.

One result of this powerful discovery was that the search for a Mental Ladder became, if not obsolete, at least a task that could be put aside for another day while attention turned to unmasking the effects of schedules of reinforcement. The question now put was: As all animals show the same kind of learning and forgetting when the environment is brought under fine control by the experimenter, are there any differences among forms of animal life in regard to how they respond to conditioning? As was true of the learning curve, the question its discovery provoked was whether all life might not behave in like fashion.

The search was a rewarding one. By midcentury, it was evident that the several ways in which reinforcement may be scheduled, in fixed intervals or variable ones, in fixed ratios of responses to reinforcements, or in variable ones, had predictable influence on living beings, regardless of species.

UNSHACKLING OF THE MISSING LINK

By the 1950s, interest in a ladder of mental life that could be erected to match the development of the Great Chain of Being had diminished. Just like skirt length and tie size, ideas, especially scientific ideas, have fashionable and unfashionable seasons and times. Thoughts had turned elsewhere after the 1920s—to the undoubted and easily demonstrated power of operant conditioning, to the ways in which instincts could be "imprinted" on an organism and, to some degree, be modified by learning. The Nobel Prize Committee would honor these approaches to the behavior of animal life by bestowing its award for medicine on three persons who both spaded and fertilized the ground for this century's ways of thinking about animal life. The attempt to build the Mental Ladder became dormant for a reason other than fashion alone: it was based on a false understanding of how evolution works. Any structure, no matter how elaborate, carefully planned, or well pieced together, will not stand if the foundation is not prepared to hold it. And the ground prepared for the erection of this ladder, if not unprepared, was surely uneven.

The well-read, educated, and intelligent mind of the late nineteenth century, as well as many minds imagining today, seemed to have viewed natural selection and evolution as containing inherently a moral for human society. The notion that emerges supposes that evolution works for the *better;* that the most suited, the best equipped, survive and reproduce these aspects of themselves that are also so equipped. (The implicit self-flattery should not escape us.) As is true of all double exposures, the image is not untrue, but it obscured the fact that evolution cares not one bit for "best," for it has no idea what the best is. Evolution, as it operates by natural selection on living forms, leads to diversity among and within species. "If," as Leonard Trelawney Hobhouse so neatly put it writing in 1926, "the struggle for existence has produced the wisdom of man, it has also sharpened the tiger's claw and poised the cobra's fang."[36]

To Victorian understanding, and perhaps still to ours, evolution appeared to move toward perfection, to lead necessarily and inevitably toward better equipped beings. If so, the concept of a ladder was a helpful one. Society could be constructed within the framework suggested by evolution and national selection would improve society if left alone to operate in its remorseless but, in the long run, successful fashion. Pure capitalism, for example, can be seen as a social and political way to give the evolution of social perfection a boost, or, at least, not to interfere with the natural workings of natural selection. Natural selection could be seen to work to eliminate the "worst" and to reproduce the "best."

Consider the search for the missing link, whether it be the fossil remains to show the physical link between humankind now and ancestor then, or the mental link that Peter, Moses, and their primate colleagues might fill. We have both reason and need to search for a link only if there is something to be linked, only if, for example, it has been decided that there is a space to be found that separates chimpanzees from human beings. The Mental Ladder was the chain, of course, just as the fossil record has come to be the supposed chain that links all living beings, past and present, extinct or breathing on today's ladder. Each new fossil find of a supposed hominid fossil brings new guesses as to the nature of the link in structure and behavior between ourselves and those very remote genetic ancestors.

We may continue work toward constructing a useful ladder: it does no harm to investigate the mental abilities of human and animal life. The harm comes when the ladder is used for inhumane reasons, inhumane either to animals or to human beings; namely, when it is used to build a society that categorizes its individuals in order to set constraints on their potential and their ability to perform. The ladder, whether under construction or finished, is then but a trophy to be admired, a sculpture that should remind the living of the wonderful beings around them now and of those shaped before our times. It belongs in the town square where it can be cleaned from time to time, taken for granted by those who see it every day, an object of admiration for newcomers, and a statement, if ignored now and then because of familiarity, that we consider the wonders of being alive to be important and worthy of respect.

The search for the missing link, the discovery of the humanness of monkeys, the opportunities to examine apes and monkeys by mental tests— these goals and happenings shaped the question that was to occupy the remainder of the century: Can people and other primates communicate? If the Mental Ladder was eventually stored haphazardly, like a once-necessary crutch stored in an attic, such was done in part because the apes and monkeys were so spectacularly human in character that people ceased to regard them as a missing link, but as beasts with something to say to us.

The next section of this book investigates the nature of attempts at communication between people and apes; it is the search for the mental missing link performed in modern dress.

People and Apes Communicating

"It is a great baboon, but so much like a man in most things, that though they say there is a species of them, yet I cannot believe but that it is a monster got of a man and a she-baboon. I do believe that it already understands much English, and I am of the mind it might be taught to speak or make signs."

Samuel Pepys, 1661

"Among animals, some learn to speak and sing; they remember tunes, and strike the notes as exactly as a musician. Others, for instance the ape, show more intelligence, and yet can not learn music. What is the reason for this, except some defect in the organs of speech? But is this defect so essential to the structure that it could never be remedied? In a word, would it be absolutely impossible to teach an ape a language? I do not think so."

Julien Offray de la Mettrie, 1748

Raising Human Babies with Chimps: Donald, Gua, and Viki **10**

T HE NOTION that we human beings ought to be able to communicate with the apes, and they with us, is recorded at least since Samuel Pepys described in his diary his visit to the docks of London to see an ape freshly arrived. In our own times, as we have already observed, the wish to communicate with the apes is illustrated by Garner's recording primate sounds on the then newly invented phonograph in order to observe how the animals responded to the played-back sounds. Later, in West Africa, he had the idea of training chimpanzees to "talk" by forming, with his hands, their mouth into words in four languages. Witmer noted the chimpanzee Peter's inability to talk and accepted the anthropological view of the day that the inability was due to a cerebral lack of the motor ability needed to produce sounds. The notion that animals might communicate with human beings was also popular in a certain genre of English literature—for example, in the stories regarding Tarzan's ability to communicate with animals and by the animal companions of Dr. Dolittle, whose language he could understand.

Ape-human communication is a longstanding hope, although it remained for our times to take seriously Pepys's idea that such communication might be studied by teaching the animal a language to be communicated by hand motions. Can the mind of an animal be released from its silence by teaching it to motion a language that is human-designed? The importance and romance of the question may blind us to the fact that the question merely reestablishes the technique Itard wanted to use to investigate Victor's mind. To be sure, Itard wanted to know about Victor's mind what was innate and what was not, while we appear to have no such question about the minds of the animals with whom we wish to communicate. But our times have succeeded in teaching animals to make signs reliably. What these signs represent, and what they might tell us about the thinking of the animal, is another matter.

SPEECH AND MEANING

Speech is a behavior, a set of motor responses; it is we who assign meaning and language to these muscle responses. While Itard and, later, Feuerbach appeared to consider the languages of the wild-boy of Aveyron and Kaspar Hauser, respectively, as more or less motor means of expression, just as swinging a bat might be regarded as the motor means of batting a baseball, in our times people have come to consider speech an expression of feelings, thoughts, and observations represented by the motor skills. Presumably, one might learn the motor ability to make understandable sounds. This ability might be taught, yet the speech might fail to have meaning or feeling. The motor ability alone does not provide meaning, and much of the problem in deciding whether animals have language or speech is in truth an argument over whether their motor responses have meaning to us. At the onset, let us separate two distinct questions: Do animals have language, and are animals able to express the contents of their mind? We can imagine a gorilla or chimpanzee capable of meaningful language who nonetheless has nothing to say about its mind, just as we can imagine a like animal whose mind is replete with ideas and understandings who nonetheless has no means to communicate these to human people or, even if it does, whose language is not understood by us.

Once we believe that speech is not merely the motor component of ideas, or ideas the exact images represented by speech, we also believe that speech is not merely a motor skill that permits the ideas of the mind to be transmitted to others. Speech is both a given and an acquired behavior. It is the very behavior that Itard and others wanted to study in order to separate its "given" from its "learned" aspect. Investigators of the last two centuries apparently understood speech as a pure production of the concepts of the mind. At times, as with some of the children described in this book, speech failed. When this occurred, the reason was thought to be some form of reduced mental ability or, perhaps, an inability to make the appropriate motor responses themselves because of organic reasons.

The trainers and admirers of Clever Hans, Lady, Van, Roger, and Peter, along with the cats, dogs, chickens, and monkeys tested during Thorndike's era, all wanted to bypass language in studying the animals. This was not because speech was unimportant, but because its presence in animal forms was suspect and because such verbal communication as there was seemed more like that of the feral children than of a mind at work. These investigators wanted to leap around language by substituting some other motor response (a bar-press, the bar of a puzzle box, or selecting a card, for example). Their path to the mind was to substitute study of some other motor pattern for speech. By so doing, they tacitly accepted the view that language is chiefly a motor response. This view remains when human beings train an animal in some form of sign language. The assumption is that the use of the signs, these being motor behavior, avoids the need for the animal to use speech, which itself is a motor behavior. The idea is that, by so substituting a relatively easy motor response for a hard one, the substitute motor response takes on the qualities and characteristics of speech itself. The logic is worthy of our interest and concern, for it may not be necessarily so that all motor behaviors have the mental characteristics represented by speech.

The idea that ape-human communication could be arranged took hold in the early years of this century, reappearing in the 1970s. Remember that Witmer predicted in 1909, only five years before the publication of Edgar Rice Burroughs's fiction, *Tarzan of the Apes*,[1] that "within a few years chimpanzees will be taken early in life and subjected for purposes of scientific investigation to a course or procedure more closely resembling that which is accorded the human child."[2] His prophecy came true. The circumstances were the pioneer attempt to reach the animal mind by the intermediary of speech.

DONALD AND GUA

Nearly twenty years and a world war were to pass before Witmer's suggested experiment occurred. During this period, Robert Yerkes suggested that one way to communicate with apes was through the use of sign language, but no one seemed to have jumped at the idea, an idea originally put forth three hundred years earlier by Samuel Pepys.

In 1927 W. N. Kellogg and L. A. Kellogg devised a plan, the execution of which would take many years and an unimaginable amount of work, emotional strain, and thought: "Had we at that time any knowledge of the personal deprivation to be demanded by the undertaking, it is doubtful if we would have persisted further in the endeavor to bring it about," wrote the Kelloggs.[3] The Kelloggs's plan was to raise an infant chimpanzee while rearing their own child. They secured the financial backing of the Social Science Research Council and the vital cooperation of Robert Yerkes, who was then directing the Yale Anthropoid Experiment Station at Orange Park, Florida. This facility would later be renamed the Yerkes Primate Center and moved to Georgia.[4] Among Yerkes' contributions was providing space at the station for the Kelloggs and, critically, the loan of the infant chimpanzee, Gua.

Gua was born in the Abreu Colony in Cuba[5] on November 15, 1930, and delivered to the Kelloggs seven and one-half months later, presumably in June 1931. Donald, son of the Kelloggs, was born on August 31, 1930. Until March 18, 1932, a period of nine months, Donald and Gua were raised by the Kelloggs "as nearly alike as it was possible to make them."[6] Figure 10.1 shows Gua and Donald, at 16 and 18½ months, respectively, ready for bed. If the Kelloggs were not yet as knowledgeable as they would become about what the emotion and commitment involved in raising a human child, not to mention the investment in-

Figure 10.1   Gua (16 months) and Donald (18½ months), ready for bed.

volved in rearing a chimpanzee companion, they were at least well versed in the history of the problem they wished to consider.

"Let us suppose," they write, to open their book-length account of the rearing of the two infants, "that by some queer accident a human infant, the child of civilized parents, were abandoned in the woods or jungle where it has had as companions only wild animals. [The comparison to the fictional Tarzan is evident.] Suppose, further, that by some miraculous combination of circumstances it did not die, but survived babyhood and early childhood, and grew up in these surroundings. What would be the nature of the resulting individual who had matured under such unusual conditions, without clothing, without human language, and without association with others of its kind? That this is not so fanciful a conception as to lie altogether outside the realm of possibility is attested by the fact that about a dozen instances of "wild" foundlings of this sort are known to history. To be sure the reports about them are in many cases so garbled and distorted that the true facts are hard to shift out. In some, however, the accuracy of the accounts is well established."[7]

The Kelloggs then describe, very briefly, their knowledge of the wild-boy of Aveyron, of Kaspar Hauser, and of the wolf-children of India, although they wrote thirty years before Singh's diaries were published and expanded by Zingg (see Chapter 2). Their willingness to undertake the experiment of rearing their child with a similar-aged chimpanzee infant is based on a continuance of the motives that stimulated Itard to undertake the examination and education of Victor. Referring to the feral children whose lives have formed the first chapters of this book, Kellogg and Kellogg wrote:

"The customary way of explaining the fact that a human being of this sort does not respond well to the efforts of those who would civilize and educate it, is to say that it is feeble-minded, that it is mentally deficient, or that it is congenitally lacking in the ability to learn and adapt to its new surroundings. Even had such children lived under civilized conditions, they would still have failed to duplicate the accomplishments of normal individuals. The opportunities enjoyed by the average child would have left them little better in their ability to react than they were when they were found. This reasoning carries with it the assumption that because these children were not up to average for their ages when their reeducation was discontinued, there must have been something wrong with them before they were placed in the jungle or prison surroundings. That they were unable to adapt completely to civilizing influences is taken as proof of an original deficiency. In fact, going one step further, it is often argued that the 'wild' children were probably abandoned in the first place because they displayed idiotic or imbecilic tendencies at a very young age. . . .

"But there is a second way of accounting for the behavior of the 'wild' children, according to the theory of external or environmental influences. It would be quite possible according to the latter view to take the child of criminal delinquents, provided he was normal at birth, and by giving him the proper training, to make him a great religious or moral leader. Conversely it would be possible to take the child of gifted and upright parents and by placing him in a suitable environment, to produce a criminal of the lowest order. Heredity, in this view, assumes a secondary role and education or training becomes the important item.

"Instead of supposing that the 'wild' children were inherently feeble-minded, as is usually done, the proponent of the environmental doctrine would hold that originally such children were probably normal. . . . Those placed with animals may actually have learned, in a literal sense of the word, to be wild themselves, in the same way that a Caucasian child reared among the Chinese grows into the Chinese customs and language, or a baby that has been kidnapped by Gypsies knows in later years only the Gypsy manner of living."[8]

The Kelloggs's suggestion as to how to evaluate the investigations of feral children well reflect that belief in modern times, especially in times beginning around 1920 and extending until the very recent past, that environment outweighs heredity. Their analysis brings modern concepts to our understanding of why the wild-boy or why Kamala and Amala failed to reach any meaningful level of acculturation and civility, while Kaspar Hauser did. The Kelloggs present the case for the interaction of environment on heredity. Because environment but not heredity can be altered by the actions of society, it is reasonable to expect society itself to undertake the understanding and reform that aids its own cause. It can do so through its legal and, especially, its educational systems.

A powerful concept to be seen in the Kelloggs's analysis is the belief offered by John B. Watson in his 1919 book *Psychology, from the Standpoint of a Behaviorist* and his 1925 book *Behaviorism* that environment so influences heredity that any child, if environment can be controlled, can be shaped to become in later life whatever the arranger wants. This hyperbolic statement was clearly meant to show that, just as there is no heredity unless the environment acts on it, so there is no environment unless there is some heredity to work on. As heredity seemed to be beyond the control of society, the pragmatic approach was for society to recognize and assume its power to control behavior.

Not all people accepted this view. Those in the United States who favored the importance of heredity over environment were instrumental in passing laws and procedures for the castration of feeble-minded males and the sterilization of feeble-minded females to avoid continued heredity, while in Germany the attempt to control heredity led to a catastrophic example of humankind's occasional attempts to regulate its genetic composition.

Political application of the nature-nurture issue does not merely swing, pendulumlike, between one pole and the other. At times the pendulum swings faster, further, and more powerfully in both directions. The first half of the nineteenth century in the United States appears to have been a period during which time both poles were touched: some advocated that environment was all and heredity unimportant; others that heredity should be shaped in order to create society. The issue has not disappeared, nor will it. The issue remains among the most profound issues for all of human society, for the answer believed determines the social structure, the judicial structure, the concept of justice, and, of course, our educational system, which decides the answer for us by its philosophy as to who is taught what by which methods and by whom. The Kelloggs grasped the nature of the two poles tightly and well, and they offered reinterpretations to us of the stories regarding feral children. They saw the alternatives as offering a technical possibility:

"Here, then, are two complete but entirely distinct methods of accounting for the same phenomena—the one according to the influence of inborn factors, and the other as a matter of environment. There remains the possibility of a middle-of-the-road interpretation which would adopt some features from each of the extreme views. The supporter of either doctrine as a means of accounting for the condition of the 'wild' children would no doubt admit that opposing influences played an important part in the development of these children even though they might not in his eyes play the principal part. . . .

"Without doubt, one of the most significant tests which could be applied to a problem of this nature would be to put to rigid experimental proof the stories of the 'wild' children themselves. To accomplish this end it would be necessary to place a normal human infant in uncivilized surroundings and to observe and record its development *as it grew up* in this environment. . . . Yet, obviously, in spite of all of the scientific zeal which could be brought to bear upon an undertaking of this kind, it would be both legally dangerous and morally outrageous to carry out. . . .

". . . Instead of placing a child in a typical animal environment, why not place an animal in a typical human environment? Why not give one of the higher primates exactly the environmental advantages which a young child enjoys and then study the development of the resulting organism? The plan is in fact similar to that suggested by Professor Lightner Witmer, who predicted in 1909:

"I venture to predict that within a few years chimpanzees will be taken early in life and subjected for purposes of scientific investigation to a course of procedure more closely resembling that which is accorded the human child."[9]

The Kelloggs's report follows a pattern not unlike the one Itard adopted in his reports on the wild-boy of Aveyron. The sections of the Kelloggs's book are separated into abilities such as eating and sleeping, dexterity, the senses, play, social and affectionate behavior, emotional behavior (this being the topic that gave Itard so much trouble, both in observing and recording), memory, along with faculties more familiar to the contemporary study of the mind and behavior, learning, intelligent behavior, problem-solving, communication, and language.

THE COURSE OF DEVELOPMENT

Measurements of height, weight, and blood pressure—standard pediatric measurements—were made with the intention of measuring the course of Donald's and Gua's development. They showed that Gua, while at first weighing less than Donald, was slowly reducing the difference. Height, muscle tone, and especially the way in which muscles are used, stretched, and contracted while sitting, standing, and lying, are shown in Figure 10.2. Gua's bone structure was rated to be that of a human child, one twice her age. Gua had 16 of her 20 "baby teeth" at the time she was separated from the mother, while Donald had two at the age of ten months. Blood pressure, both systolic and diastolic, was higher in Donald than in Gua. Gua's ability to grasp, for example, to grasp the Kelloggs's arms by hand, greatly exceeded Donald's ability. For Gua, at one-year of age:

"At present one might describe Gua by saying that she possesses the learning and mental capacity of a year old child, the agility of a four-year old, and strength which in some ways probably surpasses that of an 8-year old."[10]

Here is a sample of the daily schedule:

A.M. 7:00   Reveille.
     7:30   Breakfast.
     8:00–8:30 Sit in high chair while adults breakfast.

Figure 10.2   Donald and Gua, sitting, standing, and lying down
(photographed from above). The background squares provide information on
size, as they are 20 cm to a side. Notice the orientation of the arms.

9:00–11:30	One or more of the following:
	Ride in perambulator.
	Physiological measures.
	Observation of special behavior.
	Automobile ride to Experiment Station (for weighing).
	Photographs.
	Outdoor or indoor play.
	Experiments, tests, or measurements.
12:00	Lunch.
P.M. 12:15–1:30	Nap.
11:30–2:00	Walk or play out-of-doors for Gua (Donald still sleeps).
2:00–2:30	Bath.
3:00	Snack of 6–8 oz. milk.

         3:00–4:00   One or more of the following:
                            Outdoor or indoor play.
                            Observation.
                            Photographs.
                            Experiments, tests, or measurements.
         6:00   Supper.
         6:30   Retire.

And, here is a sample of the food diet:

Breakfast
    Warm milk (8 oz.) Canned evaporated unsweetened milk, mixed
       half and half with water.
    Cooked cereal (2 oz.).
    Cooked fruit (2 oz.).
    Rarely a soft-boiled egg, substituted for cereal.
    Cracker.
Mid-morning
    2 oz. orange juice (with 1 teaspoon cod liver oil).
Lunch
    Warm milk (8 oz.).
    One or two warmed cooked vegetables (2 to 4 oz.). Vegetable soup,
       spinach, mashed green beans, boiled cabbage, or mashed carrots.
    Occasionally 2 oz. of beef broth.
    Prepared dessert (2 oz. Junket, Jell-O, custard, tapioca pudding).
    Occasionally a little raw vegetable. Lettuce leaf, piece of celery, or
       raw cabbage leaf.
Mid-afternoon
    2 oz. orange juice (with 1 teaspoon cod liver oil in cold weather or
       dark days).
Supper
    Warm milk (8 oz.).
    Cooked cereal.
    Cooked fruit.
    Cookie, cracker, or zweiback.

    Donald did not care for the raw vegetables, while Gua "relished" flow-
ers, the leaves of some plants, and the soft bark of young trees. Both
showed aversions to new foods, and both showed temporary aversions
to certain foods, with the exception of fruits. The techniques known to
parents for getting infants to accept rejected food were used with success,
especially masking the food with another. Gua's meat-eating experiences recall those of the wild-boy and Kaspar Hauser: she picked off
crickets and small insects and appeared to chew these for the juices before
rejecting them. She once pounced on a lizard and placed it in her mouth,
but the parents retrieved it so promptly that Gua's complete reaction is
not known. After five months, Gua showed no interest in animals as

food. Meat, when offered, was tried and rejected, although the beef broth was acceptable. Once she screamed for sausages on a plate intended for human guests, but when offered them, she tasted and rejected them. Gua showed herself to be seemingly in constant thirst, and she would take cupfuls of water when offered them. In time, she learned the connection between water spigots and water, and became a frequent user of such faucets.

Gua enjoyed her bath from the first; however, she also demonstrated a preference for eating first the soap bubbles and then the soap. (The brand is not reported.) At two months, Gua was introduced to a human-like bed, with clean linen and blankets. She had been sleeping in a crib with a mattress. She liked the bed and responded to it with pleasure. One day when it was removed for housekeeping reasons, she cried until it was returned. Although she slept longer at night than did Donald, Donald took a longer nap in the afternoon. The Kelloggs report that both slept alike in terms of posture, noises, and wakening during the night. In the forest, chimpanzees, as well as gorillas, are known to build nests each night that are used for sleeping. Gua rearranged her bedclothes at night, an act that might remind some of nest-building, but after she was corrected for this behavior, it ended. Nor were other objects left around that chimpanzees used for nests. Thus, the question as to whether or not the nest-building of the forest is learned remains unanswered from the data supplied by Gua and the Kelloggs.

The way in which anthropoids, including human beings, grasp objects is one of the key measures by which anthropoid evolution is assumed. An "opposable" thumb, for example, permits grasping. When attempting to grasp a morsel of food, for example, Gua used her lips, for her thumb and finger grasp did not make the pincer movement requisite for grasping. When holding an item, such as a ball, the thumb was not much used; rather, the mobile wrist acted to cup the ball. When using building blocks, Gua stacked well enough, but experienced difficulty in adding more of them, since the precision required, especially when removing the hand quickly with the placement of the added block, was difficult. As a test, objects were presented to both Donald and Gua when they were seated in their high chairs. Gua was not permitted to use her mouth. The items used were a letter-sized envelope, a wire hairpin, a dime, and a flat nail file. Donald used the pincerlike movement made possible by his using the thumb and finger to form a pincerlike apparatus. Gua achieved some success, but she was not generally able to use the thumb and finger to grasp small objects. Figure 10.3 shows Donald and Gua grasping. Although the picture was taken to illustrate an experiment in handedness, it also shows that Gua's technique for holding objects was different from Donald's.

Although apes are capable of walking on two feet, and will surely do so when it is important to them (as when the hands are to be kept free for holding or carrying some object, such as food), they tend to use all four appendages. Donald used a walker and, it is reported, was even

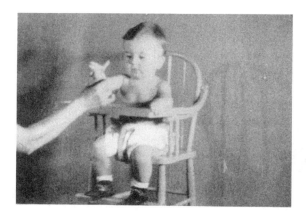

Figure 10.3   To test the ability of ape or child to locate, grasp, and move an object, a toy or spoon is held in front. Both Donald and Gua reach with the right hand. Note the position of the thumbs.

"reckless" in his attempts to move around. "If placed in the walker at the age of 13½ months or older, he would promptly get out and walk by himself, even though the getting out required a rather difficult maneuver."[11]

A walker suitable to her dimensions was built for Gua. She did not use it for walking, although she was pleased to be seated in it if someone pushed it around. She did learn to walk, and only then did she return to the walker, using it as a toy rather than as a means of assisting locomotion. Gua's walking eventually matched the speed of a human adult taking a brisk walk. A preferred way to move was to grasp the legs or pantleg of a human who was walking and "tag" along. When Gua was around eleven months, she is reported to have spent most of her time walking. Figure 10.4 shows Gua at this age. Note that the arms are held out: they are not used to regulate balance when walking, as human beings do with their arms.

Figure 10.4   Gua learns to walk. Toward the end of the experiment, Gua can lean over to pick up an object with one hand without losing her upright stance. She keeps her arms to her sides, as does a human being using them for balance while walking.

We may summarize the physical and motor differences between Donald and Gua by noting that each showed physical growth similar to its species, while differences appeared in the use of the hands, both in using the fingers and thumb to grasp objects and, when walking, in the way the arms were held in reference to the body.

THE SENSES OF DONALD AND GUA

An examination of the sensory systems revealed differences, although their significance is unclear. Gua was startled by a bright light to which Donald adapted; Gua saw and found small objects long before Donald did, Gua noticed new objects in the environment more readily than Donald; Gua reached for dust particles illuminated by sunlight; Gua showed an interest in her own image in a mirror at 8 months; at 10½ months she climbed on her bed seemingly to see herself in the mirror above it.

When presented at 14½ months with a child's alphabet book, Donald (unlike Tarzan) wanted to rotate the book, tear it, or put his thumb between the pages. He was not interested in the visual content. A month later he stretched his hand toward the pictures and colors. At 18½ months he pointed to a picture that a human being had just pointed to. Gua's response to the book was the reverse. At first, she appeared to be examining the pictures or colors; later, she wanted to manipulate the book. She touched the points with her lips or rubbed them with her fingers. At 16 months of age, in the Kelloggs's opinion, she tried to lift the pictures from the page, suggesting well-developed depth perception.

Gua's hearing appeared to be outstanding compared to the adults' as well as to Donald's. She seemed to respond to sounds (the newsboy's coming) before her human companions heard them. When she heard a voice from the radio, she wandered the room looking for the source, but eventually localized the sound and looked into the speaker. Human beings have difficulty localizing sounds that emanate either from directly in front or directly in back of the head. At the time the Kelloggs wrote, it was thought that sound localization was performed by the brain calculating the differences in time when the sound reached the two ears. A sound from directly in front or in back would reach both ears simultaneously and give no indication of origin. This is so: try it by snapping your fingers in these several locations around a person whose eyes are closed. Sounds emanating along the axis directly in front to directly behind are localized in bizarre and unexpected places. Gua and Donald showed the same responses to such sounds: they failed to localize those along the axes described, but did well at localizing those away from the axis.

Responses to taste were measured by examining the facial responses to tastes of various concentrations. Donald smiled at sweet but "made a face" at bitter. His facial expressions, as well as Gua's, are shown in Figure 10.5. The Kelloggs suggest that the same was true for Gua. It may be easier for human beings to perceive different facial expressions from other human beings than from those of another species.

"In observing behavior related to the *sense of smell* we find a strong suggestion that Gua employs olfactory stimuli for the identification of objects and individuals, in a manner quite different from that of humans."[12] For example, when she first was at home with the Kelloggs, she climbed into their arms and smelled their bodies and clothes, as if she was using this information for identification. When a perfume of extract of rose was presented to her, she promptly avoided it. She did the same six months later, although the smell had not been presented to her between the times.

PLAY AND EMOTION

However difficult it may be to define "play," much of the time both Donald and Gua were occupied in play with both objects and other beings.

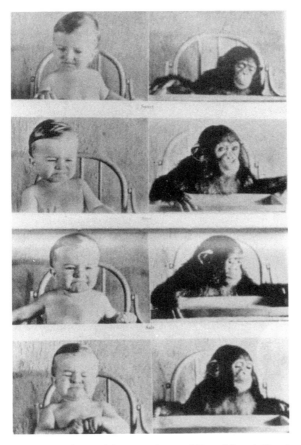

Figure 10.5   Facial expressions of Donald and Gua in response to novel tastes. The upper set shows the response to sweet; the second to sour; the third to salty; and the bottom to sour. Are the facial expressions similar?

One of Gua's favorite activities was placing her hands in the mouth of others. She also enjoyed placing small objects, such as insects and dirt specks, in her own mouth. Some of her play was remarkably like that of Donald and other human children: she would get into bed and pull the covers over her eyes, waiting for someone to play peek-a-boo. She liked to be chased by a companion. She would play ball with Donald by rolling it between them. The presence of someone outside the house caused both to climb on a chair to see out a window. Imitation between Donald and Gua, especially of vocal sounds, became common. Gua was a frequent "kisser." Both Donald and Gua showed signs of understanding social situations. For example, Donald, when cautioned not to pick up a pen, put it back, but when the adult turned away, picked it up again, only to put it down when the person turned and looked toward him. Gua made the same social deductions.

Neither Donald nor Gua showed any fear of strangers, whether animal or human. A neighbor's cat elicited from Gua a wish to put her fingers in its mouth. When introduced to a puppy, however, Gua gave chase until the puppy, cornered, snapped. Gua screamed, sought human companionship, and retained an avoidance of dogs, as well as other animals. She did not care to be led by the hand toward chickens after the experience with the puppy. When first introduced to white rats, Donald showed fear, cried, and tried to move away. Gua, too, screamed at the sight of them. For both, this view of rats was their first. Gua appeared to have four fears that were both powerful and indestructible. One was "loss of support"—that is, the human being pretending to drop her. Another powerful fear was of being left alone. As a third, she did not care for sounds that came from far overhead, such as those from airplanes or the whir of passing birds. A fourth fear, substantiated by sudden and explosive defecation, was toward toadstools. Once, toadstools were wrapped in paper and handed to her: She showed no fear and unwrapped the toadstools. After she saw the contents, she put them away but did not show the strong reaction she gave to toadstools in the wild. She found fire interesting but came to avoid touching it. Both Donald and Gua displayed "anxious behavior" when the adults were preparing to leave the house. Gua had tantrums, but only occasionally. The Kelloggs believe that these occurred at times of jealousy, when someone else was getting attention or when Donald was given a ride and she had to wait her turn.

"With reference now to Gua's emotional background," write the Kelloggs, "to her attitude, or to her disposition and temperament, it may be said that almost from the start she seemed willing and ready to do whatever was required of her. Aside from her early negativism regarding new toys, her dislike of new foods, and her uncontrollable fears, there was no evidence of her opposition or deliberate resistance to any part of the training. . . . One would say that in everyday langauge she was very good-natured. . . . There can be little doubt in this connection that her ruling emotion was fear. In those rare instances when she did object to the treatment accorded her it seemed usually because she was afraid." [13]

Among the qualities of chimpanzees that make them so attractive to people is their ability to learn from human beings and to imitate their actions. Gua was no exception. She acquiesced to being dressed in human clothes during her first two weeks with the Kelloggs. She soon submitted to having her nails clipped and to the toothbrush. She learned to use and to unlatch doors and to do the same with windows. She learned the connection between light switches and lighting. Toilet training during the nine months of training showed the following results: Of 6,000 responses by Gua, 1,000 were "errors." For Donald, the figures were 4,700 situations and 750 errors. The curve shows more errors by Gua at the onset of training, but after thirty days, the error rate was similar for the two. Gua ate with a spoon several months before Donald did so. She learned to drink from a glass.

SOLVING PROBLEMS

When problems were presented that required novel solutions, as when a cookie was hung from a string out of reach, thereby requiring Donald or Gua to get on a chair or stool, climb on it, and then swing the cookie until enough force was reached for the cookie to swing within reach, both Gua and Donald solved the problem. Indeed, on some tasks, Gua performed better at the outset than did Donald. The Kelloggs concluded their analyses of Donald and Gua's learning abilities by pointing to Gua's imitativeness: "As a distinct method by which the behavior of those in the immediate environment is accurately learned and reproduced, imitation is of tremendous importance," they write. "We are accustomed, in this connection, to regard the chimpanzee as a splendid imitator, so much so that the very name 'ape' implies its capacities. Yet the child is a more versatile and continuous imitator than the animal, as we have already seen."

Memory was tested by using several techniques, including a delayed reaction method whose assumptions and technique were those used by Hamilton when he tested a variety of animals and human beings (see pp. 254–260). For Donald and Gua, doors in the house were used, with incentives placed behind them. The point of the technique was to delay the subject from approaching any door for differing periods of time in order to determine, presumably, how long the nature of the correct door could be remembered. Donald was using his walker, and Gua was held by one of the adult human beings, for the different time periods. Donald was able to remember for five minutes, but not beyond; at ten minutes, only half his choices were correct. Gua reached this 50 percent correct point at 1 hour, a span impressively longer than Donald's. Yet at five minutes, this being Donald's best time, Gua was successful on only 70 percent of the trials compared to his 90 percent. We might conclude that Gua's memory was longer, if not always as accurate as Donald's.

To measure mental ability, the Gesell Tests for Pre-School Children were given.[14] These tests use both motor and mental ability to examine readiness for the tasks given in school. The test was first given when Donald was 10½ months and Gua 6 months. Ninety tasks were used. In all but seven, Donald and Gua were alike, Donald was superior in three, and Gua in four. One is significant: Gua displayed a more active interest in her image in a mirror than did Donald. Some modern work uses the ability to recognize one's reflection as a measure of self-consciousness or, in less conservative cases, of consciousness. Human beings and chimps have the ability, but establishing such an ability in gorillas remains elusive.

At the time of the second set of tests, and again at the third when Donald was 12½ months old, Donald scored higher than Gua. More and more often, he passed new tests that Gua failed. One of these, "Show me your nose!" is shown in Figure 10.6. Here the Kelloggs asked the

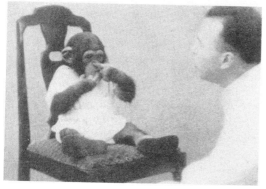

Figure 10.6 "Show me your nose!" In the top photograph, Donald responds to L. A. Kellogg; in the bottom, Gua responds to W. N. Kellogg. Does Gua's response represent the possession of language?

subject to point toward the nose and both did, although using different techniques.

"Shall we conclude from these figures that the ape is about two-thirds 'as intelligent' as a human child 2½ months her senior?" ask the Kelloggs,[15] thereby coming directly to the crux of the matter. "No," they say, because differences in maturation, in motivation, in the satisfaction felt for completing the task—all make such a conclusion uncertain. Had they written the book in modern times, they would have pointed out that another factor interefered: the tasks were developed by human beings and standardized on human beings. The intention of the Gesell Tests was to separate human beings into two groups: those ready for formal schooling and those not yet ready. The test was not devised by chimpanzees to measure what a chimpanzee needs to know in order to survive and prosper in the chimpanzee society or its natural environment. This is not to say that the test is an inappropriate test of chimpanzee ability: the issue is for whom the test predicts accurately (slum children? chimpanzees?)

and what it may predict about them (problem-solving ability, ability to survive in a forest, ability to survive in a human home?).

GUA SPEAKS

Without doubt, the aspect of Gua's development that attained most interest at the time of the study was the question of whether Gua could learn to talk. "Almost from the beginning of her human training, Gua seemed to possess a rudimentary, nonvocal form of communication by means of which her impending actions could be predicted by those who knew her well."[16] But body movement and facial expression are, of course, not verbal language, and many observers during the last three hundred years have focused on the ability of the feral child or animal to communicate by means of spoken language.

Gua used a sort of sign language to communicate wants and intentions: she would fall to the floor seemingly to indicate a wish to go to bed, grab water containers to show her wish for water, grab her genitals to indicate the need to urinate, and take the hand of a human being to indicate her wish for the person to lift a bottle to Gua's mouth. But these are not verbal communications. Here is the comparison the Kelloggs make: "despite the child's lack of progress in the acquisition of human language responses, it was precisely in the development of *articulate* sounds in which he significantly outshown the ape. There was no attempt on Gua's part to use her lips, tongue, teeth, and mouth cavity in the production of new utterances; while in the case of the human subject a continuous vocalized play was apparent from the earliest months."[17]

Gua, it seems, had her own articulations. The articulations were (to the hearing of the human listener) the bark, the food-bark, the screech or scream, and the oo-oo cry. The Kelloggs attempted to condition associations between the call and the event. For example, the presentation of the orange juice was accompanied by the grunt. The human being would say "orange," and after thirty such repetitions, "she began to bark, "uh uh . . ."[18] Gua learned to make generalized verbal responses to questions requiring a "Yes" or "No" answer: for example, "Do you want to go out?" "Do you want apple-sauce?" An attempt was made to teach the word "pa-pa." The technique was that which a human being would use with a child: saying the word, pointing, the person being the human male each time. An attempt was made to initiate imitation by moving the parts of the mouth needed to articulate "pa-pa." But nothing came of this effort. Gua attended and would put her fingers in her mouth, but nothing like "pa-pa" was ever heard by the Kelloggs.

The prominent view that chimpanzees could never learn to articulate because they lacked the necessary assemblies of cells in the brain was based on this analogy. The area of the brain implicated in human speech is lacking in the chimpanzee brain; therefore (goes the faulty logic), chimpanzees cannot have speech. The Kelloggs were skeptical. They believed that the chimpanzee possessed the motor and physical ability to permit articulation. Yerkes believed this as well, and the Kelloggs cite him as authority on this matter. Gua's inability to articulate English words must have been a great disappointment, for if such learning were but a matter of an appropriate environment acting on the necessary physiological tools, Gua should have come to articulate human language. She did not.

The Kelloggs's disappointment is evident. They were aware that some have argued that the area of the chimpanzee brain that corresponds to the area used by human beings for speech is missing or deficient in chimpanzees. Yet they were also aware that Gua had heard ample human speech from the day of her introduction to the family. Certainly, she had heard as much as had Donald. The Kelloggs were not yet convinced that articulation was impossible; but, as they tell us, it appeared unlikely to them that the chimpanzee could learn more than six words. This estimate leaves aside the question of whether the animal comprehends the meaning or merely utters the words, thereby showing the conditionability of verbal behavior. Did Gua learn to respond correctly to human commands? That is, did she show evidence of understanding human speech?

The answer, as is true of so many answers when we attempt to enter the silent mind, is both "Yes" and "No," depending on what we understand the word "speech" to mean. The Kelloggs anticipated the complexity of this issue in their 1933 work, for however full of minutiae the text may be, these data lead to clear and temperate analysis. As we will soon read, the question whose answer so divides modern investigations of ape-human communication is that of *comprehension*—whether the ape is making a motor response, whether by gesticulating or moving the mouth through which air is produced into sound, or whether the animal is using the gesticulation and speech to convey meaningful information. This is the issue, and in the long run, there is no other. Read, now, the Kelloggs' warning:

"Let us consider finally the ability of the subjects [apes] in the *comprehension of language*. We are here faced, as in many earlier instances, with the problem of inner mental processes, for we cannot tell in a strict usage of the term whether the subjects introspectively *comprehend* what is said about them or not. All we can do is to observe whether they are able to react distinctively and individually to separate words and phrases. This, then, must serve as our criterion of 'comprehension.' "[19]

RETROSPECTIVE

When read with the attitudes of our times, the Kelloggs's book says more than it intended both about the psychology of the time and about the authors. The tone of the book well suits the scientificism that had captured the study of the mind in the 1930s: opinion is minimal and is always backed by data. The amount of data is overwhelming, as the reader may have discovered from the small sample of it presented in this chapter. Much of the data appears to have been presented merely because they were collected, not because they speak to some useful interpretation. Obeying the rule that data should be presented so that anyone may recalculate or reinterpret them, the Kelloggs, for modern readers, tell us far more than we want to know. We do not doubt that the Kelloggs understood themselves to be scientific and were understood as such. Granting this, there is something missing from the book, something that leaves us hungry and unsatisfied. I suspect it is that the scientific method removes from our view what we want to know: namely, something about the interaction among the four. News of the interactions is to be found, but we sense that it is there only when it can be measured and the results tabularized. What we miss is knowing how the Kelloggs felt toward their subjects. We long for news of how Donald and Gua regarded one another; or of how the Kelloggs felt when Gua left them and Donald.

Itard becomes human to us when he allows us to know and share his feelings for Victor. We then trust him, for we recognize him as being like ourselves, the interposition of a few centuries notwithstanding. The Kelloggs do not share this humanness with us, either because of the commitment to scientificism in psychology characteristic of their time or because it is not there.

VIKI

The Kelloggs succeeded in forging a link between human and chimpanzee by demonstrating both developmental differences and likenesses. Perhaps the best remembered aspect of their courageous study was the finding that Gua did not learn to speak, even though she was raised in an environment rich in sounds and their use. Toward the end of the 1940s, Catherine and Keith Hayes undertook a seemingly similar study under similar auspices. Keith Hayes took a position at the Center in Orange Park, later Yerkes Field Station, where he and Catherine Hayes decided to rear a chimpanzee at home. The chimp, Viki, unlike Gua, had no human companion of like age and level of development, but Viki had many human visitors. Indeed, she was taken to parties and was a companion on a long automobile trip the adults took from Florida to the northeastern United States, where she was displayed to professional groups of psychologists and psychiatrists and where she was introduced as part of a speech given to civic groups by Catherine Hayes.

Catherine Hayes writes: "One of the oldest questions of human psychology is the nature-nurture controversy: Does a child mature into intelligence bit by bit as decreed by his heredity? Or does he *acquire* intelligence through experience in his social and physical environment? The answer is usually sought by comparing two children or two groups of children. We believed that by substituting a home-raised ape for one of the subjects and by equating its environment with the child's we would see the workings of a greater difference in heredity than exists between any two normal children. We have shown in actual fact that Viki's education caused her to resemble the child considerably at eighteen months. Those ways in which she differed could now be fairly ascribed to her anthropoid heredity."[20] The report of this association, one that was to last until Viki's third birthday, is of such a different nature from that of the Kelloggs that it is difficult to make comparisons between Gua and Viki.

While the Kelloggs's report is so careful and conservative, full of measurements of anything that might be of the remotest interest to the reader, Catherine Hayes's report is seemingly meant to entertain. Stories of Viki being slipped into motels, finding for her a root-beer float in New York City, and the reactions of service-station helpers and schoolboys are told with some glee, while data and measurement are, if taken at all, given little consideration. If there was scientific merit to their idea for taking Viki into their home, it was soon lost as the chimpanzee became an object of amusement and, to my mind, exploitation. The description of the chimpanzee's development soon gives up any pretense of serious research and becomes a matter of whether the next chapter can find a better story to tell of how amusing it is to be seen with a baby chimpanzee. If Garner's tragic flaw was pride, Catherine Hayes's flaw, at least as expressed in the book, was exploiting the animal for her own social ends.

"The road became narrow, winding through many small hamlets and even barnyards, and we kept getting lost in detours. We were impressed by the number of cemeteries here. Viki seized me in panic when tombstones sometimes appeared at the very edge of the highway on both sides. Keith [Hayes] reassured her. 'It's the live ones you have to worry about, Viki.'

"Darkness found us somewhere in New Jersey, hungry but with no drive-in in sight. Since our home for the night was hours away, in a friend's apartment north of New York City, we decided to break our own rule by taking Viki into a diner. As we sat down in a booth, the proprietors grinned at us, but quickly stiffened their faces in the blasé manner of this metropolitan area. They served us without comment. With her big eyes just peeking over the table top, Viki stared at them, also without comment. Many customers came and went, never glancing our way, nor did the owners call attention to us. This pleasant anonymity continued until all the French fries and hamburgers were eaten, and we called for dessert. It was ice-cream. One look, and Viki burst into such loud food barks as only ice cream can evoke. Heads straightened and turned all over the diner. Our privacy was at an end."[21]

As the Kelloggs's and Hayes's purposes for undertaking the rearing experience are different, it is most apt to say that Gua and Viki's careers were truly "incomparable," but let us determine what we can recover about Viki. Several pictures and stories about Viki tell tales that remain of value to those who would understand the animal mind. First, it is claimed that Viki was able to use three words, "cup," "mama," and "papa." Although we are told relatively little about how these words came to be learned, the report that Viki learned three words places her linguistic ability well within the range of six words predicted by the Kelloggs. It is evident from the texts that the Kelloggs were very conservative in all claims regarding Gua and that they were especially careful about claiming specific articulations from Gua. Their evident disappointment in Gua's linguistic achievements points toward their hopes and their judgment of the reality of Gua's ability to communicate orally.

We have the reverse situation with the Hayeses, for there is evident happiness at Viki's achievement. They, too, however, are suspicious. In planning what is to be done with Viki, Catherine Hayes cautions: "The significance of Viki's speech training lies not in the fact that she has learned a few words, but rather in her great difficulty in doing so, and in keeping them straight afterward. We are beginning to suspect that this tendency to confuse the words she knows may merely be the most obvious result of a more general inability to retain large numbers of arbitrary associations without conflict." [22]

Alas, the vocalizations of neither chimp are available to us for thorough analysis. [23] Nonetheless, it is instructive, that no one since the Hayeses has made any serious and verifiable claim that a chimpanzee has been taught to articulate human words. Whether the limit is one, three, or six words seems to be insignificant from our hindsight: no chimpanzee, regardless of the extent of training, comes to articulate human words in an amount anywhere near that achieved by the wild children described. In a publication issued twenty years after the Hayes's life with Viki, Keith Hayes [24] reported that after a total of seven years of training and listening, Viki had been able to master four words. About the same time, Winthrop Kellogg [25] reviewed four incidences of human beings' raising chimps and found that, although the chimps displayed much domestication and even some human-type civility, they never demonstrated the ability to articulate words in a human language.

A second achievement of the Hayeses, one whose importance was not appreciated at the time of their work, was the report that Viki was much interested in her reflection in a mirror. Viki's reaction is shown in Figure 10.7. In our time, an animal's ability to discriminate itself from its surroundings, thereby showing a concept of self as distinct from others, demonstrated by placing a red dot on the forehead of chimps and other apes under anesthesia. Upon regaining full consciousness, the animals would examine the dot as reflected in the mirror. From these studies, it seems evident that chimps, but probably not other species, react to the dot. Most pet owners have watched their companion animal appear to examine a reflection in a mirror: often the reflection is their own, and at times such pets appear to move while watching as if interested in the movement of the reflection. But when the hypothesis that there is some awareness of the reflection is examined with care, the notion is not supported. Gua had some interest in the mirror and was willing to climb on

Figure 10.7 Viki and a mirror. In modern times, the ability to use the reflection from a mirror to guide one's hand, or to recognize some alteration on one's body, has been considered evidence of self-awareness.

her bed in order to examine it. Viki, too, showed interest and evidently spent more time, and devoted more interest to it, than did Gua. Nonetheless, neither study tells us whether the chimp was merely reacting to light or whether there was an awareness of the nature of the image reflected.

THE ACHIEVEMENTS

If the Kelloggs failed to answer for all time the relationship between heredity and environment, they nonetheless showed, once again, that environment provides the opportunity for abilities to be shown. But because we do not know much about how a chimpanzee raised among chimpanzees in its natural state might learn, emote, solve problems, or speak—not to mention acquire the human habits of toilet training and eating at table with utensils—we cannot know whether Gua learned more or less. We know only that she learned different things from her feral equal. The Kelloggs demonstrated the chimpanzee's remarkable ability to learn and to imitate. The suggestion that the chimpanzee *comprehends* human speech is remarkable in itself, although at the time of the publication of the work, it would appear that disappointment regarding Gua's ability to articulate human sounds overshadowed the issue of comprehension.

We are reminded that speech is a motor response, at least at one level, one in the same league as winking or picking up objects with thumb and finger. Nonetheless, these or like motor behaviors *may* permit us to communicate with unresponsive minds. We do not need to think of gesticulation as anything more than inarticulate motor responses in order to assume that these responses represent, signify, and stand for ideas and perceptions. We now understand that we need not think of speech as necessary for communication. While the Kelloggs and Hayeses (and Garner) strove to communicate with the mind of the apes through speech by shaping the mouth or rewarding some sounds, modern investigators used gesticulation itself as the mode of speech, and others were to teach the apes motor skills (such as tapping computer keys or touching images on a video-screen).

    If only we could find a way to talk directly to the animals, so goes the idea, especially to our near-neighbors and likely relatives, the great apes. As they do not seem to be able to articulate English because of physical inheritance, because the neurological connections are not as ours, perhaps we should have the same insight that Lubbock had when he recognized that the technique used to educate Laura Bridgman could be extended to animal life by teaching his dog, Van. If Laura Bridgman, who reportedly had no sensory systems other than the cutaneous, could learn to articulate and to understand through the use of her skin, perhaps the ability to articulate sounds by the mouth was unnecessary. Perhaps the ability to gesticulate—something that Gua did very competently—could be used as a means to correspond and communicate with the animal mind. Perhaps Pepys's suggestion was the right one.

Human and Ape Communication: Washoe, Koko, and Nim

11

BORN IN West Africa, most probably in September 1965, a female chimpanzee was destined to arrive in Reno, Nevada, on June 21, 1966. There she would learn to use one language of the deaf, American Sign Language (ASL), with the hope that she might communicate to human beings, and that human beings might communicate to her. The idea can be traced to Pepys in the seventeenth century and the first execution of it to Itard in the eighteenth. It may be recalled that Itard and Victor had their first meeting in the Luxembourg Gardens, the Paris park from which the Institute for the Deaf may be seen, a place from which Itard developed a means to communicate with the nonhearing. The notion that the noncommunicating person can be taught was developed further by the work of Samuel Howe with Laura Bridgman and, later, by Lubbock and the dog, Van.

One of the major tasks given to both the chimpanzees Gua and Viki was to speak so that they could communicate with hearing human beings. The hope of establishing communication between silent minds and human minds had been prominent for centuries, but such work had become arid, perhaps because Gua and Viki were so disappointing in their command of speech. For the female chimpanzee who would make her way within the year to Reno, Nevada, the means of communication would be, not articulated sounds, but the movement of the hands. The task of teaching Washoe to use hand-signals, the reliability of which would convey meaning to human beings, was undertaken by Beatrix and R. Allen Gardner and their associates. Thus was Samual Pepys's suggestion of 1661 finally attended to, almost exactly three hundred years later.

According to the Gardners's description,[1] Washoe was raised in their home in Reno, Nevada, a home with a garage and a garden. Washoe lived in a trailer on the property. They describe the upbringing of Washoe as like that of a human child living in the same neighborhood. In her first years, the Gardners' write, Washoe learned to use a cup and to use eating utensils, to clear the table, and wash up, more or less; to dress, use toys and household tools, and to examine books. The Gardners' goal was to teach Washoe the gestures of American Sign Language (ASL). Their goals seem not unlike those of Itard, who at the onset of his work with the child Victor, wanted to teach Victor a language that could be used to permit communication between the two. Like Itard, the Gardners's work was supported by grants of taxpayers' money made through federal granting agencies. In both situations, although two centuries apart, society had decided that it had an interest in the outcome of an experiment and provided the funding to see the research forward.

After four years of training, the Gardners and their associates reported that Washoe was able to give 132 hand signs. We may remember to compare the total to the four "words" that Viki, the Hayses' chimpanzee, was heard by them to say. It was reported as well that Washoe combined signs (such as those for YOU, ME and HIDE) in ways interpretable by human beings as sentences, commands, and requests.

The trailer was Washoe's home: here she slept, played, socialized, and learned. The Gardners believe that such socialization is critical for communication to occur, even though some loss of control over the variables used in the experimental work is expected. We are reminded of the Mills-Thorndike controversy described on pp. 243–246. By keeping Washoe in this way, the Gardners may give up some control over the conditions: yet without a natural environment, one rich with complex sounds and activities, Washoe may have never shown an ability to learn the motions of ASL. After all, language recognizable by human beings has never been noted in feral apes.

THE CHIMP FAMILY EXPANDS

After five years with the Gardners, in October 1970, during which time the Gardners's and Washoe's achievements had become well known and other investigators had begun to examine the possibility of human-great ape communication, Washoe moved with one of the Gardners's colleagues, Roger Fouts, to the University of Oklahoma. (Later, Fouts, Washoe, and certain other chimpanzees from the Oklahoma facility would move to the state of Washington.) As is true of all pioneering work, the Gardners's findings raised important questions that needed answering. For example, Washoe was almost a year old when the fostering began: what would have happened if Washoe had been fostered by human beings from her birth onward? And Washoe's apparent success in communicating to her human family (and the reverse) prompted questions about the technique, questions about the meaning of the word "communicate," questions about the generality of the observations (Would other chimpanzees do the same?), as well as some additional questions whose answers would inspire the attentive mind. For example, could Washoe now teach other chimpanzees what she knew of gestural communication?

The Gardners acquired four chimpanzees, two females and two males, from 1972 to 1976 for continued study conducted in Reno. These chimps were, in order of service, Moja (1972), who arrived in Reno at the age of one day, Pili (1973–1975), who died from leukemia, Tatu (1975), who arrived at the age of three days, and Dar (1976), who arrived at the age of four days. The availability of more than a single chimpanzee provided the Gardners with the possibility of providing different kinds and degrees of chimpanzee-family stimulation and of studying the interaction among the animals. On the human side, the availability also meant that different kinds of human foster-people could be included, such as persons who themselves used ASL as their chosen language.

The precise techniques used in teaching Washoe are pioneering in nature, and therefore of much interest to anyone who wants to understand the nature of Washoe's achievements, just as it would be invaluable for us to be able to hear the cadence of Hans's tapping or the sounds produced by Garner's chimps in zoos and in West Africa. For Hans, this information was never collected, and therefore remains known to us only in the "translation" of Hans's actions into recorded data. Garner's phonograph recordings are lost, and we must note that only he heard the words spoken by the apes. The possibility of an independent judgment is gone. One reason we should like such information is to evaluate the possibility that an animal is responding in a pattern to be expected from operant conditioning.[2]

The objections and qualifications most often raised regarding Washoe's achievements are (1) whether the gesticulations are language and (2) whether the movements are but operantly conditioned. Although important issues, neither addresses itself to the main issue; namely, whether the signs have meaning. "Meaning" itself is a word, a word like "operant" or "language," that requires careful definition. That a chimpanzee can be trained to move its hands in ways that resemble sign language is demonstrable: that one way to learn these signs is by operant conditioning is also demonstrable. What remains to be learned is whether these signs represent anything and whether they convey meaning: in short, whether the silent mind is also one capable of ascribing meaning to its intentions and communications.

Washoe and the Gardners's success prompted immediate attention to two other aspects of ape-human communication: whether other genera demonstrated like behavior and whether the gesticulations were "nothing but" operant conditioning.

KOKO

The Gardners's work with Washoe became well known through both the publications and the talks they gave. The writer Emily Hahn heard one such lecture in Washington, D.C., and used information from it to form the prescient final chapter of her 1971 book *On the Side of the Apes,* a review of the newly founded U.S. federal primate centers. Another person who attended a talk by the Gardners, this one at a western U.S.

university, also in 1971, was F. Patterson. She writes, "As the Gardners described how they got the idea to teach sign language, their search by trial and error for a proper teaching method, the elaborate controls they developed to ensure that their data were reliable, and finally Washoe's willing response to their efforts to teach her language, I felt increasing excitement."[3] Patterson enrolled in a course in ASL, and, in the same month as she heard the Gardners's talk, she made an overture to the San Francisco Zoo with the hope of obtaining a gorilla with whom Washoe-like work could be done.

Patterson's attention was drawn to a young female gorilla, born on July 4, 1971, named Hanabi-Ko (Japanese for "firework's child," we are told). At the zoo, Hanabi-Ko was on and off display, in part because she was ill, and, a year later, in July 1972, Patterson was given permission to work in the zoo with the year-old gorilla, now renamed Koko.

Patterson wanted to determine if a member of another family of great apes, a gorilla, could learn to communicate by means of ASL in the way Washoe had learned. As Patterson states, and as Hahn quotes, it was Robert M. Yerkes, the pioneer investigator of primates in the United States, who pointed out the likely utility of sign language in his book *Almost Human* (1925): "I am inclined to conclude from the various evidences that the great apes have plenty to talk about, but no gift for the use of sounds to represent individual, as contrasted with racial, feelings or ideas. Perhaps they can be taught to use their fingers, somewhat as does the deaf and dumb person, and thus helped to acquire a simple nonvocal 'sign language.' "[4] As noted earlier, Samuel Pepys, the seventeenth-century English diarist, records having seen a live ape, most probably a baboon, perhaps a chimpanzee. Pepys writes, "August 24, 1661: At the office all morning and did business; by and by we are called to Sir W. Batten's to see the strange creature that Captain Holmes hath brought home with him from Guiny; it is a great baboon, but so much like a man in most things, that though they say there is a species of them, yet I cannot believe but that it is a monster got of a man and a she-baboon. I do believe that it already understands much English, and I am of the mind it might be taught to speak or make signs."[5]

The idea of an ape learning sign language is there for all to read, yet human society did not take the suggestion seriously until the twentieth century. What we are taught about human culture does indeed shape the kinds of questions we think we can ask and, often enough, stops us from asking wise and revealing questions. The view that great apes lacked the anatomical parts and brain centers needed for speech somehow led to the view that apes could not communicate, so Pepys's and Yerkes's leads went unheeded.

Yerkes appears to be suggesting the use of finger signing rather than of sign language insofar as the distinction mattered to him. In finger signing, the fingers are used to spell out English words: the technique is that of signing a language already known, such as English. In sign language, gestures are used to signify thoughts, objects, and ideas. Sign language is, then, a new and different language. The distinction is important, of course, when one attempts to translate between English, or the language we speak, and a "signed," as opposed to a gestured, language.

Patterson has thought much about the nature of language. The book she wrote describing her work with Koko opens with a presentation and analysis of a work by the novelist Walker Percy. When writing a novel twenty years earlier, Percy raised the issue of the nature of being human, with special reference to the value and importance of language. A few years after the publication of Patterson and Linden's account of Koko, Percy would publish another novel *The Thanatos Syndrome*, which examined directly the relationship between humankind and a nonhuman primate, in a way that I would regard as phenomenological.[6] Patterson and Linden are well aware of the relevance of linguistic communication to a definition of humanness and animalness.

At first, Patterson worked with Koko in the zoo setting. Since Koko was on display to the public, Koko had both a rich environment of people moving and talking around her and the attention of Patterson who was teaching her ASL. Of course, the richness of the environment also meant a loss of control as to what Koko saw and heard in her environment, but physical isolation was neither possible nor perhaps experimentally wise. Later, Koko would be moved to a trailer that she used as her home, nursery, and school. Even later, Koko's trailer would itself be moved, and she would stay in a private home while recuperating from an illness. Koko was to have a varied environment in terms of the objects and people moving through it, people both speaking and behaving. Let Patterson tell us of Koko's progress:

"Koko first began to show signs she understood the significance of the strange gestures she was constantly witnessing as early as the second week of Project Koko [as Patterson named the project]. On July 25 [1972], before Koko had been taught any signs through molding [this being the Gardners's technique of shaping and molding the hands to represent the sign desired, one not unlike what Richard Garner did in the Gabon forest] the volunteers reported that she made gestures that resembled the FOOD and DRINK signs several times during the morning before I arrived.[7]

"Over the next two weeks, Koko continued her spontaneous approximations of signs, but to me they seemed coincidental, random, unintentional. With all her fidgeting, I wondered whether any of our intent was getting through. On August 7, we began a formal routine of active instruction. My assistants and I used every opportunity that arose during the day to teach Koko FOOD, DRINK, and MORE. Rather than hand her her bottle as a matter of course, we would first hold it up and let her see it. If she responded by signing DRINK, we'd give her the bottle. If she made no response, we'd sign WHAT'S THIS? If that still elicited no response, we'd mold her hand into the sign for DRINK."[8]

The training moved along in this fashion by using a combination of

operant conditioning (defined by Patterson as withholding the milk until
the desired gesture was given) and molding of the hand into the wanted
gesture. Koko showed some interesting and perhaps revealing use of
concepts. Eventually, for example, the gesture for DIRTY, first used to
refer to her feces, was used to refer to people and events. As we shall
read in the next section, the chimpanzee Nim appears to have made the
same generality.

A year into the project, a difference of opinion arose between Patterson and the zoo administration regarding Koko's long-term care. According to Patterson, the zoo wished to continue its policy of using endangered and threatened species for breeding purposes, and the zoo was
worried that Koko was becoming less a gorilla than a human-tamed animal. A neighboring gorilla, Kong, was introduced, and visits were arranged. As both animals were far from the age of sexual maturity, the
meetings were probably useful to aid socialization but uneventful for
breeding. As a result of the disagreement about Koko's future between
the zoo and Patterson, Koko and her trailer were moved to the grounds
of Stanford University in California, where Patterson was studying for a
graduate degree, and where, presumably, training could continue without the responsibility of placing Koko on display or following zoo dictates regarding her care and breeding.

One requirement imposed by the zoo was that, if Koko was to be
taken elsewhere, she had to be replaced by another female gorilla of like
age and condition. The requirement presented a serious problem to the
investigators. Gorillas, now protected to some degree under the Endangered Species Act, are difficult to procure and translocate, as the act works
to conserve rare animal species in their natural habitat. Patterson solved
the problem by arranging to purchase from a dealer in animals in Vienna, Austria, a young female and a young male for $28,000. The funds
were raised in part from the investigators' funds and in part through a
local public campaign that solicited donations for the project. Without
soliciting funds from the public, Patterson argues, the project would have
to be abandoned and Koko's future would be "insecure."

The purchased infant gorillas, King Kong and B.B., arrived in the
autumn of 1976. King Kong was renamed Michael, but B.B. was not to
be renamed, for "The rigors of her travels proved to be too much for her
frail constitution, and in spite of our desperate efforts to nurse her back
to health, she died of pneumonia within a month of her arrival."[9] We
are reminded of Garner's experience with moving the chimpanzees under
his care from Gabon to England.

Michael became part of the family and part of the training program.

"If I was exhilarated with Koko's first words, I soon began to encounter the frustrations of trying to document something as elusive as language. I had hoped and expected that language would become a part of
Koko's life and that she would use it for *her* own purposes as she learned
about the sign language we were teaching her. . . . Instead, from the
first months of the experiment Koko introduced her own variations and

novelties into her signing, . . . it was her errors that gave us elusive glimpses of language games extraordinarily more sophisticated than the simple vocabulary growth we were looking for. . . .

"By the third month of the program, my schedule had moved up from five hours a day, seven days a week, to eight hours a day, seven days a week. I was assisted by two native signers who worked with Koko two or three afternoons a week. . . . I maintained several different systems to monitor Koko's language use. Periodically, we would take a one-hour sample of all the signs and activities Koko was exposed to as well as all of her responses. This involved logging the statements of her human companions as well as keeping tabs on Koko's sign use. I maintained a daily sign checklist on which were noted all the signs Koko had made during the day, the combinations in which they occurred, the number of times she repeated each sign, and anything unusual that might have oc-curred during signing. As Project Koko developed, I instituted monthly filming sessions, videotaping, and tape recordings."[10]

Patterson is aware of what she calls the Clever Hans phenomenon, for, as she writes, "one of the various criticisms leveled against the chimp experiments from various quarters was that they were not learning a lan-guage at all, but were in fact demonstrating the 'Clever Hans' phenom-enon. Clever Hans [she continues] was a wonder horse in Germany who in the late eighteenth century confounded everybody, including his owner, with the ability to do math problems. People tried all sorts of ways to ensure that the owner was not wittingly or unwittingly supplying the proper answer. Just at the point when the last skeptic was silenced, some-one had the idea of seeing whether Hans could solve the problem blind-folded. The owner, not understanding the origin of Hans's genius any better than the spectators, readily agreed. Blindfolded, Hans was a dolt. . . . And so the criticism of Washoe and other chimps was that what they do may look like language, but their teachers, however sincere they are, must be giving them inadvertent cues."[11] (Compare Patterson's ver-sion of Clever Hans with the analysis in Chapters 5 and 6.)

Patterson took care to check the procedure for this version of the "Clever Hans effect," an effect that Patterson understands to be inadver-tent cueing by the investigator. "To ensure against inadvertent cueing, I have checked my findings through double-blind testing, in which the ape can see the test object to be identified, but not the tester, and the tester can see the ape's response, but not the object. For instance, I'll put a toothbrush into a plywood box with a Plexiglass front, cover the box, and then leave the room. Koko will enter from another room and sit down in front of the box. Standing behind the box, unable to see its contents, is an assistant who asks Koko, "WHAT DO YOU SEE IN THE BOX?" or "WHAT'S THAT?" and writes down her response. Koko then leaves, I return with another object, and we repeat the procedure. At no time does Koko see me, or the assistant see the object. This elim-inates any possibility of cueing, and random order changes in the order of the objects presented for identification prevent the ape from using a strategy like memorization to come up with the correct answers. Thus I can be reasonably sure that when Koko makes a sign she is referring to the object presented for identification."[12]

Patterson and Linden are interested in the "errors" Koko makes, for these presumed distortions can tell us much about the thinking process, perhaps about the mind, of Koko the gorilla. It is explained: "critics miss a fundamental aspect of ape sign-language performance in dismissing it as a Clever Hans phenomenon. The horse Clever Hans had merely to look for cues that would tell him whether or not to continue tapping his foot: a go, no-go decision. Koko's options are hundreds of signs or thousands of sign combinations, and she frequently violated our expectations as to what her response will be to a particular question. Once I asked her how she slept, expecting an answer like FINE. Instead, Koko signed BLANKET THERE, pointing to the floor. Anthropologist Suzanne Chevalier-Skolnikoff [writes Patterson and Linden] supports the conclusion that Clever Hans is not the most conservative or economical explanation of ape language use. 'Apes manifest advanced cognitive processes nonlinguistically,' she writes, 'and since they appear to manifest them in signing, it is illogical to attribute their signing to simpler cueing and Clever Hans.'

"When Koko made the DRINK sign in her ear, it would seem that she made an error. After all, she did not make the sign correctly. Yet, just as a standard test may be incapable of measuring the abilities of the gifted but unmotivated child, a strict interpretation of Koko's response would suggest that Koko did not know what she was talking about and was merely randomly generating gestures. However, the context makes it clear that Koko knew what she was doing, but decided to be uncooperative. Countless other examples of similar negativity underscore her intent on such occasions."[13]

Patterson and Linden's interest in language is evident from their account in book form of Koko and Michael. By July 1977, five years into the training process, the number of spontaneous signs given by Koko is listed as six hundred, although the number would be somewhat less than two hundred if the more rigid criterion offered by the Gardners was used.[14] Given this accomplishment, it was now thought possible to use this basic vocabulary to ask questions about the nature of ape language. Compared to Washoe's achievements, Koko acquired signs far more rapidly. After three years, Washoe's vocabulary was 85 signs, Koko's 127: 46 were signs commonly used by both animals. The speed of development is not consistent; nor is it for children. Koko, at four and one-half years of age, showed a decrease in rate of responding with signs, bit a teacher, and, in the following month, showed a decrease in the number of signs used.

A set of skeptical comments regarding the likelihood of ape language comes from those who wonder why aspects of grammar, such as the interrogative, are not to be found in Koko's signing. Neither Washoe nor Koko seems to have learned the use of a question mark as a sign.[15]

It is difficult to know whether some of the seemingly clever combinations of signs reflect error, imagination, or perhaps the presence of different categories of association. For example, Koko called a "green pig" a GRASS PIG. Does this imply that she confused the sign for "grass" and "green," that she was being inventive (Patterson suggests that she became bored with repetitive tasks and invents, just as children do) or that her categories of perception overlapped "grass," "green," and, no doubt, other a posteriori names into a category grasped by the gorilla but not by the human mind?

The issue is of the greatest importance to our understanding of human-animal communication because before we can ask whether an ape communicates by language, we must know rather precisely how language is to be defined. Even if we are able to define language in a way that satisfies human beings, we may be foolish to search for these humanlike aspects of grammar in animal signs and sounds. Animals may not think, perceive, or ideate as do human beings. There is no reason to suppose that animal language has the same characteristics as does human language; that the grammar, word order, notion of a complete sentence, or the interrogative or command, are used.

If we believe that animal communication exists, then the first step toward understanding it is to determine the grammar being used. Much investigation, teaching, and assessing of animal gestures is presently aimed toward demonstrating that gestural language is a meaningful language to human beings. The question that should prompt research, however, is whether a grammar can be found. Unless we learn the grammar, we cannot know the meanings and intentions of the animal mind.

That Koko might learn the rudiments of English usage and grammar occurred to Patterson and Linden. "At the beginning of the project," they write, "there was little evidence to suggest that a gorilla could learn to understand English, and several scientists believed that if an animal could not generate spoken words, it did not have the necessary equipment to understand them either. This is a rough description of the motor theory of speech perception.

"Still, the principal benefits of learning English would be to Project Koko. If Koko could translate from English to sign language it would prove that she understood the symbolic nature of language. For instance, if Koko made a gestural rhyme on a word in sign language, we could then see whether she understood the concept of rhyming by asking her to sign a word that sounds like another spoken word. If Koko showed that without prompting she could associate words by gestural similarity, that would show that her sign homonyms were not merely mistakes, but rather evidence of a sophisticated understanding of the underlying structure of language."[16]

Here is one example of Koko's achievement. Patterson turned to spelling words that they did not want Koko to understand, much as human parents take to spelling words in the presence of pre-spelling children, especially those children at the stage of development marked by continuous repetition anywhere of whatever word is heard. This was done, Patterson reports, because "somehow Koko has figured out that "c-a-n-d-y" spells one of her favorite treats, so that now we have to use even more artful subterfuges when discussing such highly charged topics."[17] If the letters are spoken, as the passage implies, then the interpretation is sensible only if it is believed that Koko had come to understand how in English spoken letters of the alphabet form words. Another explanation for Koko's ability would invoke her receiving of physical cues, as documented by the stories of Clever Hans, Van, Roger, and the other clever animals that populate this book.

Another example of Koko's now remarkable understanding and ability is the question of whether animals, including great apes, understand the concept of "death" as do human beings. "Koko takes death quite literally. Once while she was idly playing in the room I was complaining to Maureen about the rigors of giving lectures. At one point I said, 'I can't go to L.A. [Los Angeles] every month—it would kill me.' I looked over at Koko, and saw that she was signing FROWN."[18] This example suggests that Koko understands the human concept of death and that Koko apparently understood spoken English as well as English spelling.

Patterson and Linden use other examples to demonstrate Koko's abilities in linguistic humor, insults, and imaginings. Koko was thought to have used the sign for "alligator" to prompt fear in her human companions. (It was, of course, Koko who showed fear of this animal.) In regard to Koko's ability to rhyme (in sign language) words she used and comprehended in English, consider the following exchange between Koko and one of her teachers, B. Hiller, which occurred while Koko faced an array of toys:

> *Hiller:* Which animal rhymes with hat?
> *Koko:* CAT
> *Hiller:* Which rhymes with big?
> *Koko:* PIG THERE (She points to the pig)
> *Hiller:* What rhymes with hair?
> *Koko:* THAT (She points to the bear)[19]

Why, and what, was she signifying by pointing? Was this behavior akin to Van's, the dog, indicating the "correct" card by fetching it (presumably as Lubbock looked toward it) or Roger, also a dog, solving problems in three languages? How was Koko able to rhyme in a language, English, that, presumably, she did not know, unless it is claimed that she had learned this language from those human companions around her? Are we dealing with Koko's ability to do "language games," as Patterson and Linden suggest to us, or, to be harsh, would a thorough analysis of the data regarding Koko, as the one done by Pfungst on Clever Hans, offer suggestions as to the need for experimental controls and designs beyond those applied?

Patterson and Linden write: "According to one's perspective Koko is either a dolt who has only a shaky hold on a basic vocabulary, or a bright, playful, creative creature capable of quite sophisticated innovation." Nor is this difference in opinion merely a question of empirical method versus anecdotal material. Underpinning the two perspectives are two conflicting views of the human's place in nature, or possibly, a reminder offered by the Mills-Thorndike controversy.

"For my part [continue Patterson and Linden], I find it more comfortable and nourishing to live in a world in which I can see and acknowledge elements of my behavior in the creatures around me, in which I can identify and communicate with a close relative with whom man has been out of touch for the past million years. . . .

"Still, however close Koko and I have become, I have not lost sight of the scientific importance of her achievements. I still maintain the daily checklists and logs, the videotaping session, and other means of monitoring her performance. Indeed, Project Koko is the only one of the ape language-experiments in which there is a constant, uninterrupted record of progress. Koko is the only language-using ape who has received continuous instruction by the same teacher. She is the only language-using ape who has received continuous instruction by the same teacher. She is the only language-using ape who has received nine years of language instruction. These circumstances give Koko added importance, especially in light of the new assaults that have been launched against the credibility of these experiments during the last two years. The suggestion by critics that imitation and prompting play a large role in ape conversation is based primarily on the performance of a chimp, Nim, who had many changes of environment and personnel during the four years of experiment. I am currently analyzing video-tapes and gathering other evidence to see whether Nim's was in fact a case of arrested development in language acquisition, a circumstance brought on by the environmental and methodological oddities of the experiment."[20]

Study of the training of the chimpanzee Nim, mentioned by Patterson, is not related neatly in time to the work with Washoe and Koko. The work with Nim followed the lead supplied by the Gardners in 1966, but itself took place from 1973 to 1977 while that with the gorilla Koko began in 1972. As different workers published their findings at different times, each had available different sets of information. The study of Nim was finished before the publication of Patterson and Linden's book on Koko, the work with Nim was itself published in 1979, and the Gardners's most updated account of their work is 1989, fifteen years after Nim and nearly ten after the report on Koko. Therefore, the most suitable way to present the information in a way that permits the reader to judge is to turn now to what we know of Nim. Indeed, the history of the attempt to demonstrate human-ape communication in modern times is not unlike the argument that took place between Mills and Thorndike nearly a century before. It is one that extends into our own times.

RETHINKING SPEECH

The Gardners's work, as well as that of the colleagues who now joined them in their attempt to establish human-chimpanzee communication, led to like attempts elsewhere. The pattern is like the one we witnessed when Thorndike's efforts to erect a comparative psychology began a new kind of search by Haggerty and Hamilton. Science, especially behavioral science, moves in just such fits and starts. In reaction to questions and criticisms, now to be recounted, the Gardners's own work was expanded and developed. Later work in which chimpanzees used ASL will mean more to us if we first learn about the nature of the reaction to the Gardners's reports on Washoe's abilities.[21]

Think, for a moment, about two ways in which we might ascertain the physical nature of our planet. One way is to take a sample of aspects that seem to us to have something in common and to analyze these aspects for their chemical composition. We might, to illustrate, select a car junkyard, drill into the earth's core, take a sample of houses, and examine bushes. Upon analysis, the junk yard would lead us to identify certain metals; the core would give us a variety of carbon compounds; the houses would tell us of the many different compounds, organic and inorganic, used to supply humankind with shelter; and the bushes would yield their own basic qualities, along with instructive distinctions. It is in this way that much of social science goes about its business: it first takes samples, and then examines the common elements that are to be found. Attempts can thus be made to generalize the findings to all like situations. It is a science of averages, deviations, and assessment of the degree of acceptable generalities. The current state of statistical mathematics makes possible analyses of remarkable sophistication.

The second path is to concentrate on a single living being by bringing under control a specific behavior, with the expectation that the same behavior can clearly be seen in all other beings. For example, examination has shown that the effects of schedules of reinforcement are alike for all species and all members of them. A rat or pigeon is trained to press a bar following a certain schedule, not because the results are expected to distinguish rats from pigeons. The search, rather, is for a universal statement about how all living beings react under a particular "schedule." The experimental situation is refined to the point that the rat and pigeon, for example, react alike, thereby yielding the atom of behavior—the unit to which all behavior can be reduced. We may liken this approach to putting a tiny sample of our planet under such intense magnification and scrutiny that our investigation reveals something that is universally true of all such samples. Behavioristic thinking greatly favors the approach of detailed control. The philosophic and empirical development of this approach is properly credited to B. F. Skinner and his many colleagues who have pursued this path with gusto, although its origins are to be found in the work of Thorndike and Watson.

The nature of these two paths to our understanding is important to the discussion of Washoe and her successors. The work with Washoe is in the form of a one-subject, indeed, a unique-subject, study. This study superficially appears to be similar to the observations designed by behaviorists interested in studying one being in detail rather than taking averages of larger populations of individuals.

The work that followed Washoe's and the Gardners's achievements attempted to determine whether the gesticulations formed a grammar which was judged a characteristic of a truly communicative language. The work also attempted to bypass the question by teaching the apes a novel language, one presumably without the defining characteristics of languages used by human beings. One change in procedure included the removal of the human from the teaching situation. Pfungst would have approved.

NIM

The nature of the quest to understand the minds of animals was now changed when the supposition took hold that the mind has something to say if only it can be reached. The supposition represents a major shift in thinking about the workings of the animal mind, because in both theory and practice, it restores the notion that there *is* an animal mind worthy of contacting. It also uses the principal tool of behaviorism, the learned operant response, as a means of contact through the training and teaching of a sign language.

The person who best understood the importance of the operant technique when applied to human-ape communication was H. S. Terrace, whose comprehensive account of Nim, the chimpanzee, is to be found in the book of that name.[22] Terrace begins his analysis with his interpretation of the Hayes's work on the language of Viki, the chimp, and of the then progressing work by the Gardners. As for Viki, he notes that the four words she spoke were produced only by the most intense and extravagant of techniques, including the physical molding of Viki's mouth by the Hayeses and the constant providing of a reward. There was no evidence, he concludes, that the production of the words offered any meaning. "Vicki[23] would imitate each of these words if she was given a food or drink reward. What the Hayeses showed [argued Terrace] was that a chimpanzee could learn some unnatural tricks in order to obtain a reward."[24] Washoe, to the contrary, Terrace writes, "did learn to use words of a human language."[25]

The issue Terrace addresses is how to interpret the meaning that such words might have, especially when two or more words are put together in a way that the human listener might see as the combination of different or novel ideas reflective of the chimpanzee's mind and perceptions. For example, does TICKLE ME have a different, or the same, meaning as ME TICKLE? Or, consider one of the more famous of Washoe's utterances, the signing of WATER BIRD when asked to identify a swan.

"It is difficult," Terrace suggests, "for an English-speaking person not to interpret *water bird* as Washoe's combining an adjective and a noun to create a special meaning. A moment's thought, however, reveals a number of simpler explanations. Washoe had a long history of being asked *What's that?* in the presence of various objects, including bodies of water and birds. Thus, there is no way to tell whether she was signing about a body of water and a bird or a 'bird that inhabits water.' Even if Washoe was trying to qualify the meaning of 'bird,' it is important to know whether she favored constructions of the adjective + noun or noun + adjective type. In the absence of such information, reports that Washoe 'created' such combinations as *water bird* are of no more significance than reports that a chimpanzee generated a line from a Shakespearean sonnet by banging on a machine whose keys happen to produce different words. The significance of such anecdotes is determined not by the aptness of a few meaningful combinations but by the relative frequency of such combinations among all the utterances that have been observed." [26]

One way to clarify the issue Terrace raises is to teach animals to learn unfamiliar symbols as replacements to the spoken sound. As we shall read in the next chapter, David Premack trained the chimpanzee Sarah to use plastic shapes to represent "meanings," and Duane Rumbaugh used computer keys that contained lexigrams—combinations of colors and shapes that were given different human meanings.

But, Terrace points out, "The closer I looked, [at the Gardners's descriptions of data from Washoe] the more I regarded many reported instances of language as elaborate tricks for obtaining rewards. When Washoe signed *time,* did she do so out of a sense of time, or had she simply learned a gesture to request food, as in the sequence *time eat? Time* was never contrasted with other related signs, such as *now, later, before, soon,* and so on. As far as I could tell, if Washoe had a sense of time, she was not expressing it when she signed *time.* Instead, she seemed to be imitating her teacher, who had just asked her, *time eat?* Adding a meaningless sign for *eat* is hardly the same as saying 'It is now time to eat.'" [27]

This criticism and caution are not intended solely to apply to Washoe, for whenever a substitute language is learned, the question arises, *who* is providing the meaning to the symbols, whether plastic shapes, lexigrams, or gestural signs? Terrace's point is that a string of words or signs, however produced, is not necessarily sentences that contain reliable meanings. The signs may be learned chains of responses with no more meaning to the animal than Hans's taps or Van's retrieving cards. A rat may be trained to press a bar, pull a chain, scratch his back, and twitch his whiskers in that order to obtain a reward, but the signs do not necessarily convey meaning. Similarly, my giving you a set of signals and your reacting in a way I wish does not necessarily demonstrate your understanding a sentence. A baseball coach may give a sequence of signs to which a player responds in the way wished, but the sequence itself does not have meaning.

Terrace continues, "the rich linguistic knowledge implied when a child

utters a four-word sentence such as 'mommy give me milk' is apparent
when we consider other sentences the child is able to generate: 'Sally give
me soda,' 'Daddy give cat milk,' 'I throw dog ball,' and so on. In any of
the four positions of the original sentence, the child can substitute a
variety of words, each appropriate to the circumstances at hand. Such
sophistication was not demonstrated by chimpanzees."[28] A child learns
that a structure, an order, allows substitution. The power of language
lies not in imitation but in manipulating the substitutions to achieve different meanings.

BRUNO AND NIM

Along with many students of behavior, Terrace had heard of the Gardners's success in teaching sign language to the chimpanzee Washoe. In
1968 he adopted a newly born male named Bruno, also from Oklahoma,
where William Lemmon conducted a primate facility to study chimpanzee development. Not incidentally to our story, Lemmon attempted to
retrain and resocialize chimps who had been raised as pets by human
beings so that they might adopt again chimplike behavior. Terrace was
interested in the effects of the environment on the development of the
chimpanzee as well as in several of Lemmon's own findings such as, for
example, that chimps raised by human parents developed physically more
rapidly than did those raised by chimpanzees. Bruno lived with Stephanie Lee, a student of Terrace's, for fourteen months. From watching
this interaction, Terrace developed an interest in the possibility of teaching sign language to a chimpanzee and examining the course of its learning.

Terrace's way of studying behavior had been firmly in the behavioristic mode, with energetic attention given to using control measurements
to reduce the kind and number of possible explanations of the behavior
observed. Although first both the Gardners and, later, Terrace accepted
the challenge of teaching ASL to individual chimps, differences in how
they preferred to conduct and interpret research projects would yield to
disagreement that would eventually help our understanding of what the
chimps' behavior represents. If a Great Ladder of Mentation is to be
constructed, the ability to *move* between the mind of the human and the
mind of the animal will be a great leap forward. The ladder will be very
different from one constructed on the basis of puzzle-box solving or like
activities. If cross-species communication can occur, if one can learn from
the mind of an animal to the same degree that one human can learn from
the mind of another, the Ladder of Mentation is apt to look very different from the one we now imagine. A ladder that can be based on cross-
communication is certain to have rungs representing species located very
differently from one based on a single kind of behavior, such as solving
a specific problem set by human beings.

When a male was born in Lemmon's laboratory on November 22,

1973, the chimp was offered to Terrace for a projected study on the development of linguistic competence. The infant, later to be named Nim, was the half-brother of Bruno. Both had the same father. Although few preparations had been made, (the birth was unexpected), Stephanie Lee collected Nim four days later when the mother and the baby chimp were separated. Nim was taken to New York City, where he was to reside with Lee, who since she had raised Bruno, was by now an experienced mother of chimpanzees. Since Nim was, as Terrace puts it, an "unplanned baby," much had to be improvised, including how and where Nim would be raised, who would sign to him and under what conditions, what kind of data would be kept and how they would be recorded, and, precisely, what the aims of the research were to be.

Several criticisms and skepticisms had been voiced about the Gardners's work with Washoe. One question was whether the signals, the hand motions, had meaning to the animal, or whether the signs were merely learned motor responses, responses in the same category as a rat's bar-pressing and a pigeon's pecking. A second question concerned how to assess what the human hearers understood the signs to be. As we know from the accounts in this book, human beings' wishes and motives can shape animals' perceptions. This skepticism is potentially answerable. Using a technique similar to the one Pfungst used when investigating Clever Hans's abilities, we might remove the human beings from the situation, or we might employ people unknown to the animal (for example, people who use ASL, but who have not met Washoe) to see if the effect—the hand signs—remains or is diluted. We might ask if Washoe could use this form of communication to sign to other chimpanzees. The Gardners and their colleagues would undertake ascertaining answers to this question. But the first question—What do the signs mean?—admits to no simple experiment. The reason is that the question itself is based on meaning, on a matter of human definition—namely, what do we mean by "meaning." Do the two words ME TICKLE convey meaning? Is the meaning the same as TICKLE ME? Can a chimp form a sentence, or is the idea of a sentence a measure of human thought alone? Does a string such as ME GIVE TAKE HUG BIRD HURRY convey a meaning that requires all of the words, or is it but a string of signs, only one of which conveys meaning? "Meaning" is, after all, a human concept. What "meaning" does "meaning" have when applied to nonhuman beings?

Terrace described the goals he had for Nim: "I wanted to socialize a chimpanzee so that he would be just as concerned about his status in the eyes of his caretakers as he would about the food and drink they had the power to dispense. By making our feelings and reactions a source of concern to Nim, I felt that we could motivate him to use sign language, not just to demand things, but also to describe his feelings and to tell us about his views of people and objects. I wanted to see what combinations of signs Nim would produce without special training, that is, with no more encouragement than the praise that a child received from his parents. I especially wanted to find out whether these combinations would be similar to human sentences in the sense that they were generated by some grammatical rule."[29] The difference between the work with Bruno and the goals for Nim, Terrace tells us, was the difference between "mere baby-sitting and a scientific project aimed at collecting scientifically useful information about a chimpanzee's use of sign language."[30]

Think of the arrangements necessary to undertake sign-language training with an ape, or with a human child for that matter. While, for example, we credit Itard for his work with Victor, it was in fact Madame Guérin, the housekeeper and Victor's caretaker, who cared for the boy's needs while Itard appeared from time to time to arrange and conduct the investigations. This is not to make light of Itard's contribution, but as intense as it was, his interaction with the boy comprised only a small part of the day. Someone else was responsible for the boy's food, clothing, personal care, and maintenance when he wasn't under Itard's immediate supervision. Feuerbach's detailed observations on Kaspar Hauser reveal that Kaspar was raised by a hard-working family and that he was attended by a schoolmaster. Feuerbach himself had nothing to do with Kaspar's daily needs. Reverend Singh appeared to rely on his wife and members of the orphanage staff for the maintenance of Amala and Kamala, while Freud had little or no direct interaction with Little Hans but interpreted his behavior from the letters sent by the father. Not only is someone else providing most of that care, but that someone else is seen only as a shadow to the main event, even though it is clearly these (to our knowledge) shadowy people who have the most interaction with the subject.

Such help is not free. The situations involving children, from Victor to Kamala, all required financial commitment—commitment from an institute for the deaf in Itard's and Victor's situation and the evangelical aspect of a church that supported Singh's orphanage and, thereby, Amala and Kamala. Feuerbach and Freud supported their work by paid work in other areas—Freud as a writer and his practice as a physician and Feuerbach as a judicial appointee of the state. In our times, scientific research is paid for by the state and, at times, through private foundations, through any of a number of organizations established for that purpose. These make monies available, either from the citizens' treasury or from private wealth, to individuals who propose projects deemed worthy of support. As Terrace makes clear in his book *Nim,* the gathering of such support can well occupy more than half of the investigator's time. To require this much time is not necessarily wasteful: It might be well advised when it assists proposers to think and rethink their ideas, thus increasing the probability of success. But for projects such as the Gardners's and Terrace's, where at the onset success cannot be predicted and where the course of the work may run into unexpected blind ends, it becomes impossible to predict, not merely outcomes, but the experiments. The nature of the research question guarantees gaps between what we would do in the ideal and what we may do in practice. The problems of beginning a new way of inquiry, as exemplified by ape-human communication, involves unpredictability. If the events to come were predictable, what would be the sense of doing it? At the same time, funds must be found for a project with an unpredictable outcome. These funds must be larger for situations in which full-time maintenance is required. In practice, this means that the objectives of any such project are almost certain to be compromised or constrained by the financial costs of the project.

The Gardners learned a great deal about what they might have done with Washoe as their study progressed, just as Terrace learned a lot about the problems of establishing a reliable environment of the sort he originally envisioned for Nim. Nim's childhood was marked, as is that of most children, by changes of environments and personnel. The problems of housing an infant chimpanzee for several years, along with those of acquiring financial support, led to several physical moves within the New York area and what, in retrospect, looks like a parade of caretakers and teachers. Such variations made for a rich social environment and thereby provided some fascinating information, for example, that to be found on Nim's emotions at seeing a teacher who had been away for a year or seeing a house and surroundings not lived in for a long time. But these variations also made it likely that there would be alternate explanations of what was seen and that skepticism might be avoided if the requisite experimental controls were in place. The price of control is a reduced, sometimes greatly reduced, environment; the price of no control is uninterpretability.

We work with what we have, not always with what we want, just as did Itard, Pfungst, Freud, and Singh. All the same, it does not restrict our understanding to remember that each observer and investigator works in the context of a time, a time interested in some questions and not in others; within a society that has its own preferences for spending funds for scientific or humanitarian research; and within it a set of procedures as to how information is to be collected, analyzed, and circulated to the society that paid for it.

NIM'S PROGRESS

Because Nim was at first raised among Stephanie Lee's family, he heard English spoken in rich and various contexts. Nonetheless, none of the human beings around him was a fluent signer. Such teaching was begun at about two months of age by molding Nim's hands into the shape of the signs. At four months of age, signs were being given for DRINK, UP, SWEET, GIVE, and MORE.[31] The first "spontaneous" sign, given at ten weeks, was of a drink sign that was repeated. Terrace stipulated two criteria for accepting a sign as spontaneous: three independent observers had to observe it and its spontaneity, and the sign had to appear on five successive days. Using these stringent criteria, during the next six weeks Nim gave the signs for UP, SWEET, GIVE, and MORE.

At this time, Nim was moved to a beach house on Long Island, New York, and the speed with which he added signs slowed appreciably. With the change in surroundings, coupled perhaps with the developmental age, he became interested in the natural surroundings. This interest now took precedence over learning to sign. At this time, Stephanie Lee decided that her own career was to be elsewhere, and she was replaced as caretaker-teacher by Carol Stewart. At the end of the summer, at the onset of the school year, Nim was removed to Columbia University in New York City where he would live and go to school in a newly equipped nursery school built for him. A set of teachers was hired to provide instruction: some were fervent about using the techniques of operant conditioning; others were less doctrinaire. They appeared to have espoused the variety of approaches to teaching that one finds among teachers in any elementary school.

As all teachers know, or eventually find out, teaching the subject matter is one thing, but teaching deportment is another. Socialization is essential, not merely for the moment, but also for the teaching that will occur in the future. Some teachers, especially those who teach the very young, believe that controlling deportment is primary because the class is a group in which no one can learn without rules of conduct. Some would argue that good deportment comes before learning. Others believe that socialization occurs naturally and that learning is best when rules are minimal and creativity and imagination are rewarded. These differences in viewpoint were expressed by Nim's teachers, and the intensity with which viewpoints were believed appear to have led to a long-functioning informal symposium on the issue of the role of socialization in learning, or, as it sometimes is expressed, whether learning occurs as a result of human kindness or rigid drill. Carol Stewart's approach led to the addition of two signs, HUG and CLEAN, within a month; the year progressed with the additions of DOG, DOWN, OPEN, WATER, LISTEN, and GO. Some signs were learned in the nursery classroom, and some at home. Figure 11.1 shows Nim and a caretaker, Joyce Butler, in the process of signing to one another. These pictures were taken from videotapes that were used to assess and reassess the course of learning. The use of videotape allowed later observers to quantify the behaviors with far more precision than was possible by unchecked observation. One wonders how Pfungst might have used the invention.

Nim's full vocabulary is shown in Figure 11.2. In this figure, the words indicate the equivalents of the signs learned, and the curve shows the cumulative frequency of signs-words learned. A cumulative frequency is useful to show the course of learning. We will recall that Thorndike (pp. 243–246) drew curves to show the course of learning demonstrated by

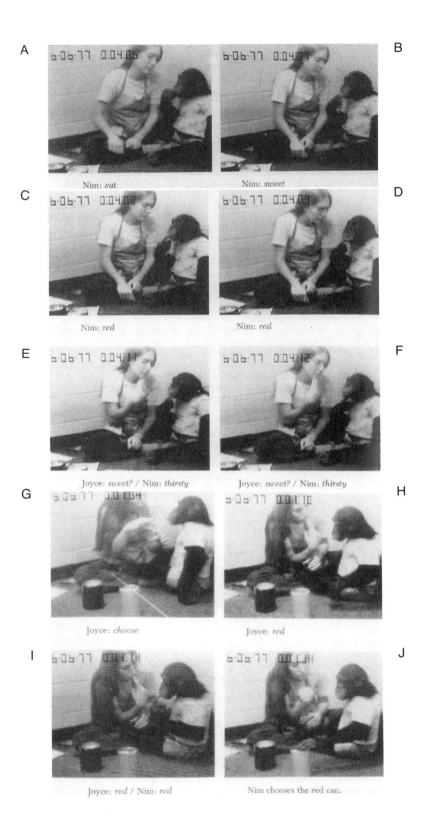

A   Nim: *eat*

B   Nim: *sweet*

C   Nim: *red*

D   Nim: *red*

E   Joyce: *sweet?* / Nim: *thirsty*

F   Joyce: *sweet?* / Nim: *thirsty*

G   Joyce: *choose*

H   Joyce: *red*

I   Joyce: *red* / Nim: *red*

J   Nim chooses the red can.

animals in the puzzle box, although these did not accumulate the data. The cumulative frequency is a favored representation of learning by behaviorists, for it shows a like function for all species. In Figure 11.2, we note that the cumulative record is "negatively accelerated," that is, that Nim is ever so slightly learning more and more signs as time passes, but at a slower and slower rate.

Nim's next home, one again in the vicinity of New York City, was a handsome house and estate grounds north of the city. This estate belonged to Columbia University at the time and offered a seemingly ideal location, at least by human standards. Because the property was found only just before the lack of a home dictated Nim's return to Oklahoma, its availability provided additional time for study under yet another type of housing and social surrounding. Laura Pettito now became the chief teacher, and Terrace received his first full grant to sustain the work.

Although Nim's signing ability was of greatest interest, his social and emotional development were of much practical interest. Several events testify, if indirectly, to such development in terms understandable to people. Among these was the practice of deceit. On one occasion, one teacher-caretaker tried to collect a bowl of cereal that had been prepared. It was missing. Nim, the only other being around, was where he had been left, on a table having his diapers changed. "It was Nim's exaggerated look of innocence that prompted Laura to grab him and to bring him to the office, signing sternly, WHERE BOWL! Nim gave Laura a puzzled look but nevertheless showed that he understood Laura's question by looking around the office as if to help her locate the bowl. Laura became more and more angry. Finally, she exploded, grabbed Nim by the hand and pulled him up on the changing table. With one hand raised as if to strike him, she again demanded to know where the bowl was. Nim must have sensed that he was in trouble. Whimpering, he took Laura's hand and led her to the sink immediately adjacent to the changing table. In a corner of this fairly deep sink, Laura saw the half-finished bowl of cereal."[32]

Figure 11.1 A–J "Dialogue between Nim and Joyce [Butler] about candy. Nim wanted Joyce to give him some red candy, hence his spontaneous sequence: *eat sweet red*. Joyce was trying to clarify what Nim signed by checking to see if he actually wanted a sweet. Nim responded by continuing the sign for *red* (the index finger touching the lower lip) and then touching his throat as if to sign *thirsty*. Joyce concluded the Nim's *thirsty* sign was a topographical [typographical?] error. The difference between *thirsty* and *sweet* has mainly to do with where the fingers touch the body. In the case of *sweet*, the index and middle fingers touch the lower lip. *Thirsty* is signed by moving the index finger downward while touching the throat . . . Joyce is telling Nim to choose the red can. Nim signs red and then chooses the red can. In this example, Nim's sign was imitative of Joyce's. Next we see an example of partial overlap between Nim's signing and the teacher's signing." (Terrace, p. 216–217)

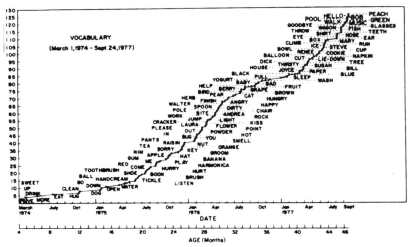

Figure 11.2   The new signs learned by Nim, placed in chronological order.
The slope may be considered a measure of the learning-curve.

Laura Pettito's work with Nim was, Terrace tells us, of a qualitatively
different nature from that of Carol Stewart's. "No longer," writes Terrace, "was Nim treated as an institutionalized child who had to demonstrate various behaviors or else. It was not that Laura made no demands
on Nim. Indeed her demands were as strong or stronger than those of
any of the other teachers, including Carol. But Laura's demands were
tempered with a combination of praise and play that seemed much more
human than the strict approach advocated by Carol. What Laura sacri-
ficed in the kinds of consistency that concerned Carol, she more than
made up for in capturing Nim's interest and respect, two attitudes essential for successful teaching."[33] But not everyone had the time or competence to work in this fashion. As any teacher or parent should know,
before you can teach anything, you have to have something to teach, and
you have to have a partly attending being to teach it to. Terrace remarks
later, "Only a small fraction of Nim's teachers and caretakers got to know
his inquisitive and pensive side. Most teachers barely developed enough
control over him to insure that he didn't act like a brat. Most of their
time was spent feeding him and playing with him because they did not
have the control needed to direct his attention to such challenging activities as looking at pictures, drawing, or playing with toys."[34] The sentence, I think, could as well have been written by Itard, Madame Guérin,
or Dr. Arbuthnot.

If interest and respect are not attitudes essential for successful teaching, and if they are more important than "demonstrating various behaviors" or drills, it would be a worthwhile exercise to review the procedures
used by Itard, Guérin, the Singhs, Garner, Lubbock, von Osten, Krall,
Witmer, Freud, Haggerty, James, Buytendijk, Thorndike, Howe, Mills,
the Kelloggs, the Hayeses, the Gardners, and the other teachers whose

work fills this book, to determine the origins of their philosophies as teachers. Some, it seems, played different roles at different times in their lives (as did Garner), but for others there seems to have been a common way of teaching, a way based largely on their sense of what it means to be a human being and on consideration of our responsibility to our own and other species. One's intellectual disposition to accept behaviorism, phenomenology, or psychoanalysis must dictate what each of us thinks of as good or as bad teaching. As is true of most of us, Nim appears to have had examples of each of these perspectives.

NIM'S ACHIEVEMENT

The list of words and signs Nim acquired tells us much, but we also want to know how the signs and words were used. Consider this exchange between Nim and a teacher, Mary (Wambach), who lost her hearing at the age of thirteen and who was a skilled user of ASL:

> *Nim:* [Looking at a magazine] TOOTHBRUSH THERE, ME TOOTHBRUSH.
> *Mary:* LATER BRUSH TEETH.
> *Nim:* SLEEP TOOTHBRUSH.
> *Mary:* LATER . . . NOW SIT RELAX.
> *Nim:* [Seeing a picture of a tomato]. THERE EAT. RED ME EAT.
> *Mary:* THERE MORE EAT! WHAT THAT?
> *Nim:* BERRY, GIVE ME, EAT BERRY.
> *Mary:* GOOD EAT. YOU HAVE BERRY IN HOUSE.
> *Nim:* COME . . . THERE.
> *Mary:* WHAT THERE? [Nim leads me into the house]
> *Nim:* GIVE EAT THERE, MARY, ME EAT. [At refrigerator]
> *Mary:* WHAT EAT?
> *Nim:* GIVE ME BERRY.

The questions remain before us: Can a chimpanzee form a sentence? Does an ape have a grammar? Do its communications have meaning? Terrace was not merely interested in the number of sign-words that Nim could produce, or in their spontaneity, as opposed to their being given in reaction to the questioner. Rather, he was curious as to how the signs were put together, whether they were used to form different strings, and whether sentences or their equivalent could be found. Nim learned 125 signs during the first forty-four months.[35] He learned to sign the names of his teachers, and he would use the signs when he saw them after even a year of separation. Nim developed an interest in drawing, using crayon and paints, and he came to sign the names of the colors he wanted. Several signs came to have seemingly unique usages. BITE and ANGRY were seemingly used as threats, and DIRTY was applied to various objects and people, more as a pejorative than as a description, at least to the human eye. He used CAT and a seeming variation to refer to HERB

(Terrace), presumably because the signs for both are similar. In this instructive example, note that CAT is the human understanding of the sign, whereas for Nim HERB and CAT "sound" alike in that the sign used for both are alike. Over the last two-year period of the project, Nim's teachers recorded twenty thousand[36] combinations of two or more signs. But the question Terrace set for himself remains: Can a chimpanzee form a meaningful sentence?

As Terrace points out, human parents are excited by the baby's first word and remember it as a great watershed, but the uttering of the first sentence is taken for granted. Its composition is unremarkable, and it does not become part of the family's collective memory. This is odd, because construction of a sentence is a difficult activity. That chimpanzees can form or remember long strings of signs does not tell us anything about their ability to form a sentence, and, it may be added, the ability to respond to a human sentence composed of a string of signs tells us nothing about the chimp's abilities to grasp the meaning of the sentence.

Terrace and his associates now undertook an analysis of the sequences using the videotapes. This form of analysis went an important step beyond the work with Washoe, at least in the earliest stages, for it accepted the signing as being a reliable ability of the chimpanzee. But it now asked, Do these signs contain and signify grammatical meaning? Terrace analyzed the videotapes into the kinds of sequences that were observed. For example, a "homogeneous" sequence consisted of EAT EAT EAT; a "heterogeneous" sequence of BANANA ME ME ME EAT (translated for analysis as BANANA ME EAT). The analytic question was whether these sequences were used in reliable ways that suggested a systematic variation in the use of signs. A further issue was context, specifically, whether the context of the signing signified something about the situation. The analysis of the videotapes led Terrace to conclude:

"As far as I can tell, the distributional, statistical, and semantic analyses of Nim's utterances represent the most intense and systematic effort ever made to evaluate a chimpanzee's ability to create a sentence. At the very least, these data invalidate two simpler interpretations of word sequences emitted by a chimpanzee: that they were learned by rote or that they were random combinations of signs, each relevant to a particular context but unrelated to one another.

"When we finally completed them during the summer of 1977, the tedious statistical analyses of Nim's signing behavior appeared to give us just what we had hoped for—a solid basis for demonstrating that a chimpanzee can create a sentence."

"But scarcely another month had passed before I realized that my satisfaction may have been premature. Contrary to what I had thought, the statistical analyses were not definitive at all. Painstaking examination of videotapes of Nim's conversations with his teachers unexpectedly revealed an aspect of his signing behavior that I—and other experimenters—had hitherto overlooked. Ironically, the only reason I found the time to study the video-tapes was a sad one: it was no longer possible to keep Project Nim going, and Nim himself had to be returned to Oklahoma."[37]

NIM MEETS WASHOE

For a variety of reasons—the constant need to find financing, the need for teachers and caretakers, the approaching departure of several teachers with whom Nim had been working, the sheer cost in human time of maintaining the work—it was decided that the time had come for Nim to return to Lemmon's laboratory in Oklahoma. Oklahoma was, of course, not only Nim's birthplace, but also the place where Roger Fouts, who had worked tirelessly with Washoe when she was under the care of the Gardners, lived, along with several other chimpanzees who had received some training in ASL. Fouts had been conducting his own research program by investigating whether chimpanzees would signal among themselves, whether they would teach ASL to one another, and whether human speech retarded the learning of ASL. Washoe, and others, were carrying on their lives and careers in Norman, Oklahoma.

On September 23, 1977, a party was scheduled for Nim to be held on the Columbia estate grounds to which Nim's teachers were invited. As the time for the party approached, Terrace took Nim to the greenhouse, a place previously off-bounds to Nim for obvious reasons. When everyone had arrived, "I carried Nim out of the greenhouse. Even then I prevented him from seeing his teachers standing in front of the . . . mansion by holding him below the umbrella that shielded us from the drizzle. When he finally saw his old teachers, Nim's first reaction was to cling to me and hoot softly as if he was both curious and afraid of the crowd he saw. Then he climbed down from my arms and urged me to proceed toward them as quickly as possible.

"To insure as formal a picture as possible, I held Nim in my arms for the first few shots, but it seemed unfair to restrain him from being reunited with his old teachers. . . . Once I released Nim, he bounded back and forth hugging almost every one of his teachers who had come to say good-by to him. Only two days earlier Nim had been ecstatic about seeing Stephanie [Lee] and her family for the first time in two years. Now his teachers were sadly bidding farewell to a chimpanzee they all adored."[38] Figure 11.3 shows the formal picture of the gathering. Nim is held by Terrace.

The next day, Nim was flown from New York City to Norman, Oklahoma. As Nim had been sedated for the flight, his reactions to the new, caged environment at Fouts's compound were not to be predicted. Terrace stayed with him the first night, but sleep came to neither. Nim acted as an uproarious child throughout most of the night. The next morning, Nim was removed from the area before breakfast, because this was a noisy time in the compound and it was feared that the shouts and screams of the other animals would affect Nim and his eventual acceptance of the

Figure 11.3   The farewell party of Nim, September 23, 1977. The photograph of the group shows some of the teachers in attendance, and features Terrace (center, front) holding Nim. The photograph gives indirect evidence of the large number of people whose work is required on such projects.

new community. Terrace took him to a wooded area, and it was here that Nim saw, for the first time since being separated from his own mother, another chimpanzee.

"One of the cages we had to pass housed Washoe, the first chimpanzee to learn some rudiments of sign language, and three other females. It was impossible for Nim not to notice Washoe and the three other chimps as we walked by. He looked in their direction momentarily but showed no reaction. He then looked away as if he had seen nothing strange. Having known Nim for almost four years, I wasn't fooled. Many times I had seen him pretend not to notice something that absolutely fascinated him. Later, at a time that apparently seemed safe to him, he would direct his full attention to the object he had pretended to ignore.

"That time arrived soon after I put Nim down in the wooded area. I had expected him to curl up in my lap and go back to sleep. Instead he took my hand and began to lead me through a grove of apple trees situated in front of Washoe's cage. Under the cover of a tree, he stared directly at Washoe, grunted softly, and pointed in her direction. His gaze was intent. Nim held on to the tree with one hand and to my hand with the other, staring and grunting for about five minutes. He was clearly intrigued by the sight of Washoe. . . . The surveillance of Washoe continued for another half hour. . . . He clearly preferred to satisfy his curiosity about Washoe from a distance."[39]

Fouts suggested that Nim might like to meet Mack, a younger animal who knew twenty signs that were within Nim's signed vocabulary. When Fouts brought Mack to Nim and Terrace, Nim seemed uncomfortable and signed DIRTY repeatedly. A second meeting later in the day was equally unsatisfying from the human point of view. On the next day, the following dialogue took place:

> *Fouts* [to Mack]: SIGN
>   *Mack:* COME
>   *Terrace:* [To Nim] HUG?
>   *Nim:* HUG [repeatedly]

Terrace writes, "I was also able to get Nim to sign HAT by placing a baseball cap on Mack's head and restraining him. Without any prompting or signing on my part, Nim signed HAT. When I released Nim, he jumped at Mack and began an unsuccessful attempt to wrest the hat away from him. Everyone was thrilled at what seemed to be two clear instances of one chimpanzee signing to another. While I was encouraged by what I saw, I remained somewhat dubious. I felt that the real test of communication between the two chimpanzees had to be made without any human beings present."[40]

It was time for Terrace to go. Nim would no longer have the opportunity, as Terrace pointed out, of being signed to for sixteen hours a day. Nim would have signing time with Fouts and, perhaps, with other signing chimpanzees, but it would not be time of the quality that had been offered during the first four years. Terrace took his leave. In a passage that may remind us of the thoughts and words that occurred to Richard Lynch Garner, then saying farewell to Elisheba in Liverpool eighty years before,

"I had to acknowledge that I was abandoning a wonderful little creature of high intelligence to an institutional existence. Whatever potential Nim had to develop his intelligence and to learn more language would almost certainly never be fulfilled. It was with real sadness that we drove away from the center. Nim was with Roger [Fouts] outside his cage when we left. He did not appear to be particularly upset. It would probably take him a while to understand how profoundly his life had changed."[41]

Terrace records a day spent with Nim a year later. During this time, Nim had lived with a group of nine chimpanzees on a human-made island. Lemmon had Nim taken off the island and later brought to an area where Terrace waited with a film crew. "As soon as I turned to greet him and he realized who I was, Nim's mood changed instantly and dramatically. He let out a wonderful shriek, leaped into my arms, signed HUG, and gave me a tremendous chimp embrace."[42] During the half-hour visit with Terrace and the film crew, Nim made nineteen signs.

SAY WHAT NIM?

While Nim spent his first year in Oklahoma, the New York City crew examined the teachers' reports. The use of the sign DIRTY toward objects and events seemed to Terrace to mark the first time an animal had shown the ability to "use an arbitrary symbol to misrepresent a bodily state" (p. 209). But Terrace was troubled about the findings. For one thing, there had been no change in the length of Nim's signings. Odd, for human beings surely increase the lengths of their sentence structures, sometimes to seemingly unmanageable, if grammatical, strings. The average was between 1.1 and 1.6 signs, and many of the symbols were repetitions. The search for understanding led to a review of the tapes:

"It was while looking at the playbacks of the videotapes that I realized I had missed an important aspect of the context of Nim's signs. When the other teachers and I were working with Nim and recording what he signed, our attention had always been riveted to his signing. We had paid too little attention to what *we* signed to Nim. . . . We did not ask systematically how often Nim's utterances contained signs that we had just signed to him. . . . Our initial analysis of the relationship between Nim's utterances and those of his teachers showed that Nim's were more dependent upon what his teachers signed than a child's are dependent on the words of its parents."[43]

When the content of the videotape was examined, it appeared that 10 percent of the signs made during the last year were spontaneous, while 40 percent were determined to be imitations of the sign made by the human signer. When compared to the behavior of human children, it was also noted that Nim's "conversations" with teachers were different. Nim interrupted far more often, suggesting either that he was not "listening" or that there was not a two-way conversation. When Terrace examined the films produced by the Gardners of Washoe (although these were commercial films, heavily edited, presumably, and not "data"), "nothing in them [the films of Washoe] suggests anything other than a consistent tendency for the teacher to initiate the signing and for the signing of the ape to interrupt and mirror the teacher."[44]

Terrace concludes: "in this light, the distributional regularities observed in the corpus of Nim's two-sign combinations cannot be considered solid evidence of Nim's ability to combine two or more words to create specific meanings. . . . I must therefore conclude—though reluctantly—that until it is possible to defeat *all* plausible explanations short of the intellectual capacity to arrange words according to a grammatical rule, it would be premature to conclude that a chimpanzee's combinations show the same structure evident in the sentences of a child."[45]

Washoe and Nim learned to sign. They learned to gesticulate in ways that reliably conveyed signs to human watchers. These gesticulations were considered by the human observers to represent words. The issue remains as to whether these signs or words, as perceived by the human observer, correspond reliably to the mind of the animal, whether they indicate interspecies and intraspecies communication, and whether they are meaningful—and, therefore, language.[46] The question is whether anything of meaning or feeling is being communicated.

Do we now have evidence of ape-human communication? Certainly, we may, depending on our definition of communication. That human observers saw gesticulations used reliably is unquestioned. That both the articulated sounds and hand movements of apes (and human beings) could be shaped by operant conditioning is equally evident. That specific gesticulations could be produced by human speech seems undeniable.

Through such interaction, do we human beings know anything about the mind of the ape? Does the ape know anything about the mind of the human being? Who is seeing what in the mirror of our minds?

Language and Meaning: Sarah and Lana, Sherman and Austin, Kanzi and Ai   **12**

T HAT ANIMALS, including human beings, can be taught to make gestures is not astonishing. Hans's taps, Van's fetches, Peter the chimp's writing—all these are examples of the animal's ability to invent or modify motor actions. The motor actions, the gesticulations, and the signs do not necessarily imply meaning or the transfer of information. Let us speak to a horse through a telephone (as did Krall), the receiver of which is placed to the horse's ear. Let us say to the horse "What is seven plus three?" Or, if we wish, "Was ist seben und zwei?" or "¿Cuanto es siete y tres?"

The horse taps the hoof ten times.

What meaning has been transferred to the horse (or, alternatively, to the child, the dog, the chimpanzee)? Not necessarily any, but to understand why this is so we return to the distinction made in Chapter 3 between sensation and perception. What the horse senses, as far as we know, is a set of electrical impulses transferred to the acoustic center of the brain. These impulses arise from the discharge of neurochemicals and therefore electrical exchanges in the neuron and axon. What the brain hears is not "What is seven plus three?" but an electronic code something like

.. ..

Presumably, the code will be a little different if the question is asked in German or Spanish, merely because the frequencies and intensities of the neurons discharged by the sound will be somewhat different. The principle remains, however, that what is sensed is the unit and patterns presented by the electronic code. We presume that it is the pattern of the code that comes to be attached to the motor response. The horse now taps ten times because the particular code is associated with that precise motor response. The horse, like the human being, we presume, has a brain that translated the sensation into a perception. We do not know what that perception is like to the horse (it is to this question that phenomenology addresses itself), but to ourselves we "hear" the code as a perception, as a question in English or perhaps in another language. If we are speakers of English we answer the question verbally by saying "ten."

Is this sound "ten" anything other than an articulation of the learned response to the perception—one that sounds to the hearer as "ten?" Or is there some additional meaning? If the question "What is three plus seven?" is asked, as an adult I am able to answer it, and I know that it makes no difference to the answer as to the order in which the numbers are put. Have I always known that as some sort of a priori knowledge? Or is it knowledge that I gained through training and have learned to make general to all such problems?

The chief point to be made is that the horse's tapping the "correct" answer can be done without the horse placing any meaning to either the question or the answer. That the tap expresses something meaningful to the human watcher says nothing about the meaningfulness of the question or answer to the horse. The horse is indeed clever to be able to decode the impulses (but all animals do this decoding constantly—it is the way the sensory systems interact with the brain). It is perhaps even more clever to learn to associate a motor response with a perception, but much of the cleverness we perceive is our astonishment at the animal showing human skills.

We do know from the training of Washoe, Nim, and a number of great apes that gesticulation can be taught and learned; that these motions represent, to the human being, signs and words; and that the human being attaches meaning to these signs. But we do not yet know, and perhaps never can know, the nature of the language being used to express the meaning we presume to hear. The Gardners and their colleagues appear to believe that the signs signify and indicate: they point to the meaning. Signifying and indicating are certainly rudimentary forms of communication used by human beings. Who among us has not been reduced to some such use of gesture when trying to make ourselves understood when in a culture whose language we do not understand? Terrace, for example, was concerned with the grammatical structure of language. He overlaid the human use of sentence structure on his analyses and found that Nim was not like a characteristic human child in language usage because of the grammatical structure. So, the question remains: Are great apes capable of using a language that conveys meaning? But now we can see that the question is no question at all, at least not a question capable of being answered, unless we know what is meant by the words "speech" and "language." Washoe and Nim's achievements prompt consideration of these questions by bringing to the fore a second generation of perhaps literate chimpanzees.

SARAH

Sarah was one among a set of chimpanzees who, as David and Ann Premack describe it, were "students of invented language" roughly during the same years that featured work with Washoe, Koko, and Nim.[1] The use of "invented language" permits the animal to demonstrate its own sense of grammar and naming. Of the nine chimpanzees under the Premacks' care, Sarah was evidently the genius, at least when it came to performing the motions required to work within an invented language. We rarely hear about apes who don't learn signing or gesticulating, but there must be some. It is helpful to learn, for example, that the chimpanzee Gussie, although given the same training as Sarah, failed to learn a single word of the invented language. All were born in Africa and reared by the Premacks and their associates, but not as children might be in the human home, as was true of the Kelloggs and Hayeses. The Premacks' opinion is that, by raising a chimpanzee in the home, "its every act will be anthropomorphized, undermining the scientific value of the animal."[2] The echo of the Thorndike-Mills dispute resounds. What some see as a vapid environment incapable of eliciting the natural behavior of the animal, others see as a control essential to the interpretation of any data.

The chimpanzees had various homes, ranging from Missouri, to California, to Pennsylvania, where they had their own rooms and nurseries, and where they were introduced to the outside, walked, and played. In the early years of training, an adult human worked with Sarah, sitting beside her at a work table. When Sarah reached sexual maturity, she, like all other so-housed chimpanzees, became raucous at times, and now the trainer worked with her through the grillwork of her cage. The tasks set for Sarah were very different from those set for others, a chief reason being that Premack was asking a different question: namely, could a chimpanzee learn an invented language and, if so, could human beings come to understand the grammar of the language? One goal was to learn about language itself, and a second was to inquire as to how the ape understands the world. As we can see, the study of ape-human communication has now shifted to fresh ground, to the issue of meaning, and, perhaps, to the standard phenomenological question—What is the nature of the animal's umwelt? It is a question that could not have been asked without the success of the first generation of studies on ape-human communication. At the same time, it is a question that treats gesticulation as a way to communicate, not as an answer regarding the nature of the communication.

David Premack had shown his interest in great ape communication in 1954 when, as a fellow at the Yerkes Laboratories in Orange Park, Florida, he received permission to work with a chimpanzee, Sally. Thus, his work there coincides with the time during which Viki was undergoing her home-rearing. But, by the time the requisite funding was received, Sally had been assigned elsewhere and Premack moved on to Minnesota. That Premack's interest was in the study of language is shown by his own analysis of what he set out to accomplish: "I sought to operationally analyze language, to decompose it into atomic constituents and to provide training procedures for each constituent. The tedium of interminably designing and applying training programs was relieved by two factors: first, by the operational analysis itself, and second, by the challenge of designing nonverbal procedures for assessing the conceptual structure that underlies language."[3]

The Premacks began with the notion that during language training, the most obvious aspects are the relationship of student and teacher. For this reason, the first "words" learned should be those that relate to both: the name, actions (give), food, the pupil's name. The words, however, were not to be spoken words, for the interest was in how language develops, not whether a great ape can cross-translate meaningfully two languages, such as ASL and English. The "words" were symbols of differing shapes and colors that could be attached to a magnetic board by the pupil. Figure 12.1 shows Elizabeth, the chimpanzee, "attaching" or "writing" a request: The top "word," hidden, is "Debbie," the name of the teacher. (Debbie, the teacher wears her name symbol on the pendant around her neck.) The next two symbols are "give" and "Elizabeth." The message is not compete, however, without the direct object ("apple" or "banana").

The training was tedious, in part because the Premacks' interest in language development led them to include concepts such as "if——then," the notion of "some," the idea of "all." All of these concepts are difficult even for human beings to learn. They report that Gussie learned not a word, and other chimpanzees learned their first word only after a thousand trials. The other chimpanzees involved in language training were Peony, Walnut, and Elizabeth, who is shown in Figure 12.1. Other kinds of animals, notably pigeons, have little trouble learning to associate a

Figure 12.1   One of the animals worked with by the Premacks, Elizabeth the chimpanzee, writes to Debby, the human trainer. The top symbol is blocked from our view, but it is "Debby." The next words are "give" and "Elizabeth." Elizabeth may add a word, such as "apple" to write the sentence, "Debby, give Elizabeth apple." To aid recognition, Debby wears a neck chain with the symbol for her name on it.

symbol with a particular behavior, the particular behavior being the amount of pecking at a disk or the symbol itself. Why, then, is the clever and imitative chimpanzee such a slow learner when required to associate a like symbol with the simple motion of placing the symbol on a magnetic board? The question teases, and undoubtedly speaks to the complexity of language and perhaps to the artificiality of our techniques.

In order to examine the chimpanzees' ability to develop such generalities, the Premacks' work involved teaching the chimp to apply concepts, such as "same-different" (Are these two words [symbols] the same, or are they different?); interrogatives (Yes or No?); relationships ("What is the relation between one clothespin and another? Between a clothespin and a pair of scissors?); categories of objects (Is an apple red or round?); imperatives (Is this true?); analogies (If A = B, C = ?, where, for example, *A* = half a banana, *B* = whole banana, *C* = half an apple, *D* = ?) For example, examine Figure 12.2. Here Sarah is being asked whether the two symbols at the top are different or the same. Notice that the symbols for "same" and "different" have some aspects in common, some not. In C and D, Sarah is being asked "What object is the same as [or different from] the object that is shown?" (The technique is a variation on the problem in which the odd member of three is to be chosen, as used by Hamilton and described on pp. 254–261.)

Experiments of this nature permit analysis of *what* is being learned as well as *how* it is being learned. Returning to the question of why the pigeon appears to learn the symbol association more easily and rapidly than does the great ape, the Premacks tell us that "Learning occurs on many levels. An individual's resources must be divided over all the levels,

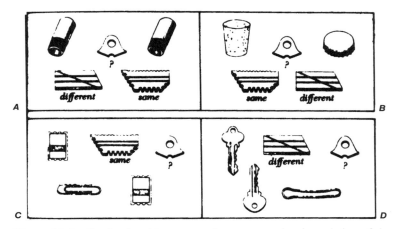

Figure 12.2 Sarah, the chimpanzee, demonstrates her knowledge of the concepts "same" and "different." Quadrants A and B ask, "Are these two objects the same or different?" Sarah decides whether they are alike. In C and D, the question posed is, "What object is the same as (or different from) the object that is shown?" Again, Sarah is to select the object that matches.

and the division need not be equal. With the pigeon, the division favors less abstract levels, so the bird learns mainly to approach color (or whatever else is used in training). Hence matching is learned slowly, and its transfer is weak. The ape, in contrast, devotes its resources to the more abstract levels, and thus learns matching quickly, and its transfer is stronger. The ape not only divided its psychological pie differently, it probably has a larger pie to divide."[4]

The Clever Hans question rushes forward, as it always must when human beings and animals presume to communicate. Premack takes on the question: "Were Sarah and the other subjects merely clever Hanses, or Gretals as the case may be, different only in sex or species. In its response to social cues, the chimpanzee is not remotely like a horse. Clever Han's [sic] trainer was apparently totally unaware that he was cueing the horse. When the chimpanzee tried to use facial cues, however, only the most naive trainer failed to observe it. Chimpanzees and children are alike in this regard: when they do not know the answer to a question and decide to use social cues, they peer visibly into the trainer's face. . . . Of the more than 20 trainers we have had occasion to use in the past 10 years, we only once found a beginning trainer so apparently absorbed in setting out the test items, following the data sheet—and doing all of this without being bitten—that he did not sense the brown eyes intently watching his every move. The first suggestion that something was awry came from the subject's abruptly improved performance. Peony suddenly performed well above chance—on words she failed with the experienced trainers. . . . There is no question but that both chimpanzees and children try to use social cues on occasion. However, I have never seen either do so with sufficient stealth to go undetected by an experienced trainer, and especially not by a second trainer who observes the first trainer and subject."[5]

Schillings could have written the same analysis of his interaction with the horse, Clever Hans. Freud might have said the same about Little Hans, had he ever observed him during the time of the neurosis.

To avoid what he thought of as the Clever Hans problem, Premack devised a "dumb" trainer technique in which information from both the chimpanzee and the human trainer were relayed by line to the third person who, in effect, conducted the session. The reason for using the "dumb trainer" was that Sarah, as is true of other animals described in this book, was interested chiefly in working with trainers she knew. The observation that apes, in particular, are not responsive to strangers in signing experiments may be regarded as a corollary to the Thorndike-Mills controversy regarding the nature of the appropriate environment.

The symbols Sarah learned are shown, as projected on a TV screen, in Figure 12.3. These constitute the lexicon of 130 symbol-words she acquired by 1976. In addition to having Sarah place the symbols on a magnetic board in a grammatical sequence of left-right, up-down, so that sentences were written on the vertical, Premack and his colleagues used several other techniques to show the reproduction of the invented lan-

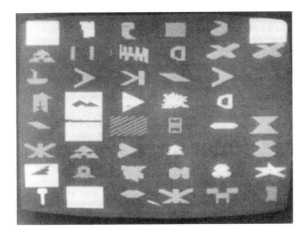

Figure 12.3   Part of the lexicon of 130 words. These simulations of the plastic shapes are photographed from a TV monitor: they are not the plastic shapes themselves.

guage. A joy-stick was employed, a technique used later by other investigators to provide the animal with the opportunity to guide its responses, rather than to merely choose among alternatives. A typewriter—a word-processing computer interface—was also used. The chief form of analysis consisted of measuring correct placements and analyzing usage, such as the number and kind of "sentences" produced. Such matters as the order of words and of sentences and the usage of prepositions show that at least some of the animals exhibited reliable choices. The issue of reliability is critical, for without consistency in choices, the experiments cannot lead to conclusions. The finding of reliability shows that the chimpanzees were making certain discriminations and certain choices in a consistent fashion.

The Premacks' interest in chimpanzee linguistics now shifted to an issue that students of great ape behavior had never before confronted directly; namely, can the animal's linguistic competence be used to tell human beings something about how the chimpanzee understands its world? The question is, of course, a benchmark of the phenomenologist's way of thinking about psychology. What may describe the umwelt? Can we learn anything about the nature of the animal mind by describing how it senses and how it perceives? What can we learn about what is innate and what is given? The Premacks ask by inquiring as to Sarah's sense of intentionality.

SARAH'S INTENTIONS, SARAH'S LIES

The word "intentionality" may be confusing, for it is used in a technical sense in fields of inquiry as diverse as philosophy and ethology. By the word the Premacks mean nothing more unusual than what you and I mean by the term: the attributing of motives, wishes, and desires to others. I may think that Sarah has an intention when she uses the symbols to write "Give me banana." Sarah may think that I have intentions as well; namely, whether or not to give her the banana. Or I may in my mind, or in hers, have the intention of giving her an apple instead. Or I may intend to give her nothing. And she may intend that I intend to do so or not do so. As the Premacks point out, intentionality is not a concept dear to the behavioristic way of thinking. It is, in fact, a notion that the behaviorist approach sees as belonging to that ever-murky soup called The Mind. Intentionality cannot be measured or counted, argues the behaviorist; therefore, it is just as well to discard it, for it is not a useful concept for erecting a science of behavior. It is not, says the behaviorist, that there is no concept called intention any more or less so than there is a concept called ghosts, but to use these words as explanations, is to miss the path, not to mention the highways, to a useful analysis of behavior.

One level of intention may refer to whether a sequence of acts containing a goal is understood. For example, a videotape "depicted a human actor in a cage-like setting, attracted by food [a bunch of bananas] that was out of reach. In one, the actor was shown jumping up and down in a futile attempt to reach bananas suspended overhead; in another, stretching to reach bananas outside the cage on the ground; or, in pushing a box that impeded his path; and last, pushing a box that not only impeded his path but also was laden with heavy cement blocks. These videotaped cases were comparable to actual problems that Köhler[6] gave his apes to solve, except that Sarah was not asked to solve these as her very own problems. She was being asked how her *trainer* could solve these problems when confronted with them."[7]

Sarah's task was to select an ending to a sequence when the videotape was stopped, for she was then shown two photographs. One showed the actor stepping up on the chair, using a rake to capture the errant bananas; or another his taking the cement blocks out of the box. The trainer left Sarah with the pictures, and Sarah rang a bell to recall the trainer whenever she elected. Note that in terms of intentionality, it was not Sarah's intentions that were being investigated, but her perception of the intentions of the trainer. Sarah made two "errors" on this series of twelve videotape problems. The Premacks argue, therefore, that Sarah was able to recognize both the problem and the solution.

One frequent aspect of intentionality, at least in human society, is the combination of lying and deception. Indeed, some people argue that what is characteristic of human society is not language, but deception. Here is the Premack's description of this work: "To find out whether or not animals can lie, we designed a simple test using the four young chimpanzees. It gave them the opportunity to tell the truth in some situations [to lie in others]; to recognize between truth and lies; and to do all this flexibly—as we human beings do. We divide a room into two compartments, separated by a mesh partition, and placed two containers on one side. We put fruit under one of the containers as the chimpanzee watched, then carried the animal to the other side of the partition. Although the ape knew which container was baited, the partition prevented him from reaching it.

"On the side of the partition with the containers, across from the ape, we stationed a trainer who literally did not know which container was baited. If he was able to tell, from the animal's behavior, which container was baited, being a cooperative fellow he shared the food with the ape. These friendly trials were followed by others in which an equally uninformed, but unfriendly, trainer (wearing dark glasses, a bandit's mask, etc.) replaced the cooperative one. He, too, watched the animal: but he, upon correctly determining which container was baited, kept the food for himself, eating it greedily in the ape's presence.

"We then turned the experiment topsy-turvy, placing the chimpanzee in the trainer's position, and vice-versa. Now, it was the *ape* who did not know which container was baited. It was now the trainer who tried to inform the ape. And, while the cooperative trainer pointed to the baited container, the unfriendly trainer pointed to the *unbaited* one."[8]

Four chimpanzees were tested under these arrangements. Two operated as full-fledged liars who directed the trainer incorrectly. One animal was seemingly not able to lie by misdirecting the trainer, yet it understood that the unfriendly trainer could not be trusted to tell the truth. Another animal, though capable of misdirecting the trainer, was not itself misdirected by the trainer. One animal was capable of avoiding the information given by the unfriendly trainer, misdirecting this trainer when the animal knew the location, and using the information gained to give the friendly trainer accurate information. Evidently, this form of intentional behavior admits to a variety of intentions, and lying and deception are among the possibilities.

Among other questions that may be asked about the nature of the chimpanzee's umwelt is how items in its world are represented. It may be recalled that in one test of intentionality, Sarah was asked to examine photographs to decide which depicted appropriately the wanted outcome of action shown on the videotape. Note that, first, the investigators expected that Sarah would process videotape perceptually in a meaningful way; that the dots would be converted according to their location and speed into a sense of action; that the sensation would become a perception. Giving her photographs to select implies that Sarah was able to translate the two-dimensional universe of the flat picture into the three-dimensional universe suggested by the videotape. We recall that Viki and Gua were interested in drawings and photographs, but clear evidence was previously lacking that any ape perceived printed pictures as conveying the nature of meaningful objects.

What words represent is, of course, one way in which beings represent the inner world to the outer. The use of pictures is another form of such representation. The Premacks were, therefore, intrigued by Sarah's visual representations. Generally speaking, the chimpanzee has difficulty matching a photographed image with the original. A picture of a shoe, for example, is not easily matched to the shoe itself. While a human child can do this at 18 months, the chimpanzee is five or six years old before it can do so. Nor does film offer much better opportunities for matching. The Premacks point out that a familiar environment, such as the home or the yard, is not recognized; nor is, for example, a likeness of a well-known trainer. The Premacks suspect that the chimpanzee does not translate the two-dimensional world of the photograph to the three-dimensional world of reality with the ease that human beings appear to enjoy. They built a replica of the chimp's room with models of the furniture and the objects to be found in the original. When objects were placed in containers in the model room, and when the chimp was shown them, the chimp, nonetheless, could not translate the information into news that permitted it to find the objects in the real environment. In short, the animal did not appear to be able to take information from model representation in space to the larger, yet more familiar, template.

Surely all animals, and certainly great apes, must have some kind of spatial map; otherwise, how do they navigate their surroundings? The Premacks then arranged two rooms with identical furniture and showed the chimpanzee the location of a bait in one room, and then led it to the second room, identically arranged, where it found the bait in the identical and appropriate container. If the rooms were exactly alike, the information was transferred. But the Premacks could not be sure that visual transfer had occurred, for it remained possible that the animal merely took the rooms to be one, not two. The ape was then placed in a position to see both rooms, and when the container of one was baited, the animal found the bait in the like place in the second room, thereby suggesting that the notion of equality was present in the ape's judgment.

The size of one room was then reduced by using canvas and introducing smaller but similar furniture. When this new room now approximated the other, the animal showed itself able to transfer information about the bait, but the ability disappeared when the room angles were modified by 45 degrees. It seems likely that, however the chimpanzee builds its mental map of its universe, it is able to call on the model to locate objects to a degree near that of humankind. Nonetheless, these tests of mapping did not show how this map is developed or how it may be used. However the chimpanzee maps its universe, it does so along categories that we human beings have not yet discovered or recognized.

When chimpanzees were asked to show their sense of spatial relationships by placing, in a pictured puzzle, the eye, nose, and mouth of a chimpanzee on its photographic representation, we find the results shown in Figure 12.4 The left picture shows what is described as one of Sarah's more accurate attempts to place the three parts of the phase in position. The right side shows the more typical result from very young children and chimpanzees in which, commonly, the parts are placed along the black-white axis rather than in location on the face itself. The lack of

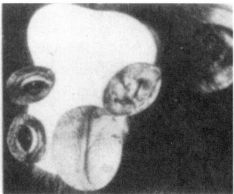

Figure 12.4   The face-puzzle. Human children do well at representing the human face by age four. The left figure shows an example of Sarah's solution of the puzzle. The figure on the right shows how younger children and juvenile chimpanzees tend to solve the problem: they arrange the pieces, seemingly, by darkness and lightness.

ability seems to be not one of perception alone, for the Premacks report that when the chimps are asked to match the facial parts with a flashlight, bell, and flower, they do so in terms of the function. That is, the eye is matched to the flashlight, the ear to the bell, and the flower to the nose.

Can language training itself increase intelligence? It is a strange concept, but one worthy of thought, especially since information gathered later from other chimpanzees suggests that those able to "learn language" are also those able to make humanlike inferences regarding the nature of their environments. It may also be that language training affects the mind in ways that encourage the organism's understanding of representations. Perhaps the training alters the synapses or axons of the brain in some unknown fashion. There is no doubt that Sarah performed far better than the nontrained chimpanzees in the family, but this difference may reflect age, or possibly Sarah's unique abilities, rather than the direct function of training versus none.

We now know from the Premacks' wish to study the development of an invented language, as opposed to the notion of using hand gestures as symbolic words, that the great apes are capable of giving meaning to nouns and gestures, and that those symbols, to a degree yet uncertain, may represent aspects of the mind and thereby of the umwelt of spatial and temporal maps. We may thank Sarah's genius and the ways in which the Premacks phrase their questions for these insights.

Alas, the human and the primate condition have much in common, as well they might. In 1983, the Premacks wrote of Sarah, "Today, at almost twenty years of age, Sarah cannot be taken outdoors for walks or into special rooms for testing. Everything is brought to her. Her cage is connected to an outdoor run that overlooks the pond (in Pennsylvania). Her trainers bring a TV set and phonograph into her room every day;

they prepare her tests and offer her several classes a day, not unlike tutors indulging a special student with private lessons five days a week. Sarah was always constitutionally short-tempered and demanding; with increasing age, she has become even more of a crank."[9]

One hopes that such is not the necessary result of language training.

LANA AND LANA

Through the LANA Project (LANguage Analogue Project) the chimpanzee, Lana, was able to use an artificial language, one that uses a computer keyboard on which lexigrams are placed on the keys. The project was begun by psychologist Duane Rumbaugh with Ernst von Glasersfeld, a psycholinguist who invented the lexigram language, called "Yerkish," Pier Pisani, who designed the computer programs, and Harold Warner, a biomedical engineer. The scope of talents indicates the difficulty inherent in reaching the goals of the project. The effort received its impetus from the Gardners's work but proceeded along different lines; namely, it used an invented language represented on a keyboard. Use of the keyboard is, of course, a modern way of representing Hans's taps; the computer keyboard is a modern reinvention of Krall's "alphabet-writing blackboard" or the horse Lady Wonder's "horsetypewriter." The system differs from earlier systems in eliminating direct human interaction between question and answer, something that was always present for Hans and Lady Wonder. Having the benefit of the Gardners's experiences, Rumbaugh stipulated the goals of project LANA as twofold: "first to determine whether or not chimpanzees could acquire and use complex grammatical linguistic skills, and second, to find out if modern technology and computer science might be advantageously employed to aid in the acquisition and maintenance of such skills."[10]

Lana was born at the Yerkes Center in 1970 and was just over two years old when she began language training. She was provided with an experimental chamber (shown in Figures 12.5 and 12.6) that contained space, a view toward windows, a keyboard, and instrumentation to allow a window view, films, slides, music, food, drink, and a hopper that could deliver objects. Figure 12.7 shows Lana at work in her chamber. In the early stages of the project, Lana had a companion, an orangutan named Biji, whom she had known previously. Biji and Lana were expected to be social companions, but when Biji proved a distraction rather than a helpmate, the companionship was terminated.

The first question, of course, was whether Lana would learn to use the computer keyboard. If she could, the second question was whether she would learn to use the lexigrams to produce what are called stock sentences—the sort that can be used to request nourishment and entertainment. Because the computer system could be in place every hour of the day, the need for attention from teachers was eliminated, thereby also eliminating the possibility of teachers and trainers giving unconscious

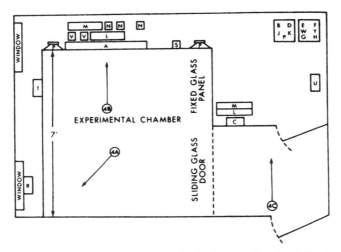

Figure 12.5   Lana's chamber. 4b the keyboards, visual display, and vending devices; 4c, the computer, rack, and the experimenter's keyboard; M, the "receive" component of the projectors; R, the window view; S, the object hopper; T, the slides; V, drink.

cues to Lana. The system appears to eliminate the human cues that the several horses used to control their taps. Yet we must remain wary about the nature of ape-human interaction, just as in other human-animal situations we have already noted, such as with von Osten's and Krall's horses, the stableboy in Krall's studies, and the use of "pencil-writing" by Lady Wonder's owner. The use of a keyboard also meant that all exchanges could be recorded for future examination. Moreover, while Lana's remarks were being recorded for future use, they were also being produced without a human in Lana's presence. Furthermore, Rumbaugh and Premack appear to think alike in terms of the Mills-Thorndike controversy: both clearly favor control.

A visual language was chosen, for this sensory system had been crucial in both Washoe's and Sarah's learning. Rumbaugh acquired a lexical language that, at the outset, used nine colors to stand for a like number of "subjects": autonomous actors, spatial objects, spatial concepts, ingestibles, parts of the body, states and conditions, activities, prepositions, affirmation, and sentential modifiers. Nine design elements were then added to yield the elements and examples of lexigrams. The mutations produced 255 individual lexigrams, known as the language Yerkish. Each has a single meaning, and each meaning corresponds to an English word. What Rumbaugh calls the conceptual lexigram classes include, for example, under the category "autonomous actors," familiar primates, unfamiliar primates, nonprimates, and inanimate actors (such as the keyboard). Under "parts of body" are included ear, eye, foot, hand, mouth, and nose. Under "edible units" are M&M candies, nuts, raisins and so on. Under

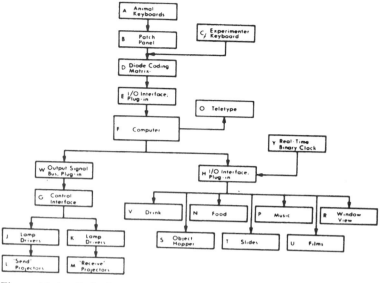

Figure 12.6   Code for the arrangement of Lana's chamber.

Figure 12.7   Lana the chimpanzee in her chamber pushing the symbols.

"drinks" are coffee, coke, juice, milk, and water. In addition, there are "directional prepositions" (out-of); "additive conjunctions"; "similarity/difference markers" (same as, different from); "quantitatives," such as all or none; "comparatives," such as less and more, and the question, a "?"

Yerkish has a grammar. It is not a language in which the human "fills" in missing words, supplies prepositions or directives, or, probably most important, judges word order to be inconsequential, as was true of earlier languages, such as ASL, taught the great apes. The same order could have different meanings, depending on intention. For example,

"Tim move into room."   = Indicative statement
"? Tim move into room."   = Interrogative (query)
"Please Tim move into room." = Imperative (request)
"No Tim move into room."   = Negation

Tim is Timothy V. Gill, initially a trainer; later, "deviser of strategies." Gill and Rumbaugh, while describing the way in which Lana learned and was instructed, tell us that it was not until after Biji was removed that Lana became interested in the slides, movies, and music. Lana "turned these on" by requesting that the appropriate window be raised to give her access.[11]

The first experiments offered Lana three-word-long sentences. Two of each trio could be valid because Lana could build on them. For example, "Please machine give" is valid because Lana could add an object to it, such as "milk" or "apple." A second valid statement was "Please machine make," this being a statement that permitted Lana to add "music" or "movie." An invalid sentence was "Please window Tim," or "Please make machine." The former was invalid because it is a random set of otherwise useful words (The resemblance to Washoe and Koko-like communication may be noted.) The latter is invalid because it is an inversion of an otherwise valid statement. Unlike the work with the great apes using ASL in which grammar is assumed by the human listener, here the grammar is given by the human experimenter and it is the ape who must acquire the human-devised grammar. Rumbaugh and Gill report that, at a 90 percent level of accuracy, Lana came to reject the invalid "sentences" by pressing the period key. Lana also learned, at the same high level of success, to recognize each of nine "stock" sentences.

"Yes" and "No" were learned by reference to specific situations. Examples were "Window open?" (Yes or No.) "Door shut?" (Yes or No.) Lana generalized Yes and No. She first used "No" when she saw someone drinking a Coca-cola. She became angry and hit the "No" key. She also used "No" to indicate to Gill when the machine failed to produce the desired reward. The first conversation was also about a drink. On this occasion, she and Gill discussed the likelihood of his sharing his drink with her.[12]

Consider the following dialogue:

TIM: ?What color of this.   [10.10 A.M.]
LANA: Color of this orange.   [10.11 A.M.]

TIM: Yes.

LANA: ?Tim give cup which-is red? [This was probably an attempt to request the orange; however, because a red cup was part of her object/color naming materials, Tim responded with the latter object.]

TIM: Yes. (Thereupon he gave her the cup, which she discarded.)   [10.14 A.M.]

LANA: ?Tim give which-is shut.   [10.16 A.M.]
?Shelley give.

TIM: No Shelley.   [10.16 A.M.]

LANA: Eye. (A frank error, probably.)   [10.16 A.M.]
?Tim give which-is orange.   [10.21 A.M.]

TIM: What which-is orange?   [10.21 A.M.]

LANA ?Tim give apple which-is green. [At this point, Lana frequently confused the keys for the colors orange and green.]

TIM: No apple which-is green.
[In other words, "I have no green apple to give."]

LANA: ?Tim give apple which-is orange. [Thereupon she bounded with apparent enthusiasm to the door to receive "the orange-colored apple."]   [10.23 A.M.]

TIM: Yes. [And he gave it to her.]   [10.23 A.M.]

Rumbaugh summarizes: "An important development in the experiments reported here was Lana's acquisition of many critical linguistic-type skills for which she had received no specific training. Unquestionably she would never have acquired these skills spontaneously had she not received prior, very specific training in certain language fundamentals; nevertheless, having received such training, she showed a readiness to expand her ability to use linguistic-type communication in a number of significant, and, to her, novel directions. . . . Lana's performances on occasion involved transferring information learned in one situation to another, and generalizing skills learned in limited contexts to deal with broader problems.

"The first instance of Lana's spontaneous acquisition of linguistic-type skills was the most crucial—her learning to identify lexigrams. . . . A second example of the spontaneous acquisition of skills on Lana's part was her use of names . . . she began spontaneously requesting the names of objects for which she had no name. Alternately, she had 'invented' her own names for objects by combining lexigrams already in her vocabulary in novel ways. In addition, she spontaneously began to use 'this' as a pronoun to refer to objects whose names she had not learned. . . . Lana moved beyond the 'stock sentence' phase and began composing entirely novel and appropriate sentences on her own. Finally, it was Lana herself who initiated the practice of conversation with us."[13]

We cannot yet say with certainty whether Lana has acquired language, in large part because we human beings are not yet certain of our definition of language. If we are egotistical enough to define language as what human beings do, not surprisingly, we will conclude that other animals do not have language. If we define language to be gestures that communicate, then the simplest movement that indicates reproduction is language. There is no language without accompanying motor movement, and it is this movement, whether it be the movement of the glottals, the motion of the hand, or the pressing of the key by chimpanzee or pigeon, that we measure and take to substitute for language.

Savage and Rumbaugh conclude: "To place the LANA Project and the other chimpanzee language studies in proper perspective, the findings should be related to primate nonverbal skills, on the one hand, and to human linguistic skills, on the other hand. We must look to see what an animal who does not typically acquire language, when given some language skills, can tell us about the relationship which exists between these two forms of communication, just as Lana, herself, has learned to span the gulf which, until now, has separated nonverbal from verbal creatures, so must we, by careful study, try to understand the nature and implications of this bridge she has structured."[14]

SHERMAN AND AUSTIN

As so often happens in research regarding living beings, the answer to the original question never quite surfaces in a "Yes" or "No" form. Rather, the answer leads to another, more refined, question, this one operating at yet a different level. Perhaps it is for this reason that research on behavior seems never-ending, in the sense that seemingly simple and straightforward questions—such as "Do apes have language?"—are never answered simply, in the way the question "At what speed does a body fall?" can be. Investigations regarding behavior seem to evolve: one idea, like a species or a specialized piece of morphology, is tried out, works for a time, and then becomes useless or of negative value. This is why there is a psychology to the questions we ask; why the nature of the questions we ask, as we have seen from Itard forward in time, tells us much about the state of the mind. The speed of falling bodies presumably does not change over years and eons; but the nature of the psychological questions we ask does, because the asker is itself a living, changing being.

Another idea, usually seen as unimportant, now suddenly seems to gain value; data thought to bear on one question are later seen to be relevant to another. This process of intellectual evolution leaves the impression of a science of fits and starts, especially when the historian tries to show the development of logical thought, of one idea leading univocally to the next. The history of behavioral science and intellectual thought cannot be understood as a lock-step process in which hypothesis leads to a test of confirmation of the results which leads, then, to a new hypothesis. Our knowledge of ourselves and the other animals of our planet has no simple trajectory: our knowledge comes through myth, through encompassing ideas that hold together the fragments of our understanding. The question "Do apes have languages?" is not forgotten, but it can now be seen to be too simplistic to be answered. It is not a question capable of scientific answer, but a question whose answer depends on the definitions of the times. All depends, of course, on what is meant by language. This is not to say that the questions of behavioral science can be reduced to questions of semantics. It does mean that many of the questions we think we are asking about behavior, though seemingly sensible, are not. They are projections of our own psychologies. The nature of the question, then, tells us much about the workings, conceptualizations, and assumptions of *our* minds.

When the ideas and findings produced by the Gardners, the Premacks, Terrace, and Rumbaugh are given time to settle and let us see more clearly what they and their chimpanzee companions can tell us, we can see that there is more order than we expected. Sue Savage-Rumbaugh, who had been involved with Project LANA, has pointed out that the chimpanzees demonstrate three initial "atomic" skills: requesting, naming, and comprehending of communication.[15] Whether through sign language or lexigram-learning, the chimpanzee first asks for a wanted item; then gives such items "names"; and finally uses the name to communicate to another being. When the question is put—Can we devise means to examine how each of these aspects is being learned?—we begin to understand what it is that the ape is able and willing to learn. Similarly, to ask the seemingly important question "Is ape language like the language of children?" leads us to search for an answer that can detract from our wish to understand ape language. Comparisons, such as that between ape and child, are of course useful: it is the practice of finding and measuring such comparisons that dominated the attempts to construct the Mental Ladder.

Whether or not ape-thinking follows the same development as child-thinking may tell us much about the rungs on the Mental Ladder, but the correct placement of these rungs does not necessarily tell us anything about the composition of each. Comparing apes to children does not always tell us anything about either, although it may reveal something about the relationship between the two. Remember that Gua, Viki, Washoe, and Nim were raised much as if they were human children. Koko, Sarah, and Lana were raised as captive animals with a large degree of human involvement in their environments.

An additional similarity, a subtle one, shapes the course of research by shaping the sort of question asked. Savage-Rumbaugh notes that "the Gardners, Premack, Rumbaugh, Patterson, and Terrace all began their work with the assumption that when an ape 'correctly' used a symbol, it had some referent clearly in mind. These researchers were of the opinion that when a chimpanzee learned an associative connection between a displayed object and a particular behavior, the chimpanzee was, in fact, *naming*. Naming was viewed as a rather simple skill and thought to be readily accounted for by known principles of conditional discrimination learning. Thus, the real focus of these projects was the combination of symbols that the apes produced and whether or not these combinations constituted evidence of syntactical competence."[16] Savage-Rumbaugh believes that naming is not a simple process and that, in fact, the key to recognizing what the ape knows is to determine how it comes to name.

"My aim," she writes later, "in comparing symbol use by children and apes is to cast the study of language acquisition in apes in a new light. At the very least, it should be clear why it is not fruitful to pursue such questions as 'Do apes have language?' and 'What do they have to say to human beings?' Instead, we must ask (a) how the processes of symbolic and nonsymbolic communication between two organisms differ; (b) whether the symbolic system appears *de nouveau* simply by learning which words 'go with' what things, or whether it somehow emerges out of the nonverbal system; (c) what actually occurs when a creature who does not engage in symbolic communication with conspecifics is taught to use symbols with members of another species; (d) whether it is possible to determine objectively what these symbols actually represent to chimpanzees; and (e) why turn-taking is important."[17]

Savage-Rumbaugh designed experiments in the hope of answering these self-imposed questions. She had the benefit of firsthand knowledge regarding the LANA Project and access to the equipment Lana used. Two chimpanzees, Sherman and Austin, born a year apart, became part of the new project, which was called the Animal Model Project. The chimpanzees are shown in Figure 12.8. The two animals are of somewhat different temperaments and, because of the age at which they began work with the study, one is larger and physically dominant. They were raised in the laboratories but interacted a great deal with both apes and people. They were included in the laboratory routine: they participated in preparing food, cleaning, and answering the phone. They watched TV, got tickled,

Figure 12.8   Austin gives a hug to a smiling Sherman.

drew and painted, walked, and explored.[18] Figure 12.9 shows Austin replacing a kitchen utensil.

The design of the keyboard used by Lana had undergone changes over the years of work, but the principle of using a lexigram to stand for a concept or noun was retained. In a separate experiment designed to determine whether the apes now understood spoken English, the apes were asked, either by lexigram or by spoken English, to select a named item. Both Sherman and Austin were correct sixteen out of sixteen occasions when the lexigrams were used, while in response to spoken English, Sherman was correct two of fourteen times, and Austin only one of fifteen. Because there were sixteen items, these frequencies reported for spoken English are at chance level, yielding a conclusion different from that offered by Patterson regarding Koko's ability to hear spelled words.

Since a first priority of the research program was to uncover how the apes learned the association of lexigram and word, the teacher sat with each subject (there were four subjects at this time), giving a food reward and praise when a single key was associated. The only task was to associate a single key with a single meaning. Then a second symbol was introduced (and varied with the first), and then a third. The introduction of the third symbol produced learning deficits; that is, the animals now had more trouble with the first and second, these being lexigrams seemingly learned already.

Savage-Rumbaugh interprets these data to mean that the learning of the second lexigram went seemingly rapidly because the ape was following a strategy; namely, if the teacher shows A, pick A (that is, *match* the stimulus); but when the teacher does anything else, select lexigram B. With the introduction of a third symbol, the strategy became inoperable, and the chimps now turned their attention from the keyboard to watching the teacher. It would seem that learning to associate a symbol or

Figure 12.9 Austin putting in place kitchen utensils.

gesture with a noun or meaning is not a simple task for a chimpanzee. If a chimpanzee does learn these associations rapidly, it would be wise, these data suggest, to look for other sources, such as the teacher, as the source of the animals' strategy. These four animals, including Sherman and Austin, became dunces after two "associations" were learned. Despite months of additional training, the degree of learning was nothing like that reported for other chimpanzees, although these chimpanzees learned associations in different ways.

Savage-Rumbaugh explains: "The answer emerged from analysis of videotapes of our training sessions. These showed that there was only one aspect of our training situation that seemed to be salient from the chimpanzee's perspective. This was the teacher's decision to give, or not to give, the food in her possession after a key was lit. Instead of attending to the relationship between the object the teacher held up and the symbol they then lit, our chimpanzees were attending to the relationship between the symbol they lit and the ensuing event—i.e., whether or not the teacher gave food to them. They treated the object display as simply a discriminative cue for the onset of the behavior of symbol selection, while they essentially ignored the type of object the teacher displayed. Their behaviors suggested that they thought the teacher's decision to call a response 'correct' might have something to do with *how* they lit the symbol, how they treated the teacher, etc. as opposed to *which* symbol they lit."[19]

There is an important lesson here to be learned and remembered. It is a lesson that would have saved Lubbock and others from false conclusions had they learned it: when an experimenter establishes a task involving operant conditioning, the experimenter must know precisely *what* is being associated with *what*. Merely saying that I want the animal to associate the key with a certain behavior does not guarantee that this is the association being made. The animal may well learn to associate a look, a movement, opened eyes, speech, a twitch. The Cemetery of Misinformation is replete with experiments and experimenters whose arrogance has lead them to believe that they, not the animal, are in charge of the associations being made. The judge is always the subject, not the experimenter or teacher. All living beings are Clever Hanses, Vans, and Rogers, for all of us are "clever" enough to come to make associations, whether or not these be the one that the experimenter working on us had in mind.

Savage-Rumbaugh's perception enabled her to redesign the work. For example, the keyboard was redesigned so that the ape now selected the reinforcement it wanted. In this way, the animal could not make an error. The same was then done with food. As a result, the animals rapidly learned to distinguish the four food symbols from one another as well as from the other lexigrams. The original, rote learning, did not work, in the sense that the apes did not learn to do what the people wanted them to learn. Savage-Rumbaugh summarizes: "An obvious implication of these failures is that associations *between particular symbols and particular objects* are not readily acquired by apes."[20] The question we are led to ask is, *What,* then, were Washoe, Koko, Nim, and perhaps Sarah and Lana, learning? *What* aspects were being associated?

It would be comforting if we knew that a definitive answer to this question was possible. Savage-Rumbaugh considers one possible answer in respect to most of the studies described, but it is useless to do more than suggest possibilities. None of the project's happenings was videotaped[21] fully from the beginning, so different projects used very different means to produce very different behavior. A gestural sign is not, after all, the same act as placing plastic on a magnetic board or punching lexigrams on a keyboard. In addition, those who invest years of their lives examining important and valuable research questions develop a loyalty to their work and to the animals and people with whom they are associated. One lesson to be learned from von Osten and Lubbock, as just two examples, is that such investment brings with it pride and a loss in objectivity. It is to Savage-Rumbaugh's credit that she understood that the *failure* to learn by rote was of greater potential importance to our understanding than redemonstrating the typical learning curve proposed by Thorndike or labeling associations as language-learning.

APES TEACHING PEOPLE

"During one training session, instead of gesturing for the experimenter to go to the refrigerator and take some food out when the bowls became empty, Erika [one of the four] lit the symbol 'orange' and stared directly at the refrigerator. Noting this, the experimenter selected orange slices from the refrigerator and loaded them into the dispenser. As soon as Erika had finished requesting and consuming the orange slices, she again looked toward the refrigerator and this time said 'banana.' Again the experimenter fulfilled the request by selecting bananas from the refrigerator and placing them in the dispenser."[22]

Notice that, although anecdotal, the suggestion is that the ape is communicating a request to a human being that involves the person's completing an action requested by the chimpanzee. The distinction was now encoded between asking an inanimate object (the machine) to provide the wanted substance and the animate object (a person). ("PLEASE MACHINE GIVE BEANCAKE," "PLEASE SUE GIVE BEANCAKE.") Once again, as was true of Clever Hans and, no doubt, of the other animals described, had we data on errors and information, that kind of error tells us something of value about the process of association. In this situation, when the ape made an error, he or she would attempt to correct it chiefly by substituting a different kind of food, rather than by correcting the name of the person or shifting from asking the machine to asking the person.

Requesting and naming, the first two of the three processes that guided these investigations, now appear to be clarified, and there was an indication that the third, comprehending, could be investigated. "We were encouraged," Savage-Rumbaugh writes, "with the progress that the chimpanzees had made in learning to distinguish the request and labeling functions of a symbol. However, we still had not achieved behaviors characteristic of true speech episodes in which a listener and a speaker use symbols to control and coordinate each other's behavior in meaningful rule-bound exchanges. Such exchanges are an essential feature of human languages." [23]

"Given the limited vocabulary available to Sherman and Austin at this point in their training (food names, tickle, out, blanket, and key), the most straightforward way of assessing receptive competence was to turn the request task around and ask them to give their teachers the items they had requested from us. . . . As far as we can tell, no one else working with apes has reported giving such a simple and straightforward test of comprehension to their subjects, although it has been successfully employed with children as young as 13 months of age." [24]

The teacher sat with three foods between person and chimp. The teacher then requested one of the foods by using the keyboard. At first, the ape looked at the teacher, repeated the request, and then waited for the food to be given. The ape did not appear to understand the human request as a request for ape-action. A coin was introduced, offered as a reward for providing the teacher with food, which the chimpanzee could use to procure something for itself: "once Sherman and Austin caught on to the basic idea, they handed us food more rapidly than we could eat it." [25]

In this way, Sherman and Austin came to satisfy the three goals of Savage-Rumbaugh's work: clarification of the acts of requesting, naming, and comprehending. The last of these, it would seem, was only in its suggestive phase: it remained to be seen whether the use of symbols would yield clearer evidence of the kind of communication possible between ape and ape, between ape and human being.

COMMUNICATION AMONG PRIMATES

It will be recalled that Roger Fouts had reported the successful use of ASL *among* chimpanzees. It would be pleasant to report that Austin and Sherman promptly began to communicate with one another, but they did not. In Savage-Rumbaugh's work, either Sherman or Austin accompanies teacher A to the refrigerator and helps to hide food inside a container. The food (or drink) was one of eleven possible substances. Teacher A then asks, say, Austin to take the container to teacher B, who does not know what food is in the container. Teacher B then takes both Austin and food to the keyboard at which awaits Sherman, who does not know the nature of the food either. On the first such occasion, Austin (the knower) went with the teacher to Sherman and the keyboard. Surprisingly, even though he had no training, Sherman used the keyboard to request the food: "During the first eight to ten trials, it seemed that the informer was recalling the food he had seen placed in the container and requesting it for himself. It did not appear that the informer knew that the listener needed his special information. The uninformed chimpanzee, however, was taking advantage of the information, which was coincidentally transmitted by the informer. Once each chimpanzee had been in both roles several times, the informer chimpanzee began to realize that the other chimpanzee needed information before he could use the keyboard and also that unless the other chimpanzee did use the keyboard, neither of them received any food. Comprehension of these contingencies was displayed by both chimpanzees during trials on which one or both of them erred."[26] However suggestive this form of communication may be of true communication between the apes, the communication is contaminated by the activity and presence of the human teacher. The question now to be asked is whether Sherman and Austin can bring such communication under their own control.

Primates, at least nonhuman primates, are not distinguished by their willingness to share food. Indeed, they are notorious for not sharing food even with offspring. Nonetheless, Sherman and Austin were taught to give food to one another. Neither they nor the teachers found this to be an easy lesson to learn or to teach. The goal, of course, was to use such food sharing when intra-ape communication was examined. When an attempt was made to encourage the two to communicate in a single room, the results were discouraging. The videotapes suggested that the animals were not attending to one another and that when a message was given, the intended receiver failed to notice the request. It was therefore decided to use two rooms. The decision was a fortunate one, for now the animals began to attend to one another through a window, as shown in Figures 12.10 and 12.11. During a seven-month period, a total of 523 exchanges of food took place. For Sherman, the percentage of cor-

Figure 12.10   "Sherman watches as Austin requests food at the keyboard."

Figure 12.11   "Austin passes the requested food through to Sherman."

rect food requests was between 84 and 100 percent during each month; for Austin, the like percentage was from 82 to 93 percent.[27]

Sherman and Austin now began to invent lexigrams for newly introduced foods, such as *strawberry drink, lemonade, pineapple,* and *pudding.* New terms, such as *bowl, box,* and *lock,* were added, and, Savage-Rumbaugh reports, as long as no more than three were introduced, the chimpanzees were able to select a new lexigram in association. The inquiry into the chimpanzees' comprehension ability now moved to the use of tools. The initial tools and their uses were a key (to unlock padlocks), money (for use in vending machines), a straw (to be used to obtain liquids), sticks (for dipping into food), sponge (to soak liquids, to be squeezed into cylinders), and wrenches (to unscrew bolts.) The area was baited with food, and when the ape was interested, the teacher demonstrated the use of the tool. The name of the tool was named on the keyboard. The apes then could use the keyboard to request the tool, thus permitting Sherman and Austin to name tools. All this activity, of course, was used to investigate how and whether the two could learn to request the tools of one another. Figure 12.12 shows one such example.

Savage-Rumbaugh reports that, along with this learning to communicate with one another, the two chimpanzees began to attend to one another more and more frequently. "Consequently," she writes, "their communication assumed an increasingly human countenance, not by virtue of specific training, but by virtue of use. Moreover (with the exception of turn-taking), we were not teaching or shaping these regulatory skills in any prescribed way; rather, they were apparently being developed because they were indispensable to the smooth flow of communication."[28] Savage-Rumbaugh suggests that to ask whether or not apes have language is to refine what the chimpanzee does have, and what it can do.

A

B

C

D

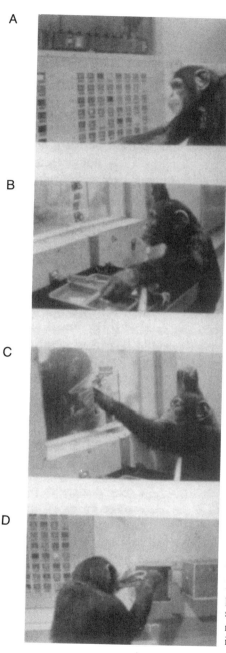

Figure 12.12 A: "Sherman says "Give key."

B: Austin, observing Sherman's request through the window, reaches for the key.

C: Sherman approaches the window, and Austin gives him the key.

D: Sherman takes the key to the padlocked food site and, holding the small padlock between the index and middle fingers of his left hand, starts to insert the key.

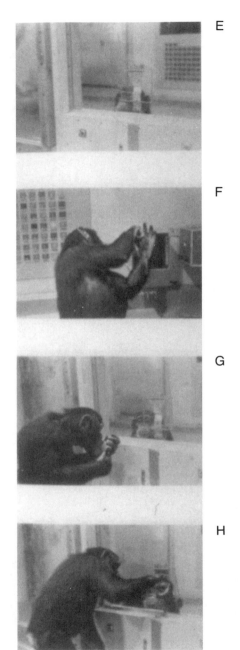

E: Austin watches while Sherman attempts to open the key site.

F: Turning himself, as well as his arm, Sherman manages to unlock the padlock.

G: Sherman brings a container of the pudding (which he found in the food site) back to Austin, sampling Austin's share along the way.

H: Sherman hands the container of pudding to Austin.

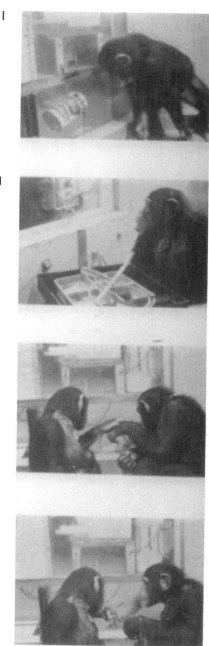

I: Austin watches as the straw site is filled with juice.

J: Austin (left) after asking for straw at the keyboard, goes over to Sherman. Sherman is still looking at the keyboard to see which tool Austin needs.

K: After giving Austin the straw, Sherman goes into Austin's room where they take turns sipping through the straw. Sherman is drinking, and Austin is gesturing for Sherman to give him a sip.

L: Sherman complies, giving Austin a turn at the juice.

M: Austin then gives Sherman another sip.

N: Austin watches as food is placed in the wrench site.

O: Austin asks for the wrench, then goes to the window where Sherman has already selected the needed tool and is preparing to pass it through the window to Austin.

P: Sherman slips the wrench to Austin.

Q

R

S

T

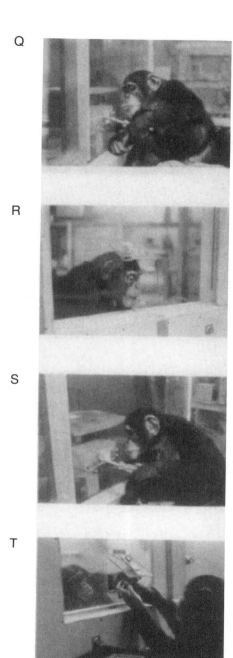

Q: Austin studiously attempts to fit it over the tight bolt.

R: Sherman watches from the side as Austin tries to open the wrench site.

S: Austin carefully turns the wrench to loosen the bolt.

T: After obtaining M&Ms and bananas from the wrench site, Austin passes an M&M through the window to Sherman.

"The crux of the question," she concludes, "is, when animals use symbols in ways similar to humans' use of words, do they know and mean what they are saying? For animals to know and to mean what they are 'saying,' the process of representation must be extant and operating. . . . Sherman and Austin gave strong behavioral evidence that, at a minimum, a significant portion of their symbols came to represent items not necessarily present."[29]

What does such "behavioral evidence" tell us? Consider another experiment. Skinner and his colleagues[30] trained two pigeons. The first was trained to peck at one of three keys marked with the letters R, Y, or G. The choice served as the indicator for the second pigeon to select, by pecking, a key that was either red, yellow, or green. (Notice the human-oriented relationship between R, Y, G and the color of the keys.) Both pigeons received a food reward when the second selected the "right" color.[30]

Is this communication?

Savage-Rumbaugh thinks not, at least not in the same sense that she expects Sherman and Austin to communicate. To thrust aside the behaviorist's well-directed digs, she is forced into an argument based on consciousness and awareness. "It is important," she writes, "to recognize that the procedure used by Epstein et al. [the Skinner experiment on pigeons] does not require that either pigeon see the other at any point or that either pigeon realize that their joint behaviors produce their 'shared' consequence. Indeed, their consequences are shared only in the sense that they have occurred simultaneously. The pigeons are not aware of the shared consequences of their joint actions or that reinforcement is linked to their inter-individual performances."[31] It is, of course, exactly this kind of defense that pleases the behavioristic mode of thinking, for it points directly to the murkiness that results when "awareness" comes to have explanatory power. The behaviorist trap springs, tautly.

Once the question of ape-human communication is phrased in a way that can be answered, as opposed to the simplistic "Do apes have language?" we are able to determine just what it is that the chimpanzee is able and willing to do. When the question is put properly, other forms of life can be asked to demonstrate their knowledge. A male pygmy chimpanzee, Kanzi, a rare species, is reported by Savage-Rumbaugh to learn far more spontaneously than the other subspecies of chimpanzees tested. Kanzi reportedly generated his own rules of grammar by combining a gesture with a lexigram by, for example, selecting the lexigram for "dog" and the gesture for "go" and then going to an area housing dogs to play with them.[32]

An initial report on chimpanzee language comprehension in 1993 was based on experimental work with Kanzi and a human child, Alia, in which the development of their language comprehension was compared. Each was tested by being asked to manipulate lexigrams in response to human speech involving 660 novel sentences ("Can you make the bunny eat the sweet potato?"). The results show differences between child and ape in syntactic structure, but also show similar development of comprehension of human speech.[33]

On another front, T. Matsuzawa, working with the chimpanzee, Ai, has shown that this ape is capable of simple arithmetic, naming colors, and combining words to form evidently new meanings. Ai came to recognize symbols, called graphemes, presented by a computer screen.[34]

The answer to whether human and ape can communicate is "Yes." In the sense that human beings are able to observe gestures reliably given and nonrandom symbol usage, we can say, by definition, that apes communicate. What these animals communicate, in the sense of the nature of the meaning and feeling that is being communicated, remains unclear and uncertain. At the outset, we expected these animals to communicate with us in our language. That the ape has no grammar that we know does not mean that no grammar exists, but that we have yet to understand the rules of expression. Similarly, that the apes can learn an artificial language, one invented by human beings, does not mean that the ape has no language of its own, but that it can learn one we invent.

We still do not know the umwelt of these animals. We have not been able to translate the ape's understanding of its environment to imagine the world of the ape. We do not know the meaning of the ape gestures or language. To so know, we must learn to ask the right questions of the animals, and they of us. Nor do we know very much about the mind of the ape, in part because we have spent our time asking whether communication is possible, not what is being communicated. We have only begun to ask the question from the phenomenological perspective; namely, what kind of world does the animal experience? Our task now is to learn the language of the animal in order to understand its perceptions. Then we may communicate.

The mind of the ape appears to be more approachable now than it has ever been before. The discovery and evaluation, the refining and discarding of ideas and procedures, inexorably lead to better methodology. If foot or hoof tapping is not the technique, sign language may be; if not, then a language of lexigrams might be. What is promising in the quest to make noisy the silent mind is the steady refinement of techniques, the progress in understanding the scientific method as applied to the mind and behavior. But method does not appear in isolation; it is always prompted by the questions one asks and wants answered. A behavioristic question is different from a phenomenological question, which is different from a psychoanalytic question. Each way of understanding has a different language, different suppositions, different measures of truth. We have not yet decided what we want to know about the mind of others. This indecision is a simple reflection of the ways in which we human beings create and thereby understand our universe.[35]

What have we learned about how we construct our universe in order to include what we believe to be in the minds of others? The next, and last, part begins the search for a way to answer that question.

Principles and Myths

PART V

"But man, proud man,
Dressed in a little brief authority,
Most ignorant of what he's most assured,
His glassy essence, like an angry ape,
Plays such fantastic tricks before high heaven
As make the angels weep."

Shakespeare, *Measure for Measure*
(II,2,117)

What Feral Children   **13**
and Clever Animals Tell Us

EACH OF OUR minds knows it is alone in a universe it creates; yet in our attempts to communicate, among ourselves and other species, we reveal a struggle to know other minds, to overcome the isolation of our own mind. This book is a chronicle of some attempts to grasp the workings of other minds. If the concept of death and the ability to make tools are not what distinguishes human beings from other beings, perhaps what does is the quest to understand the minds of others. It is a human belief that, through language, we can both transfer the contents of our mind to another and come to experience the contents of the minds of others. The examples reported in this book are among the more elaborate attempts to do so. Each attempt is prompted by a wish common to all of us, by our everyday concern to discover what goes on in the mind of another. The question is not *whether* we communicate. Of course, we do: your signs and gestures tell me much about your feelings, emotions, and intentions. But they do not tell me as much as I would like to know, so I find myself constantly searching for ways to discover the nature of your intentions and your meanings. I guess these only through my own private world, as has been documented by the accounts contained in this book. I can understand your mind only through the filters of my mind, for my mind is both the coder and decoder of your communications.

While I have been writing to you concerning a few of humankind's more notable attempts to understand the minds of others, you will have come to your own conclusions about humankind's successes and failures in regard to these attempts. Your interpretation of the contents of this book will depend much on the kinds of questions you want to hear asked. For example, if you want to know whether animals have language, you will be frustrated by the lack of a clear answer to this question, although you will understand by now that I suggest it to be the wrong question. If you entered our communicative relationship wanting evidence that horses and dogs communicate to people, you will not be disappointed, but you may be discouraged to find that clear, unambiguous evidence is yet lacking.

POSSIBLE PRINCIPLES

I wish to suggest the following notions as among the ideas to be gleaned from the examples described in this book of human attempts to make public the silent mind:

A proposed difference between physical and behavioral science.

The relationship of "experimenter" to "subject."

What we might make of the modern version of the Thorndike-Mills controversy—that the laboratory not only exploits but also dampens and extinguishes understanding of the meaningful and communicative behavior of living beings.

Whether we know if knowledge is innate or constructed from experience.

The appropriateness of comparing species and genera by the image of a Mental Ladder, in our effort to know the essence of what is "human" (and therefore what sets us apart from the animals).

There are other principles, questions, and generalities. Some the reader will have discovered independently of what I have intended.

Physical and Behavioral Science

Consider the nature of the science of behavior as understood and practiced by Itard and those who followed his example. Itard's approach to understanding Victor (1) emphasized detailed work on the individual, rather than measurement and description in terms of norms and averages; and (2) assumed that the chief way to gather information was to concentrate on what the individual could learn.

By the first, I mean that Itard's approach, this general way of doing behavioral science, gathers information by investigating unique situations. It is not interested in gathering data on samples that are then described by using deductive or inductive statistical measurements. The interest is in the unique case rather than in the mean of a sample. The approach is also characteristic of phenomenology and psychoanalysis. These practices, too, concentrate on the individual, assuming that whatever is found there may be made general to all humanity. Much of modern behaviorism also concentrates on detailed work on one individual in preference to the descriptions of groups, on the grounds that an effect, once properly demonstrated by the unique individual, will be found to be true of all. Logically, the concentration on the individual is separate from concentration on the unique case. However, Itard was not, at first, interested in Victor merely because he was unique, but because his specific kind of uniqueness encouraged experimentation and observation that could not be done on others. Victor's uniqueness was, of course, the way in which he was presumed to have been reared.

But the emphasis on unique cases leaves unresolved a potential flaw; namely, how can we know if what is found is, in fact, true of other beings? Perhaps the uniqueness is truly unique, for it produces behavior not found in other human beings. As an example, recall that Freud considered Little Hans's oedipal projections to be characteristic of all of humanity. Perhaps they are, but how can we know that they are not merely the imaginings of this one child? It may also be that Little Hans's ruminations were his alone, that they were ideas unique to him. On what grounds does the psychoanalyst believe that these projections are necessarily general, necessarily characteristic of all people?

By the second, I mean that Itard's way of describing and measuring Victor's mind was to determine what could be taught him. Itard did not merely wait and watch Victor in the expectation of seeing ideas mature and develop in him: He assumed that what was interesting to know was the nature of what Victor could learn. Itard assumed that he would come to understand Victor's mind by investigating Victor's capacities. To do this, Itard wanted to know what Victor could learn. Acceptance of a central assumption—that it is what can be learned that is important to study—characterizes most of the stories I have related. Making the same assumption about the nature of behavioral research are Singh with the wolf-girls, the teaching of Kaspar Hauser, Thorndike with his cats, dogs, and chickens, and all the work with chimpanzees and gorillas that assumes that one communicates with the mind of others by discovering what that mind can learn.

Our common way of thinking about the difference between physical and behavioral science, described in Chapter 3, is that the goal of the first is to eliminate variance, while the second accepts variance as the essential characteristic of the subject worthy of study. The physical sciences seek to eliminate variation because variation confounds accuracy of prediction. The behavioral sciences should accept variation as the essential aspect of living beings, and thereby strive to measure variance as a technique of describing the nature of life itself. We often confuse the legitimacy of these different goals, thereby leading us to the conclusion, for example, that the physical sciences are more "scientific" than the behavioral because they strive for accuracy and prediction. Some appear to think that a measure of the applicability of science is accuracy of prediction, but variance, too, is a legitimate interest of the scientific method. Science is a unique method, a method independent of what it studies. Measures of variance can be just as reliable as formulas that strive to eliminate or reduce variance. As always, the meaningful issue is what one wants to know, what one wants to accomplish through the application of the methods of science.

Let us put to rest the notion that there can be no science of living beings or that scientific procedures somehow diminish and degrade the awesomeness of life. The chief characteristic of life forms, as opposed to physical objects, is variation. It is variation that permits evolution, for without variation, there is nothing for natural selection to select. The study of variation may be done in two ways: by study of the unique or by study of the general. In this book, we have examined examples of both, although study of the unique case dominates, to be sure; but what Thorndike, Haggerty, and Hamilton contributed is the importance of general variation. Both ways must be investigated because we cannot know what is unique without knowing what is general. Behavioral science, therefore, proceeds on two fronts: the study of the unusual and the study of the variation characteristic of groups.

Subject and Investigator

The many instances of comingling of the behavior of subject and investigator recounted in this book illustrate what is self-evident: beings interact. Naming oneself "investigator" and another "subject" is a fantasy, and an arrogant one at that, as Itard appears to have come to understand about his relationship with Victor. Garner never appears to have comprehended this truth about his relationship with the chimpanzees he studied, and living experimenters may speak for themselves on this issue—or let writers of the future do so on their behalf.

Only rarely do we take seriously the truth that subject and investigator are but human designations of their own perceived role in relationships. We fail, for example, to ask the effect of Little Hans on Freud. Would Freud have found some other child whose behavior illustrated the concepts of anxiety, castration threat, and the oedipal complex? Or did Little Hans's behavior serve to incubate from Freud ideas that would never have come from him otherwise? What was the effect of Moses, Aaron, and Elishiba on Garner? The interaction is a powerful one, especially at the death of the animals. What accounts for his pathetic years that followed the fame of the West African adventures?

And what of those who would seek to comprehend the minds of apes through communication with them? What were Gua's effects on Donald and his parents? Viki on the Hayses? Washoe on the Gardners? Koko on Patterson? Sarah on the Premacks? And Lana, Sherman, Austin, and Kanzi on Rumbaugh and Savage-Rumbaugh? We have little information and we can guess, but we do not know, in part because the nature of the joint communication is unreported by the human interpreters, and, in part, because the accounts are not yet finished.

The lack of information regarding the interaction between presumed subject and putative experimenter needs attention and investigation. The view that the investigator is functioning in some all-knowing, all-seeing capacity, while the subject's actions are but the unvarying response to the investigator's machinations, misleads our interpretations. It does not seem to have occurred to Pfungst, for example, that he was a stimulus for

Hans's behavior, or to Freud that Little Hans's father was both interpreter and prompter of Little Hans, or to Thorndike that blowing smoke
at his subjects might have affected their appreciation of him when he
handled them, or to the teachers of sign language to chimps that the
animal influenced the questions posed by the investigator.

The fact that the subject and investigator themselves are worthy of
study as a behavioral phenomenon leads us to a reconsideration of the
Thorndike-Mills controversy. In its modern version, the issue of the exploitation of animal species in such a partnership rushes very much to
the forefront.

What Is Exploitation?

The Thorndike-Mills controversy focused on Mills's view that Thorndike's experiments with cats, dogs, chickens, and fish were unlikely to
discover much of significance regarding these animals' minds and behavior because the animals were tested in a situation (the puzzle box) artificial to them. Thorndike's view was that one would "eulogize" animal
behavior forever if one relied on uncontrolled studies and unverified observations; that control of confounding variables was essential; and that
the puzzle box emulated the animals' natural activities while permitting
accurate assessment of ability.

During the century that has passed since the dispute, the notion of
exploitation has been added to the core of the dispute. This notion has
developed naturally enough. Witmer's investigation of Peter the chimp,
it may be recalled, had as its origins the goal of determining the relationship of people to apes by using the method Witmer had developed to
test and measure the abilities of children. His goal was to indicate for
each child the most appropriate kind of training or education. If testing
can do so, there would be few quarrels, for providing each human being
with the kind of education from which he or she can prosper would seem
to be an important goal of a just society. But, if the tests are understood
as selecting not on the basis of ability, but on some other criterion—say,
on the basis of sex, gender, race, or economic background—we can appreciate the view that such tests are being used to exploit, to eliminate
options, to suppress people into a life that discourages development or
success.

The exploitation charge, first directed at those who would test and
measure human ability, in time would be extended to animals. By what
right, it may be asked, do people "test" animals, decide that one individual or one species is better suited for survival and success than another,
or place in experiments subjects that may be harmed or demeaned? One
aspect of this viewpoint echoes Mills's argument that an animal not comfortable with its environment will display for you nothing of value regarding its abilities. A like argument is heard, perhaps less convincingly,
from Krall, who noted that Muhamed's and Zarif's failures were due to

the horses' discomfort with the questioner rather than representative of the horses' ability to think or calculate.

The complexity of the question is due to many issues being wound about one another. Any sensible answer demands an unravelling of the several threads so bound. If our concern is solely with animal behavior, the nature of the environment is surely relevant. A tame animal is different from a wild one; a domesticated one different from its genetic parent; an animal reinforced by food different from one not so rewarded; an animal contained whether by a laboratory, a cage, or by natural boundaries, different from one unconstrained.

If we are interested in the rights to be given animals, these being rights in the same sense that society gives rights to its members, then we are concerned with an issue that is very different from that of whether the behavior is natural. The student of nature becomes involved in the discussion of rights when society believes that rights are to be given to some species and not to others, depending on whether they are more, or less, like us. We do this when we assume that, say, chimps or orangs have more humanlike rights than, say, rabbits or sheep, because they are more like us. This sort of awarding of rights requires that society be able to establish a Physical or Mental Ladder that distinguishes animal forms from one another and establishes likenesses to human beings. If this Mental Ladder is designed to include communication, as it does when apes are taught to sign in the expectation of our describing their minds, or of communicating with them, we find ourselves needing to evaluate the idea of a Mental Ladder and to determine the probable origins of our kind of consciousness.

Innate or Learned?

What engaged Itard's interest in Victor was the question that interested European society following the discovery of Wild Peter: Could study of these children reveal which ideas and capacities were innate and which were learned through socialization with human companionship? Neither European society nor its physicians and scholars were able to answer the question from observation of Peter, and while Itard's lengthy investigations of Victor tells us much about their relationship, little about the question posed comes to the fore.

The question itself is part of a more longstanding and more difficult question regarding the roles of nature and nurture, of what is innate and what is acquired through experience. We make our first mistake in answering this question if we assume that the two are alternatives, that behavior is one or the other, rather than both. We make our second mistake if we ask what amount or percentage of behavior is one, how much the other. We make our third mistake if we equate nature with genes and nurture with experience alone.

All living beings, unlike the matter of concern to the physical sciences, have a genetic base that reflects variation at the level of the gene. It is true that my body and yours, and those of both Clever and Little Hans, are a product of a pool of genes. It is true that these genes determine the limits on my behavior and capacities, and thereby set constraints on *what* I may learn. Peter the chimp will not solve differential equations, and no amount of education will help me climb or skate with the grace he displayed. What he and I may learn is acquired "on top of" the genes, so to speak, but while it is true that the inheritance of these genes sets the limits of what I can learn, limits do not dictate *what* I can learn. Witmer well understood this. His goal in designing mental tests was to assure that constraint not be confused with content. The slum child did not know Greek and Latin, to be sure, but not because of mental constraints.

When "genetics" is used as an *explanation* of mental ability, the slum child is not educated, and the lack of education is taken to reflect the validity of the theory of genetic constraint. Such would merely be a fine example of faulty logic were it not for the fact that the logic is used to direct educational and social policy, to constrain individual lives. Itard knew nothing of such constraints, so he strove mightily to show that Victor *could* learn. In like fashion, had Kaspar's teachers known of genetic constraints, they would have placed him in an asylum for safekeeping and not troubled themselves to civilize and educate him. Happily for Kaspar, they did not know of such genetic limitations. Had Laura Bridgman's family believed her sensory deficiencies to be genetic, no attempt would have been made to teach her to communicate in alternative ways and a life full of meaning would have been stunted.

Just as Peter, Victor, Amala and Kamala, and Kaspar could tell us nothing of significance about the nature of their unfettered minds, so it may be argued, by extension, neither do gesticulating apes. The teaching of a means of communication, whether it be called gesticulation or language, is a socializing experience invented by human beings. What we learn from them is limited to that which fits our own experiences, for this is all we know and all we can ever know. The question—What is innate? What learned?—is another example of a nonsense question for the behavioral sciences. It sounds like a valid question, but, as it admits to no meaningful answer, it is really no question at all.

The Mental Ladder

Concepts—whether scientific or religious, logical or emotive, historical or futuristic—are the ways in which the human mind that constructs a universe organizes its perceptions. Whether they are given to us at birth or whether we must learn them through socialization, we use them to construct our understanding of that world we presume to exist independent of our minds. These concepts may be true or false, in the sense that they may or may not match that world independent of us. And they may be dangerous if they lead us to acts unworthy of our humanity.

The concept of the Mental Ladder is an engaging one: it, along with its master, the concept of the Great Chain of Being, has been with humanity since humankind first wrote about its understanding of itself. As we have read, investigators who examined human and animal minds were at times prompted by an assumption that a ladder of mentation, rather like a phylogenetic ladder, was a useful description of the history of nature. The concept is now out of fashion, because it is believed that there is no way to erect the steps. Every animal, by definition, is now seen as a master of its own eco-niche and of its own ways of doing things. All animals are alike in that they succeed in their environments. It does not occur to us to think of threatened and endangered species as "stupid" because they are unable to accommodate to the human destruction of their environments. To say that one species is "smarter" than another is to apply obviously human values to animal forms that are themselves as successful as we are.

But we are hypocritical on this issue. Obviously, we do not value the behavior or soul of the fly or cockroach to the same degree that we do the chimpanzee or domestic dog or cat. We human beings do have a ladder representing our judgment as to "likeability," and it is not a great surprise to learn that the species placed on the top of our ladder are those genetically most like us and those we have domesticated to behave in ways we prefer. While the ladder of mentation may now be considered a suspicious concept by those investigating the minds of animals, it nonetheless continues to reflect the way we human beings think about animal life. Remember the late Victorian ladder shown in Figure 8.1, which placed the dog along with human beings at the top of the ladder of mentation. We impose the ladder on our comprehension and evaluation of other creatures. To do so is not "wrong"—but it is human. That we do so merely illustrates that we design our mental universe by the use of concepts, which suggests that our first task is to develop ways to understand our own mind by understanding the concepts it employs.

The attempts to make public the contents of the silent mind—the feral child or clever animal—do indeed illustrate futility, but the efforts appear futile only if we assume that their goal is to expose the mind of others. The attempts themselves are of the greatest importance to us because it is the nature of the attempts that illustrates the first principles of how my mind conceptualizes. The studies reported in this book have largely accepted the view (1) that behavioral science can and should imitate physical science; (2) that the investigator takes control of variables that alter the behavior of the subject; (3) that the use of human and animal subjects in this way is justified, as I noted, citing Aristotle, in the first sentence of the book, since "human beings, such being their nature, desire to know"; (4) that the human investigator can separate what is innate from what is learned by appropriate experimental design; and (5) that there are ways, perhaps a mental or phyletic ladder, that enable us to distinguish the capacities and abilities of animal life.

What if none of these assumptions is true? What if the assumptions are nothing other than false descriptions of how our minds categorize? Consider the notion that the assumptions themselves are but descriptions of how the human mind conceptualizes; that they are no more necessarily true or necessarily false than any other set of concepts we have come to use to judge our world.

THE PSYCHOLOGIES

If the task before us is to name and describe the concepts that are used to create our private universes, then we might search within the psychologies discussed. We might expect phenomenology, psychoanalysis, and behaviorism to have already searched for them. Both psychoanalysis and phenomenology, unlike behaviorism, consider perceptions to be the data of psychology. Perception is the master filter through which all information and ideas flow into the mind. When Freud reads the letters from Little Hans's father, he searches Hans's perceptions for meanings, for ways in which Hans's statements represent experiences. The substitutions and symbols that appear in the spoken word and behavior refer instead to an unconscious set of ideas that determine Hans's world. The father's moustache stands for much more than hair, just as the zoo animals come to substitute for Hans's ideas about dimly understood relationships among human beings. What Hans cannot understand, he explains by substituting concepts. The oedipal situation, after all, serves to organize the perceptions of experience and make reality reliable. Freud may have discovered it, but Little Hans invented it. It is a splendid example of a category, for it explains and organizes experiences by giving them meaning. Whether it is "true" is the wrong question to ask of concepts, for about them we ask only whether they are successful and universal.

Phenomenology is, at once, more serious than psychoanalysis about the importance of perception and less successful at determining the categories that define our perception of reality. In its attempt to describe the mind, phenomenology both melds with psychoanalysis in its use of the verbal report as the basic data and shies from it. Phenomenology wants to take on a larger task, for it wants to find the nature of the categories that limit, filter, and shape experience, whereas psychoanalysis already thinks it knows the nature of the categories. Let us examine these epistemologies more closely, for they are longstanding and compelling contributions regarding the nature of our minds.

Phenomenology proposes that we can reasonably assume that each mind is both creator and judge of its own private universe. Study of the mind begins with a description of the contents by describing the phenomena within it. Although it may appear to be easier to understand the mind of another human being than that of a nonhuman animal, this is not necessarily so. We may be more easily misled or deceived by our own species than by the analysis of another. It is true that a shared language appears to help me understand the mind of another human being, for language is among the chief ways that our species uses to provide information to one another. But language itself is an invention of that private universe, and there is no guarantee that any of us has found the code that permits us to translate correctly between our mind and that of another. Not only is there no guarantee; there is no way of knowing we have done so.

The mind is forever a private, separate universe, and our mind is forever alone in the universe because we have no way to decipher the code, goes this argument. One way to decipher is by decoding the sensory systems, to learn what they select to transfer and how they do so. Our mind is a composite of the neural connections it brings to life and those connections that develop during our lifetimes. The sensory systems—the eye, the ear, the nose, the tongue, the skin—these and more are carriers and, we suspect, powerful distorters of the world our minds try to imagine, create, and keep constant so that we may live within them. For this reason, some have regarded the study of the sensory systems as the paramount way to find the code, if one exists. The rationale is that the standard way of defining the umwelt is by analyzing the perceptions that create it. Howe's work with Laura Bridgman, Lubbock's with Van the dog, Von Uexküll's analyses of the umwelt, and Buytendijk's analysis of the ways in which dogs track—all are examples of phenomenological ways of thinking about the mind described in this book.

Laura Bridgman, bereft of all senses except sensations of the skin, would have been of commanding interest to the modern phenomenologist. Her inability to use sensory systems made her a near-perfect example of what the mind might hold uncontaminated by sensory experience. She would have been the phenomenologist's wild-boy of Aveyron, for she might have provided the answer to the question "What would happen if a human was without . . ." that so attracted Itard to study of the wild-boy. But she lived before phenomenology made its intellectual statement in our times. Instead of being the modern wild-boy of Aveyron, taught but unteachable, she profited from Dr. Howe's teaching and his painstaking, patient, and altogether remarkable colleagues who taught the little girl to communicate, not with sight and sound, but by skin sensations.

The phenomenologists' emphasis on perception stalks and underscores the stories related in this book. William James saw clearly the importance of evaluating how the senses translate information to the mind; Freud emphasized the reality of Little Hans's perceptions to Little Hans himself as the basic data on which a theory of the development of the human mind was placed; Pfungst saw immediately that he must measure what Clever Hans was able to see, so that he could measure the contributions which the animal's senses made to his tapping of the hoof. To construct the mind, we want to know, first, what is in it (in the contemporary sense of delineating the neural connections and pathways) and, second, how the senses alter these connections. Phenomenology is not alone in desiring this information.

The myths of psychoanalysis escape clear definition; otherwise, they would not need to be myths. As is true of all myths, the product is greater than the sum of the parts, for to name the elements is to destroy the design. Psychoanalysis remains a most persuasive myth, one able to capture the human mind and behavior of our times. Its concepts are prominent and perhaps dominant in its explanation of the ways people in our times imagine the minds of others to function.

If we are to have the first slight sensation of understanding our mind, we must understand psychoanalysis and come to comprehend how our own mind comes to terms with its concepts. The task is difficult and daunting, and may even be impossible, yet what we learn from the attempt is itself worthwhile. It is both an ironic and a powerful statement to write, but this is so: Psychoanalysis is a myth that attempts to describe the origin and nature of human myth itself.

Hans's father's accounts written to his friend and colleague Sigmund Freud and Freud's statement and analysis of the case are critical points in the development of psychoanalysis, for with these data Freud formed clearly and publicly the notion of the oedipal period. All of psychoanalysis pays homage to this central idea. Within this powerful myth is the lasting notion that early experiences affect later experiences, that the developing mind goes through identifiable sequences, and that the developing mind invents or re-creates myths to explain observations.

Note the similarity of this myth and view of individual development to classic ethology. The notion of the relationship of the innate releasing mechanism and the fixed action pattern, an idea fundamental to ethology, fits neatly with the psychoanalytic notion that events in early mental development become fixed and that they influence the course of one's personality ever after. That Freud and Lorenz, representing psychoanalysis and ethology, have a common intellectual source should not astonish us. Both men were of the same general culture, and both were educated and shaped by nineteenth-century philosophic and scientific views. That one view, psychoanalysis, emphasized human beings and the other, ethology, looked primarily to animals, allowed us to mislead ourselves by emphasizing the differences among species and masking the similarity of the approaches.

Both psychoanalysis and ethology, while making no great issue of it, yet consider perception and sensation to be fundamental to behavior. For this reason alone, they may be regarded as companions to the phenomenological emphasis on perceptual experience. Psychoanalysis has no mode of explanation, no myths to explain, unless these be perceptions in the human mind, perceptions that, as distinct from sensations, are representations of reality as we construct it. When Little Hans reacted to his father as if he, the father, were the black horse, then seemingly to the sight of the father's nakedness, then to the father's moustache, we see the changing perceptions in Little Hans's umwelt. The basic datum of psychoanalysis is composed of such data, the perceptions, but it adds to these data something the phenomenologist does not yet have, the concepts that give continuing structure to the perceptions.

The reason that psychoanalysis can do nothing to assist our understanding of the animal mind is that it relies on language to understand perception. We know the dangers of such reliance (think of Richard Lynch Garner's study of the language of apes), especially when language is assumed uncritically to be an accurate representation of the perceptions of the mind expressing itself. The psychoanalytic technique teases out representations to encourage the verbal behavior to follow itself, to let one idea lead to another, to determine the connections between the representations in the expectation of finding both larger and narrower representations that hold the individual representations together. The association method is a means to determine the categories of the mind, a task the philosopher Kant set for modern thinkers in the eighteenth century. Notice that it is a method whose goal is to find relationships between perceptions. The basic datum is still the perception.

However evident it may now be that the datum is the perception, if we wish to learn to decode the mind of others into representations that we can understand, the nature of the perception is yet one more example of how isolated and private our own mind is. The memory spoken of by another person may have some meaning to me. It may bring forth a memory of my own or perhaps be translatable into some situation or feeling familiar to me. Yet it is not my perception; it is yours, and I can never know when I have understood you, when I have successfully translated your memory so that I may count it among my own perceptions.

That I can never know is unnerving. I may agree with phenomenology and psychoanalysis that the datum to examine is the perception, but what if the perception is unknowable? Is it worthwhile to search for it? Or, by doing so, do I merely mistake verbal behavior for something real and reliable rather than understanding it as one more way in which motor skills are used? Without denying the importance of perceptions, I may decide that understanding the perception is not possible, that such attempts lead merely to data that themselves go nowhere, that attempting to understand the mind by listening to verbal behavior is no more sensible than trying to understand hydrology by skipping a stone into the water. Relaxing it may be, but I learn very little about the events that shape the course of the phenomena merely by watching the stone. Nor do I know much about your perception of, say, "envy," merely by hearing you say the word.

When Thorndike decided to ignore the assumed perceptions and concentrate on the motor behavior of animals, to study the action itself and not the perception on which it is presumably based, he was setting forth a new view of how to study the mind. His view would rival psychoanalysis and maintain such strength that phenomenology would never be understood, at least in English-speaking countries, to be a serious way to help communicate and translate between and among minds. His approach, that of measuring the change in motor behavior as new tasks were acquired, was oriented toward animals, not toward human beings. If his technique were a sound one, it would not have the drawbacks of psychoanalysis, since it would not depend on language and it would extend the study of the mind to all sentient beings, not merely those who speak in a language that we human beings presume to understand.

Thorndike's words contain no evidence that he understood the nature of the philosophic position he was adopting, although these ideas become most prominent later in the theories of J. B. Watson and B. F. Skinner. Their view is that the mental events known to be part of perceptions, along with supposed mental states (sadness, thinking), are impossible to evaluate and thereby poisonous to any serious attempt to understand behavior. It is not argued, as critics have mistakenly assumed, that such mental events do not exist or that the mind is without perceptions, emotions, images, and the like. Rather, since we cannot know the nature of these save through the dubious means provided by verbal behavior, such mental events, their descriptions and presumed workings, should not be the basic data of a theory of behavior. The physicist is not expected to understand extrasensory perception or emotion, although some make the attempt, for neither is the fundamental unit for understanding the subject matter.

Through the works of Thorndike, Watson, and Skinner, there evolved a way of examining behavior by ignoring (but not by denying, as some appear to think) the perceptual stuff of the mind and behavior. Concentrate on the observable, on the behavior; forgo the illusions of mental states, intervening variables, and hypothetical constructs. I may tell you that I am hungry, and you may record and measure this statement as what it is, verbal behavior. You may measure how often I said it, under what conditions, and what response from others the verbal behavior elicits. But you should not infer some mental state from the utterance. *I* may think *I* know of such a state, and *I* may describe it to you, but the description is but verbal behavior, and verbal behavior is, like any other motor behavior, to be recorded and measured. This point is so often misunderstood that it deserves repetition: Descriptions of mental states are useful verbal data, but they are like any other piece of verbal behavior. They should be treated as data to be collected and not as representations of mental states, for these states are not only unknowable, but also not a necessary part of the prediction of behavior.

When we wish to establish communications with other beings, we appear to adopt either the behaviorist or the phenomenological approach. Much of what is most unclear about the ape-human language experiments is muddy because we confuse the goals and methods of these two ways of knowing the mind of another. Let us conclude by examining this point in the hope that it may be instructive regarding the psychologies we invent.

HUMAN AND ANIMAL COMMUNICATION

Descriptions of how people have attempted to understand the human and animal mind are actually descriptions of how we view our own mind at work. In our attempts to extend our private universe to the private universe of another being, we have no choice but to translate the stuff of our own mind onto our expectations regarding the stuff we understand to reside in other minds.

The animals and people who populate these pages are real only insofar as I use my sensory system and my mind to understand them, to find some place in the concepts resident in my mind in which these dogs, horses, fish, cats, children, people of differing mental abilities and achievements, and curious investigators, may fit. Their reality now resides in my mind and, I hope, in yours. Surely I write these last words with the hope that these aspects of my mind can be translated to yours, in part because I want to share that aspect of my private universe that so admires humankind's attempts to understand its own mind and that of animals. While the behaviorist wants to measure the frequency of gesticulations, the phenomenologist wants to know what they represent, to know what, if any, meaning they might have. The issue is far more than a disagreement about technique: it is also a disagreement about goals. Our goals shape the questions we ask and the images we use. Consider, for example, the image of the Ladder of Mental Ability, that ladder created with the hope of showing the psychological relationships among all of animal life, including ourselves.

At the top of the erected ladder, holding it in place from above, being careful not to fall in the process of readying and steadying the ladder, are those whose minds favor and categorize by phenomenology, psychoanalysis, and ethology. At the bottom, steadying the ladder, keeping it erect and safe from unwanted motion, are the behaviorists, useful indeed for providing requisite steadfastness, if not the courage to lead by climbing first.

The myth of communication is a necessary myth. While the strength of the myth of evolution is that it explains and provides a story about who we are, and how we came to be, a script that gives some meaning, however sparse, to our lives and actions, the necessity of the myth of communication is that it permits us to think that we can leave our private universe: we are saved from pure solipsism. Imagine the limitations of our world if we did not try to get into the minds of other beings, if we did not accept the other people on our planet as themselves having feelings and giving meanings to that which they experience. It may be that it is the ascribing of such consciousness to others that separates human beings from other animals on the rungs of the ladder. We cannot know until we take the next step in the analysis; namely, the step of discovering how to re-create the umwelt of others. Until we learn to do this, we shall continue to misunderstand what animal and feral minds might tell us about their perceptions, and thereby their universes. Without such understanding, our grasp of communication and language is certain to be confused. Worse, it is certain to represent only what we want to believe to satisfy our own mind's boundaries and categories.

We human beings are psychologists by design. It is an aspect of our equipment to believe that others have minds, more or less similar to our own. Every now and then a philosopher or scientist tells us that we are wrong, that animals are no less or more thoughtful than an automobile, but our natural way of thinking denies these notions. That we cannot prove the existence of the minds of others to the satisfaction of the philosopher does not change our behavior: we act as if the minds of others are as real and as knowable as our own.

The "natural psychologist" feature of human beings is a salient characteristic, one well worth investigation, for by requiring us to create and recreate a universe of minds, each one filled with its unique history of experiences, sensations, perceptions, and emotions, we may learn much about the nature of human thought. As so many persons described in this book illustrate, we ignore investigation of our own motives and ways of understanding at great peril. Worse, by ignoring these, we lose sight of what is so characteristically human: the creation of other minds.

We do not know whether chimpanzees, dogs, chickens, and fish do likewise with their mental time and mental space, but do we not, in our own minds, believe that these beasts also have rich and complex beliefs about their own minds—and perhaps about ours? When my dog barks at me, do I not attribute to the dog some purpose, a purpose reflected in my reflecting about the contents of the dog-mind? In this way, the mind of my dog is inextricably bound to my mind, for I have no way to create its mind other than by applying the categories and concepts of my own.

If we do not know, just now, how to assess our communications with other minds, we should not be discouraged from continuing the attempt. The lessons from the events described in this book show us some wrong ways and point to a path that must be taken if we are to find the right way. The story of animal-human communication cannot yet be told, for it is far from finished. Our human sin is, as it so often is, that of unchecked arrogance, of the kind that "makes the angels weep" as the quotation from "Measure for Measure" that began this chapter and section warned, this being the arrogance that appears when we are uncritical of our own techniques and unmindful of the ways in which our own ways of thinking, our own categories, govern how we create other minds, and how we investigate our own.

As we cannot help being human, let us discover what it means to be so. By so learning, we can perhaps come to find more about the minds of others, but first we must deny the arrogance of thinking that we are objective and devote our attention to examining our own categories, and thereby the power and weakness of our human natures. The need for such reflection regarding the nature of being human is the moral of this account of humankind's attempts to decode the reflections from the minds of feral children and clever animals.

POSTLUDE

For that which befalleth the sons of men befalleth beasts; even one thing befalleth them: as the one dieth, so dieth the other; yea they have all one breath; so that a man hath no preeminence above a beast: for all is vanity.

All go unto one place; all are of the dust, and all turn to dust again.

Ecclesiastes

Notes

Chapter 1. Nature and Nuture: Children Without Human Parenting

1. J. A. L. Singh, and R. M. Zingg, *Wolf-children and Feral Man* (New York: Harper, 1939). The quotation is from pp. 182–183. The contents of this book require description. The first part, written by Singh, is an account of the Indian wolf-children, about whom more later. The second part, by Zingg, is an account of feral human beings. To my knowledge, it is the most thorough and accurate such work to be found, but perhaps because the book's title fails to convey the rich material to be found therein, Zingg's splendid work is rarely referenced.

   The third part, one to which the title gives only the most vague of suggestions, is a reprint in English of the descriptive part of Anselm von Feuerbach's work on Kaspar Hauser, who will also be covered in the present volume. This book thereby contains Singh's first-hand account of the wolf-children, a splendid accounting and analysis of feral human beings from classic times to the recent present, and a valuable reprint of the classic work on Kaspar Hauser. One cannot guess the contents from either the title or the table of contents.

   Much of the information contained in this section, represented either by direct quotation or discussion, comes from Zingg's contribution.

2. The earlier meaning of the word "feral" refers to the release of a domesticated or socialized being into the wild. The word has come to be used to describe any animal taken from the wild into captivity—a definition just the reverse of its earlier meaning. I use the word in its longstanding meaning. Peter is feral, as he was presumably born in a socialized state, placed in the wild, and recaptured into civilization.

3. Sources recorded at the time, however, note three aspects of Peter that belie this interpretation. For one, he had the remnant of a shirt collar around his neck. Second, his legs, but not his thighs, were tanned, suggesting that he had worn breeches without socks. Third, it was thought by a physician that his tongue was inappropriately connected, with the result that he could not speak easily.

4. Singh and Zingg, p. 193.

5. Singh and Zingg, p. 185.

6. The letter is from Countess Schaumburg-Lippe to Count Zinzendorf. The count had requested custody of Peter with the hope of learning from Peter

what sorts of knowledge Peter was born with and which he had acquired through experience. The application was declined because Peter had been given to someone else, namely, the Princess of Wales. See Singh and Zingg, p. 185.

7. Singh and Zingg, pp. 186–187.
8. Singh and Zingg, p. 187.
9. The information is taken from Singh and Zingg, pp. 184–185. Zingg's sources appear to be A. Rauber, *Homo sapiens ferus, Rauber,* 1888, and J. F. Blumenbach, *Vom Homo sapiens ferus Linn und namentlich von Hammelschen wilden Peter. Beytrage (Beitrage) zur Naturgeschichte,* Vol. 20 (Gottingen: Heinrich, 1811). On Peter, Zingg uses a translation of Blumenbach; namely, T. Bendyshe, *The Anthropological Treatises of Blumenbach* (London: Longman, Roberts, and Green, 1965).
10. The date must be wrong.
11. Singh and Zingg, p. 187.
12. Singh and Zingg, p. 187.
13. Singh and Zingg, p. 187.
14. Throughout this book, material inserted between parentheses () or brackets [] within a quotation is added to assist the reader's understanding of places, terms, and phrases. () insertions are by the author being quoted; [] insertions indicate additions made by the current author.
15. Apparently, this child, described by Buffon in French, encouraged Rousseau's views, published in 1754, that argued for an appreciation of the natural state and against the degrading results of socialization.
16. I use three sources chiefly. The first is a translation by George and Muriel Humphrey of J-M-G Itard: *The Wild Boy of Aveyron (Rapports et memoires sur le sauvage de l'Aveyron)* (New York: Century, 1932). To the translation, G. Humphrey has added an exceptional introduction that provides one of the finest interpretations to be found. The Humphreys's book is more about Itard than Victor, for obvious reasons, and we learn much about Itard's work after his work with Victor. The account of Itard's contributions to sign language and the training of the deaf are most informative, if unexpected, from the title of the book. The second is H. Lane's account of Itard and Victor, *The Wild Boy of Aveyron* (Cambridge, Mass.: Harvard University Press, 1976) (The original is "de l'education d'un homme sauvage ou des premiers developpemens physiques et moraux du jeune sauvage de l'Aveyron. Paris: Goujon, 1801). The third is from Singh and Zingg, already cited.

    An additional source is R. Shattuck, *The Forbidden Experiment, the Story of the Wild Boy of Aveyron* (New York: Washington Square Press, 1981). I have used this book for Shattuck's presentations and translations of some of the eyewitness documentation. Shattuck's purpose, however, is to use these documents to create a literary account; for this reason, I feel less secure with his account than with Lane's.
17. Rousseau suggested that orangutans and chimpanzees were human beings who, having been reared in the forests, had lost or never gained the socialization necessary to be human. See G. W. Hewes, in D. M. Rumbaugh (ed.), *Language Learning by a Chimpanzee, the LANA Project* (New York: Academic, 1977).
18. Humphrey and Humphrey, pp. vi–vii.
19. Lane uses this setting to open his book. I use the story on his authority. The date evidently is autumn-winter 1799–1800.

20. Lane, p. 4. Much of the description is taken by Lane (therefore by me) from the description provided by Pierre-Joseph Bonnaterre who wrote a "Notice" in 1800) using information taken from those who worked with Victor before his shipment to Paris. Lane, who found the long-missing "Notice," offers an English translation on pp. 33–48 of his book. Pinel's account remains missing. See Humphrey and Humphrey, note 1, p. vii.
21. Humphrey and Humphrey (trans.), p. xxi.
22. Humphrey and Humphrey, pp. xxiii–xxxiv.
23. Humphrey and Humphrey, pp. 5–6.
24. Humphrey and Humphrey, p. 7.
25. Humphrey and Humphrey, pp. 10–11.
26. Humphrey and Humphrey, p. 11.
27. Humphrey and Humphrey, pp. 21–22.
28. Humphrey and Humphrey, p. 25.
29. Humphrey and Humphrey, pp. 26–27.
30. Humphrey and Humphrey, p. 29. The "or" sound at the end of Victor would have the sound of "oh" in French. Shattuck (1980) tells us that at the time there was a long-running play in production with the title *Victor or the Forest's Child* (p. 110).
31. Humphrey and Humphrey, pp. 31–32.
32. Humphrey and Humphrey, p. 11.
33. Humphrey and Humphrey, p. 44. Later, when Mme. Guérin's husband was missing from dinner, Victor continued to set a place for him. He did so after M. Guérin died, and when he saw Mme. Guérin cry at the sight of the empty setting, he cried as well.
34. Humphrey and Humphrey, p. 51.
35. Humphrey and Humphrey, p. 53.
36. Humphrey and Humphrey, p. 59.
37. Humphrey and Humphrey, p. 60.
38. Humphrey and Humphrey, p. 63.
39. Humphrey and Humphrey, p. 68.
40. Humphrey and Humphrey, p. 73.
41. Humphrey and Humphrey, p. 86.
42. Humphrey and Humphrey, pp. 96–99.
43. The quotation is from H. Lane, p. 166–167. Lane tells us (p. 167) that the ministry granted 150 francs a year to Mme. Guérin and that Victor went to live with her nearby. One visitor, a decade later, described Victor as unchanged. Victor died in 1828, two years after Pinel's death, at the age of forty years. Itard died ten years later, in 1838. For information on Itard, and the distinguished remainder of his life and career, see Lane.

Chapter 2. *Kasper Hauser and the Wolf-Children*

1. Anselm von Feuerbach (Paul Johann Anselm, Ritter von Feuerbach) (1775) was trained in law and became known for his strong views against the use of torture. The initial publication is *Kaspar Hauser, Beispiel eines Verbrechens am Seelenleben des Menschen* (Ansbach, 1832). The publication, along with the testimony of others and notes, are collected in H. Pies, "Kaspar Hauser, Augenzeugenberichte und Selbstzeugnisse, herausgegben, eingeleitet und mit fussnoten versehen" (2 vols). Stutgart: G.m.b.h., 1928. The title of the orig-

inal (Beispiel . . .) was translated as "An instance of a Crime against the Life of the Soul (the development of all its intellectual, moral, and immortal parts) of Man." The title makes evident Feuerbach's views. The English translation of the title is "Caspar Hauser, an account of an individual kept in a dungeon separated from all communication with the world, from early childhood to about the age of seventeen" (London: Simpkin and Marshall, 1833).

The text I have used for the description presented here is the 1832 edition (with a foreword by Lieber) that is reprinted in J. A. L. Singh and R. M. Zingg, *Wolf-children and Feral Man* (New York: Harper, 1939). Once again, it is to Zingg's scholarship that we are indebted for the preservation of these documents.

2. Singh and Zingg, p. 284. The German is a phonetic rendering of what was heard. The first sentence sounds like "I want to be a horseman as my father is." In the English translation of the play *Kaspar* by Peter Handke (1969), the phrase is given as "I want to be a person like someone else once was." I am grateful to Katherine M. Faull for translating these dialects.

3. Singh and Zingg, pp. 284–285.

4. Singh and Zingg, pp. 285–286.

5. Singh and Zingg, p. 288.

6. Singh and Zingg, p. 288.

7. Singh and Zingg, p. 289.

8. Singh and Zingg, pp. 291–292.

9. Recent work on self-recognition among the great apes appears to show that only the chimpanzees, and, questionably, the gorillas and the orangutans, recognize themselves by means of a mirror (Chapters 10–12).

10. Singh and Zingg, pp. 296–297.

11. Feuerbach's concern was with whether the treatment of Kaspar constituted torture; Feuerbach was not interested directly in the psychological data to be gathered. In an important way, this lack of direct interest makes his account all the more useful, for the events are laid out carefully and are not seen as supportive of one psychology or another. In contrast, Itard always discusses like events with Victor in terms of the psychological principles involved.

12. Singh and Zingg, p. 315.

13. Singh and Zingg, pp. 317–318.

14. Singh and Zingg, p. 354.

15. Singh and Zingg, p. 341.

16. Singh and Zingg, p. 342.

17. Singh and Zingg, pp. 353–354.

18. Singh and Zingg, p. 343.

19. Singh and Zingg, pp. 346–347.

20. Singh and Zingg, p. 359.

21. Singh and Zingg, note 10, pp. 360–365.

22. The quotation is from that part of Singh and Zingg, p. xxxi, that includes the firsthand account of the discovery of the wolf-children by the Reverend Singh and his colleagues.

23. Singh and Zingg, pp. xxxi–xxxii.

24. Singh and Zingg, pp. xxxii–xxxiii.

25. Singh and Zingg, pp. 3–6.

26. Singh and Zingg, pp. 7–9.

27. Singh and Zingg, pp. 8–9.
28. Singh and Zingg, p. xv.
29. The names Midnapore and Godamuri do not exist in modern atlases. Midnapore is now a district, a county. The area is now in Orissa. See Map 29, H6, *New York Times Atlas of the World.*
30. Singh and Zingg, pp. 11–17.
31. Further evidence (in addition to suckling) that Singh knew little about the behavior of wolves.
32. Singh and Zingg, pp. 39–40.
33. Singh and Zingg, p. 40.
34. Singh and Zingg, pp. 52–54.
35. Singh and Zingg, p. 97.
36. Singh and Zingg, pp. 103–104.
37. Singh and Zingg, pp. xxvi–xxvii.
38. Most likely the forerunner of what is now the Eastern Psychological Association.
39. Singh and Zingg, p. 113.
40. The question of the veracity of the dates, ages, and times arose and will continue to arise. Zingg (p. 141), in an excellent essay on the wolf-children, points out that stories of babies being cared for by wolves are common in north India (in 1939). Zingg presents a splendid analysis of the relationship between the Hindu religion and wolves, and the loss of life in British India caused by animals.

   Arnold Gesell, the distinguished developmental psychologist and physician, who was among the preface-writers to the Singh and Zingg book, also wrote a separate book, *Wolf Child and Human Child* (New York: Harper, 1941). Not everyone has accepted the accuracy of Singh's account, and some telling points were made by W. F. Ogburn and M. K. Bose ("On the Trail of the Wolf Children," *Genetic Psychology Monographs* 60 (1959: 117–193). See also C. Maclean, *The Wolf Children* (New York: Hill and Wang, 1978).

Chapter 3. Thinking about the Mind

1. C. Darwin, *The Origin of Species* (London: John Murray, 1859). Also, *The Descent of Man and Selection in Relation to Sex* (London: John Murray, 1871).
2. Plato writes of the Great Chain of Being in the *Timaeus.*
3. A. O. Lovejoy, *The Great Chain of Being, A Study of the History of an Idea.* (Cambridge, Mass.: Harvard University Press, 1936). This book deserves rediscovering, not merely for the content, but for its style. What I write here owes a great, but not total, debt to Lovejoy's splendid scholarship. The quotation is from the unpaged Preface.
4. Lovejoy, p. 58. Lovejoy refers us to Aristotle's *De Anima.*
5. Pope, *Essay on Man.* Nonetheless, the same poem contains the caveat "The proper study of mankind is man."
6. C. Darwin, *The Expression of the Emotions in Man and Animals* (London: 1872; Chicago: University of Chicago, 1965).
7. Lovejoy's conclusion is worthy of our attention. "But the history of the idea of the Chain of Being—in so far as that idea presupposed such a complete rational intelligence of the world—is the history of a failure; more precisely and more justly, it is the record of an experiment in thought carried on for

many centuries by many great and lesser minds, which can now be seen to have had an instructive negative outcome" (p. 329). The reasons for the judgment follow in Lovejoy.

8. I believe that the term and idea of "natural psychologist" is from Nicholas Humphrey (1983).

9. The picture (1838), which was also made into a well-circulated print in black and white (1839), is of a Newfoundland dog. See Ormond, 1981 (Sir Edwin Landseer, Philadelphia Museum of Art and the Tate Gallery).

10. Morgan, *Animal Life and Intelligence* (Boston: Ginn, 1891).

11. Morgan, pp. 340–341.

12. Morgan, p. 341.

13. An engraving of *A Distinguished Member of the Humane Society* is in the possession of the British Museum. A number of engravings exist in private and museum collections.

14. William James, in the two-volume *Principles of Psychology* (New York: Henry Holt, 1890). That James truly defined psychology by this book may be seen by comparing the chapter titles with contemporary introductory textbooks of psychology.

15. I. Kant, *Critique of Pure Reason,* first published, 1781. See also Chapter 1.

16. Morgan, 1891, p. 243.

17. Most specifically, those of G. Romanes.

18. Muybridge's photographs of animals are in E. Muybridge, *Animals in Motion* (New York: Dover, 1957).

Chapter 4. The Psychology of Psychoanalysis: Freud and Little Hans

1. S. Freud, "Analysis of a phobia in a five-year-old boy." The Standard Edition of *The Complete Psychological Works of Sigmund Freud, Vol. X: Two Case Histories,* pp. 141–142. The psychiatric saying regarding the order of patients is first the child, then the mother, and then, maybe, the father. The idea is that the family problems are set on the child, the most defenseless member of the family. When the child shows the accompanying neurosis, help is sought. When the other sees the results, she decides that there are longstanding neurotic conflicts in need of clarification. The father may then seek such help either because he feels the need or he wishes not to be left out of what seems to be conspiracy between mother and child.

2. English edition, p. 5. There is a footnote in the original regarding the idea of child analysis.

3. "Wenn du das machst, lass' ich den Dr. A. kommen, der schneidet dir den Wiwimacher ab" (German edition, p. 245).

4. "Aber ich will kein Schwesterl haben!" (German edition, p. 248). I changed "baby" to "little."

5. "In der Nacht war eine grosse und eine zerwutzelte Giraffe im Zimmer, und die grosse hat geschrien, weil ich ihr die zerwutzelte weggenommen hab'. Dann hat sie aufgehört zu schreien, und dann hab' ich mich auf die zerwutzelte Giraffe draufgesetzt." (German edition, p. 272). I have altered the translation to retain the ambiguity as to the number of giraffes.

6. The other was "Why anxiety?," meaning, why should this be the body's reaction to repression? It does not appear to be an adaptive response.

7. English edition, p. 39; German edition, p. 274.

8. English edition, p. 45; German edition, p. 280. The English, but not the German, edition contains the italicized prepositions.
9. English edition, p. 47; German edition, p. 281.
10. English edition, pp. 48–49; German edition, pp. 283–284.
11. English edition, p. 51; German edition, p. 285.
12. English edition, p. 54; German edition, p. 288.
13. "Ich, ich bin ein junges Pferd!" (English edition, p. 58; German edition, p. 292). The statement is not unlike that made by the young man in the play (and, later, film) *Equus* by Peter Shaffer.
14. English edition, p. 98; German edition, p. 333. "Er ist der Installateur gekommen und hat mir mit einer Zange zuerst den Podl weggenommen und hat mire dann einen andern gegeben und dann den Wiwimacher. Er hat gesagt: Lass den Podl sehen und ich hab' mich umdrehen mussen, und er hat ihn weggenommen und dann hat er gasagt: Lass den Wiwimacher schen."
15. English edition, p. 100; German edition, p. 335. The italics are in the English edition, alone.
16. "Three Essays on the Theory of Sexuality," 1905. In English, standard edition, 7, p. 125. The German edition is "Drei Abhandlugen zur Sexualtheorie, p. 5.
17. English edition, p. 102; German edition, pp. 337–338.
18. English edition, pp. 102–103; German edition, p. 338.
19. This information comes from two sources: J. Hanks and W. G. Bringmann, "Whatever Happened to Little Hans? *History of Psychology Newsletter* 221, no. 3/4 (1989): 78–81; "Memoirs of an Invisible Man. I, II, III, IV." *Opera News* 36, February 5, 12, 19, 26, 1972. This is a magazine publication of the (New York) Metropolitan Opera House. The opera articles know nothing of the significance of Hans's background to psychology.

Chapter 5. The Psychology of Experimentation and Behaviorism: Clever Hans and Lady Wonder

1. The description is assembled from Oskar Pfungst, *Clever Hans (The Horse of Mr. von Osten)*. It is reprinted from the 1911 English edition in a book edited by R. Rosenthal, *Clever Hans, the Horse of Mr. von Osten* (New York: Holt, 1965). The Rosenthal edition contains the 1911 English translation of the Pfungst work, along with a 1965 commentary by the editor, and two 1911 commentaries by the American psychologist James Angell and the German Carl Stumpf. Stumpf is prominent in the story. The description is recreated from information given on p. 18.
2. The early reports on Lady were conducted and published by J. B. Rhine (1895–1980), the noted investigator of psychic phenomena and, especially, of extrasensory perception, and Louisa Rhine, with the assistance of William McDougall, all, at that time, at Duke University. Lady was probably born in 1924 [she was a three year old when Rhine investigated her in 1927] and died after 1952. Rhine and Rhine (1929a) gave the first analysis of Lady, finding her telepathic toward human beings. A second analysis published six months later (Rhine and Rhine, 1929b) showed Lady unable to demonstrate the telepathy. Lady's ability to divine the location of missing persons and animals was reported in *Life* magazine, vol. 33 (December 12, 1952, p. 25) and in *Newsweek*, vol. 60 (December 15, 1952, p. 21).

3. December 15, 1952.

4. *Newsweek,* vol. 41, February 16, 1953, p. 32.

5. Karl Krall, *Denkende Tiere, beitrage zur Tierseelenkunde auf grund eigener Versuche, der Kluge Hans und Meine Pferde Muhamed und Zarif* (Leipzig: Freidrich Engelmann, 1912) includes a fine discussion of Hans I, along with a photograph of his skull to show the cranial capacity. Incredibly, neither Stumpf nor Pfungst, nor Rosenthal mentions the fact that Hans was the second von Osten horse to show such talent. They appear to believe that Hans was a unique phenomenon who suddenly burst on the Berlin scene. But all three should have known better, and the first two almost certainly did. Krall's book is the most complete statement of Hans I and Hans II extant.

6. J. B. Rhine comments on the trance state, but seems to consider it as being related to the hypnotic state he associates with telepathy and prescience.

7. My colleague Achin Kopp provided me with the benefits of his scholarship and sensitivity to nuances of the German language by reading Krall's book and helping me interpret what Krall reported.

8. I have found evidence, however, that Krall and Pfungst were booked to debate the issue in various German academies after the war. Alas, I find only programs and no record of what was said.

9. Rosenthal, *Clever Hans (The Horse of Mr. von Osten)* (New York: Holt, 1965).

10. The quotation is in the text of the English edition presented by Rosenthal, p. 29; German edition, p. 25. The original statement is from Lowenfeld, *Handbuch des Hypnotismus* (Wiesbaden, 1901).

11. Pfungst tells us that a number of the experiments were conducted in the horse's stall, although the account does not always make clear which were conducted under which conditions.

12. English reprint, p. 30; German original, pp. 27–28.

13. English reprint, p. 33; German original, p. 29.

14. English reprint, p. 35; German original, pp. 30–31.

15. English reprint, p. 35; German original, p. 31.

16. English reprint, pp. 42–43; German original, pp. 33–36.

17. The data are organized in this table from information given on p. 43 of the English reprint, p. 36 of the German original.

18. English reprint, p. 47; German original, p. 41.

19. English reprint, p. 46; German original, p. 38.

20. English reprint, pp. 47–48; German original, 39–40.

21. English reprint, pp. 48–49; German original, pp. 40–41.

22. English reprint, pp. 56–57; German original, p. 45.

23. Pfungst's career with clever animals did not end with Hans. He investigated the case of Don, a German setter belonging to the Ebers family, presumably in the early 1910s. Don spoke. Pfungst made phonographic recordings of the speech and discovered that the "words" could not be recognized by people under these conditions. In other words, those who could hear words in Don's speech heard what they wanted to hear. (H. M. Johnson, "The Talking Dog." *Science* 35 (1912): 749–751.

24. English reprint, p. 102.

25. English reprint, pp. 109–110; German original, p. 84. The example is that of a man; most of the subjects were women.

26. English reprint, p. 239; German original, p. 169.

Chapter 6. Experimentation and the Experimenter: Clever Hans's Companions

1. The information about von Osten is from Krall, pp. 10–12 and 349–354. K. Krall, *Denkende Tiere* (Leipzig: Englemann, 1912).
2. German newspaper accounts of the time refer to von Osten as Hern von Osten. I wonder if writers using English, such as Yerkes, heard this orally as "Baron von Osten" and hence, in writing, ascribed the mistaken title to him. A reprint of the *Berliner Tageblatt*, August 19, 1904, shown in L. D. Fernald, *The Hans Legacy* (New York: Erlbaum, 1984), p. 29, supports this view.
3. The English-reading world knows of von Osten and Clever Hans mostly through secondary sources. Pfungst's book is available in English and German, so those who are curious will turn to it for information. A history of von Osten and his horses is supplied by Krall, pp. 275–354.
4. "Sein dauernder tiesser Hass gegen sein Pferd, dem er ein "Ende vor dem Mortelwagen" wunschte, wahrte bis zuletzt." Krall, p. 349.
5. Krall is described by much later sources (e.g., Yerkes) as being a wealthy jeweler. Fernald (1984) says the same. I have seen no evidence for this, and none against it. Krall describes his own background as that of a horse-trainer, but, of course, he may have been both.
6. Why are Rendich's and Nora's achievements not well known? Or Rendich's telling comment about Hans? One reason is that the information is available to us only from Krall through his tedious book, published only in German and now very rare. The inaccessibility of Krall's work to modern scholars may explain lack of interest, although it does not justify it. Pfungst does not mention it, although we know that his professor, Stumpf, surely not only knew of it but also visited Nora.
7. K. Krall, *Denkende Tiere*, 1912, pp. 30–31.
8. Krall, p. 488.
9. Internal evidence from Krall suggests that these tests were done in Berlin while von Osten lived, and not at Elberfeld after von Osten's death. Pfungst does not appear to have known of them.
10. It is reported that Krall selected Arabian horses on the advice of the German military. The reason for their interest in the project is not known. One can imagine any number of possible motivations in the Germany of 1910–1911, as elsewhere, when the military become involved in academic and scientific affairs.
11. Freely translated from Krall, p. 172.
12. Krall, p. 488.
13. I have edited Krall's German words in the hope of providing the flavor of the commands and requests. One can almost hear Krall's "Falsch!," rolling over the years. The horses' responses are as reported by Krall, pp. 139–141.
14. Krall, *K. Denkende Tiere*, p. 156.
15. The two papers are E. Claparède, "Les Chevaux Savants d'Elberfeld," *Archives des Psychologie* 12 (1912): 263–304, and "Encore Les Chevaux D'Elberfeld (Avec une Note de M le Dr de Modzelwski," *Archives des Psychologie* 13 (1913): 244–284. I am much indebted to Danielle Murphy of Lewisburg, Pennsylvania, for helping me to understand the nuances in these two long and repetitive papers.
16. My knowledge of Berto comes through Claparède, as I am unable to locate

Claparède's reference to the original Krall paper regarding Berto. It is "Krall, K. Sur l'instruction de Berto," a translation of "Berto, das blinde rechnende Pferd," *Gesichte fur Tierpsychologie* 1 (1913): 10.

17. I translate freely from Claparède, p. 253.

18. Claparède, p. 282, but the data are from Modzelewski.

19. H. M. Johnson, "The Talking Dog," *Science* 35 (1912): 749–751.

20. Krall published the pictures of Garner without reference. My judgment is that Garner, in these pictures, is around sixty years old, an estimate that would place the time at around 1910, a time after Garner had more or less finished his work in West Africa described in Chapter 8. I have no idea where this work occurred, as there is no reference to it among Garner's known works. I can only guess that Garner and Krall were in communication, and that the photographs and information came from Garner to Krall personally. Such would explain both the highly exaggerated claims made, and the source of the photographs. But, see Chapter 8.

Chapter 7. The Psychologies of Perceiving: Phenomenology and Ethology

1. U. Eco, *Foucault's Pendulum.* Translated by William Weaver from the Italian (New York: Harcourt Brace Jovanovich, 1989), p. 467.

2. Lord Avebury (Sir John Lubbock), *On the Senses Instincts* [sic], *and Intelligence of Animals with Special Reference to Insects* (New York: Appleton, 1908). See also Lubbock, 1884.

3. Lubbock, 1908, pp. 272–273.

4. Lubbock, p. 273.

5. Lubbock, p. 273.

6. The text of Howe's remarks is taken from Lubbock who, while placing the passages in quotation, does not write that they are from Howe. I assume it.

7. Lubbock, pp. 278–280.

8. M. L. Huggins, "Kepler, a Biography." Mentioned, but not cited by Lubbock; I have been unable to locate the original.

9. Recent editions of Lofting's works are altered to dismiss his comments on the differing intelligence of human races, for the original stories express views of nonwhite races that are unacceptable to modern readers. Yet, the original Dr. Dolittle stories well reflect the viewpoint of a hierarchy of mental abilities—of a mental ladder—as described in Chapter 9.

10. "B.B.E.," " 'Roger,' A Record of the Performances of a Remarkable Dog," *Century Magazine* 59 (1907–1908): 599–602.

11. The quotations are from B.B.E. pp. 599–601.

12. "B.B.E.," 1907–1908; the quotations are from pp. 601–602.

13. "B.B.E.," 1907–1908, p. 602.

14. I do not know the source of Yerkes's knowledge about Clever Hans, for Pfungst's work was not yet published. Yerkes writes that Pfungst "has demonstrated the form of a number of involuntary movements which well might serve as signs to Hans" and that these are "of the head and eyes of (the) trainer or questioner." As we know from Chapter 5, Pfungst did not so prove. The reference is R. M. Yerkes, "The Behavior of 'Roger,' Being comment on the foregoing article based on personal investigation of the dog," *Century Magazine* 59 (1907–1908): 602–606.

15. An early use of the term "ethology" to describe a psychology of one's character, in the moral sense of character, may be found in T. P. Bailey, *Ethology:*

Standpoint, Method, Tentative Results (Berkeley: University of California Press, 1899).

16. K. Lorenz, The classic paper is "Der Kumpan in der Umwelt des Vogels." *Journal of Ornithology* 83 (1935): 137–213. An English translation may be found in K. Lorenz, "Companionship in Bird Life, Fellow Members of the Species as Releasers of Social Behavior," in C. H. Schiller (translator and editor), *Instinctive Behavior: The Development of a Modern Concept* (New York: International Universities Press, 1957). The most useful single reference for understanding Lorenz's viewpoint is "Physiological Mechanisms in Animal Behaviour," *Symposia of the Society of Experimental Biology* (New York: Academic, 1950). It is not incidental, one supposes, that Lorenz occupied the chair at Konigsburg which Kant had held a century and a half earlier.

17. To illustrate the nature of this conceptualization of the mind, I have relied on Jacob von Uexküll, especially the 1934 work, *A Stroll Through the Worlds of Animals and Men, a Picture Book of Invisible Worlds* (Streifzuge durch die Umwelten von Tieren und Menschen) (Berlin: Springer, 1934). In English, this may be found in C. H. Schiller, *Instinctive Behavior, the Development of a Modern Concept* (New York: International Universities Press, 1957) and F. J. J. Buytendijk, *The Mind of the Dog* (Boston: Houghton Mifflin, 1936). Both writers published splendid contributions to the psychology of animal life, von Uexküll writing in German and Buytendijk in French.

18. Von Uexküll, p. 25.

19. Von Uexküll, p. 6.

20. Hubel and Weisel, "Receptive Fields, Binocular Interaction and Functional Architecture in the Cat's Visual Cortex," *Journal of Physiology* 160 (1962): 106–154.

21. Von Uexküll, pp. 13–14.

Chapter 8. Peter and Moses: Chimpanzees Who Write

1. Witmer's development of the Psychological Clinic is detailed by P. McReynolds, "Lightner Witmer: Little-known Founder of Clinical Psychology," *American Psychologist* 42, 1987:849–858. On the matter of Peter the chimpanzee, Professor McReynolds graciously shared with me newspaper articles he had found published between 1907 and 1910. I am very grateful to Professor McReynolds, professor emeritus at the University of Nevada (Reno), for his collegial assistance.

2. Author of *The Home-Life of Borneo Head-Hunters* (1909). It was from Borneo that an orangutan was procured and provided to Lightner Witmer. In "Observations on the Mentality of Chimpanzees and Orangutans," Furness (1916) had reported his work on teaching the orang to speak.

3. Lightner Witmer, "A Monkey with a Mind," *the Psychological Clinic* 3, no. 7, (December 15, 1909): 179. Witmer edited the journal which mostly reported events that occurred in the clinic that he had organized. The comment regarding Furness refers to his attempts to teach an orangutan to speak. See *Proceedings from the American Philosophical Society* 55 (1916): 281.

4. Witmer, p. 179–180.

5. Witmer, p. 179.

6. Witmer, p. 180.

7. The pictures appear in Witmer, (1909), pp. 180, 181 and following, p. 199.

Internal evidence in the paper suggests that the pictures were taken at the clinic (not at the theatre) by Witmer's academic colleague, Robert Twitmyer.

8. Witmer, p. 181.
9. Witmer, p. 182.
10. Witmer, pp. 184–185.
11. Witmer, p. 193–194.
12. Witmer, p. 194.
13. Witmer, p. 182.
14. Witmer, p. 195.
15. Witmer, p. 196.
16. Witmer, p. 197.
17. Witmer, p. 197.
18. Witmer, p. 199.
19. Witmer, p. 205. From newspaper accounts furnished me by Professor McReynolds we can read of Witmer displaying other primates, sometimes, it appears, in public halls, but I find no archival material on this matter.
20. R. L. Garner, *The Speech of Monkeys* (London: Heinemann, 1892); same title (New York: Webster, 1892).
21. R. L. Garner, *Gorillas and Chimpanzees* (London: Osgood and McIlvaine, 1896). Parts of it, along with parts of the other books, appear in a U.S. edition *Apes and Monkeys, Their Life and Language* (Boston: Ginn, 1900).
22. Garner, 1896, pp. vii–viii. The following quotes from Garner are also from this work.
23. The situation is much different today. It would be hard to imagine a contemporary zoo or an animal act that did not include gorillas or chimpanzees. Whether this state of modernization is best for either them or us is difficult to decide.
24. To the human eye, that is.
25. Garner, pp. 14–16.
26. As names have changed several times, for the location, see *The Times Atlas of the World,* plate 91, north of Catherine Iguela, marked Lag. N'Kumi. Lamberene, which was to be made famous by the work of Dr. Alfred Schweitzer, is to the northeast.
27. Who took the pictures? Moses?
28. Garner, pp. 19–21.
29. Garner, pp. 59–62.
30. Here Garner drops to a description in the past tense, suggesting to me that he had seen this event himself. Yet, nowhere does he confirm this impression directly.
31. V. Reynolds, *Budongo* (Garden City, N.Y.: Natural History Press, 1965).
32. J. Goodall, *The Chimpanzees of Gombe* (Cambridge, Mass.: Harvard University Press, 1986).
33. D. Fossey, *Gorillas in the Mist* (Boston: Houghton Mifflin, 1983).
34. A. C. Bradley, *Shakespearean Tragedy, Lectures on Hamlet, Othello, King Lear, and MacBeth.* (London: Macmillan, 1904).
35. Garner, pp. 76–82.
36. It may be recalled that the horses Muhamed and Zarif confused the German "v" and "f" when they tapped.
37. Garner, pp. 96–98.
38. Garner, p. 98.
39. Garner, pp. 108–112.

40. Garner, pp. 99–101 and 112–115. Garner tells of the death of Moses in two episodes, the first being Garner's reaction upon his return with Aaron and the second being Garner's personal reaction. I have transposed sections of each to form this single story.
41. Garner, pp. 118–121.
42. Garner, pp. 124–126.
43. Garner, p. 128.
44. Garner, pp. 142–143. The obituary is *The New York Times*, 1920, 24 January 11:5.

Chapter 9. Exploiting the Missing Link

1. Lightner Witmer, "A Monkey with a Mind," *The Psychological Clinic* 3, no. 7 (December 15, 1909): 267.
2. I rely on S. J. Gould, *On Ontogeny and Phylogeny* (Cambridge, Mass.: Harvard University Press, 1977).
3. I depend on the excellent biography of Thorndike written by G. Joncich, *The Sane Positivist: a Biography of Edward L. Thorndike* (Middletown, Conn.: Wesleyan University Press, 1968).
4. The dimensions and like information originally appeared in the *Psychological Review,* Monograph Supplement No. 8, 1898. Thorndike's collected papers on this topic appear in a later compilation (1965) called *Animal Intelligence, Experimental Studies* (a facsimile of [the] 1911 edition). The 1965 edition was published in New York by Hafner. It contains an introduction by Paul G. Roofe of the University of Kansas which unaccountably spells the name as Thorndyke.
5. Thorndike, *Animal Intelligence* (New York: Hafner, 1965), p. 22.
6. Thorndike, p. 11.
7. Thorndike, pp. 21–22. Italics in original.
8. Thorndike, p. 23.
9. Thorndike, "The Intelligence of Monkeys," *Popular Science Monthly* 59 (1901): 273–279. Quotation from p. 273.
10. Thorndike, 1965, p. 27.
11. Thorndike, p. 38.
12. Thorndike, p. 40.
13. Thorndike, p. 59.
14. Thorndike, p. 61.
15. A most reasoned commentary on the deficiencies in design and the arrogance of interpretation is to be found in L. T. Hobhouse, *Mind in Evolution* (London: Macmillan, 1926).
16. T. W. Mills, *The Nature and Development of Animal Intelligence* (New York: Macmillan, 1898).
17. T. W. Mills, "The Nature of Animal Intelligence and the Methods of Investigating It," *Psychological Review* 6 (1899): 262–274. The quote appears on p. 263.
18. Mills, "The Nature of Animal Intelligence," p. 262.
19. Mills, "The Nature of Animal Intelligence," p. 266.
20. "For the king had at sea a navy of Tharish with the navy of Huram: once in three years came the navy of Tharish bringing gold, and silver, ivory, and apes, and peacocks" (I Kings 10:22). "For the king's ships went to Tharish bringing gold, and silver, ivory, and apes, and peacocks" (2 Chronicles 9:21).

21. Thorndike, 1901, p. 274.
22. Thorndike, 1901, pp. 277–278.
23. Thorndike, 1901, pp. 277–278.
24. Thorndike, 1901, p. 279.
25. The principal data are reported in R. Kyes and D. K. Candland, *Journal of Comparative Psychology* 101 (1987): 345–348; however, the observation cited here is unique in frequency and thus not reportable. The paper describes the experimental work, however.
26. M. E. Haggerty, "Imitation in Monkeys," *Journal of Comparative Neurology and Psychology* 19 (1909): 337–455. Quote on p. 342.
27. Haggerty, p. 343.
28. Haggerty, p. 348.
29. Haggerty, p. 350.
30. The interpretation is mine. Hamilton does not mention Thorndike in this connection.
31. G. V. Hamilton, "A Study of Trial and Error Reactions in Mammals," *Journal of Animal Behaviour*, (1911): 33–66, 40–41.
32. Hamilton, p. 32.
33. G. V. Hamilton, "A Study of Sexual Tendencies in Monkeys and Baboons," *Journal of Animal Behaviour* 4 (1914): 195–318.
34. G. V. Hamilton, *A Research in Marriage* (New York: Albert & Charles Boni, 1929)—a "Kinsey Report" for the 1920s. The cover specifies that the book is only available to members of certain professional societies as, it would seem, the data on sexual behavior would be dangerous to others. Hamilton's splendid analysis of the data is far in advance of its day and of value to any day.
35. M. E. Haggerty, *Children of the Depression* (Minneapolis: University of Minnesota Press, 1933), No. 6, Day and Hour Series of the University of Minnesota. Haggerty became dean of education at the University of Minnesota.
36. L. T. Hobhouse, *Mind in Evolution* (London: Macmillan, 1926).

Chapter 10. Raising Babies with Chimps: Donald, Gua, and Viki

1. E. R. Burroughs, *Tarzan of the Apes,* Original edition (Frank A. Munsey Co., 1912). The story has been repeated for so many different purposes that its original nature is lost, just as are so many other examples recounted in this book. For example, many readers would be astonished to learn that neither Tarzan or Jane ever saw England, that Jane was from Baltimore, Maryland, or that Tarzan saved her from a forest fire—in Wisconsin. Of special interest is the way in which Tarzan learned English, as it is in some ways akin to ape-language learning. He learned the letters from the children's books left among his parents' possessions, but he could not speak the language. It was D'Arnot, the Frenchman, who taught him to speak English. D'Arnot taught him English as they conversed in French. One can only guess at the sound. I am grateful to my longtime friend Owen T. Anderson for pointing this out to me.
2. Lightner Witmer, "A Monkey with a Mind," *The Psychological Clinic* 3, no. 7 (December 15, 1909): 205.
3. W. N. Kellogg and L. A. Kellogg, *The Ape and the Child, A Study of Environmental Influence upon Early Behavior* (New York: McGraw-Hill, 1933).

4. The federal primate centers were the subject of a splendid book by E. Hahn, *On the Side of the Apes* (New York: Crowell, 1971).

5. Madame Abreu made important contributions to the study of animal behavior in the West. She collected primates at her Havana, Cuba, home; this group formed the nucleus of the group that established the Yale Station and was probably the source of animals for Hamilton's work (Chapter 8). Most of what is known about Madame Abreu and her colony is to be found in Yerkes's book *Almost Human* (New York: Century, 1925). More recently, Emily Hahn has revitalized our knowledge of the situation in her most informative book, *On the Side of the Apes* (New York: Crowell, 1971).

6. Kellogg and Kellogg, p. 16.

7. Kellogg and Kellogg, p. 3.

8. Kellogg and Kellogg, pp. 5–7.

9. Kellogg and Kellogg, pp. 9–11.

10. Kellogg and Kellogg, p. 30.

11. Kellogg and Kellogg, p. 71.

12. Kellogg and Kellogg, p. 109.

13. Kellogg and Kellogg, p. 186.

14. Arnold Gesell, creator and standardizer of the test, is the Gesell who served as one of the commentators on the book about the wolf-girls by Singh and Zingg.

15. Kellogg and Kellogg, p. 269.

16. Kellogg and Kellogg, p. 275.

17. Kellogg and Kellogg, p. 281.

18. Kellogg and Kellogg, p. 185.

19. Kellogg and Kellogg, p. 289.

20. C. Hayes, *The Ape in Our House* (New York: Random House, 1951), p. 95.

21. Hayes, pp. 154–155.

22. Hayes, pp. 240–241.

23. I am told, however, by Donald Dewsbury that some film footage of Viki making the sounds survives.

24. K. J. Hayes, and C. H. Nissen, "Higher Mental Functions of a Home-raised Chimpanzee," in A. M. Schrier, and F. Stollnitz, (eds.), *Behavior of Nonhuman Primates,* vol. 4 (New York: Academic 1971), pp. 59–115.

25. W. N. Kellogg, "Communication and Language in the Home-reared Chimpanzee," *Science* 162 (1968): 423–427.

Chapter 11. *Human Ape Communication: Washoe, Koko, and Nim*

1. The information about Washoe is from R. A. Gardner, B.T. Gardner, and T. E. Van Cantfort (eds.) "Teaching Sign Language to Chimpanzees" (Albany: State University of New York Press, 1989), Preface and Chapter 1.

2. An earlier draft of this section contained quotations regarding the care and teaching of Washoe written by Gardner, Gardner, and Van Cantfort (1989) and a picture of the Gardners and the chimpanzee Moja. As R. Allen Gardner did not agree to the use of the quotations and picture, they are not cited. The correspondence between R. Allen Gardner and me on this subject has been placed by me with the archives of the Ellen Clarke Bertrand Library of Bucknell University, Bucknell University, Lewisburg, Pa., USA 17837. The contents are unrestricted.

3. F. Patterson and E. Linden, *The Education of Koko* (New York: Holt, Rinehart and Winston, 1981), p. 7.

4. R. M. Yerkes, *Almost Human* (New York: Century, 1925).

5. The quotation is provided by E. Hahn, 1971, frontispiece. The quotation does not appear in G. Hewes.

6. W. Percy, *The Thanatos Syndrome* (New York: Farrar, Straus & Giroux, 1987). Perhaps not-coincidentally, Percy lived and wrote in the small town of Covington, Louisiana, also home to the Delta Regional Primate Center of Tulane University.

7. Patterson and Linden, p. 27. I have capitalized the presumed meanings of the signs as do the Gardners, for consistency. These authors use italics.

8 Patterson and Linden, pp. 27–28.

9. Patterson and Linden, p. 63.

10. Patterson and Linden, pp. 72–73.

11. Patterson and Linden, pp. 77–78. The quotation would seem to show some serious misunderstandings regarding von Osten, Oskar Pfungst, and Clever Hans, among them being a two-century error in the dating of the event.

12. Patterson and Linden, p. 78.

13. Patterson and Linden, pp.79–80. The quotation in their text from Chevalier-Skolnikoff is not cited. For information on the views of those who consider the Clever Hans case to be relevant, see T. Sebeok and U. A. Sebeok, *Speaking of Apes* (New York: Plenum, 1980) and J. Umiker-Sebeok, and T. Sebeok, "Clever Hans and Smart Simians." *Anthropos*, 1981, 76, 89–165.

14. Patterson and Linden, Figure 1 (following p. 83).

15. Patterson and Linden, p. 92.

16. Patterson and Linden, pp. 99–100.

17. Patterson and Linden, p. 101.

18. Patterson and Linden, p. 191.

19. Patterson and Linden, p. 140.

20. Patterson and Linden, pp. 209–210.

21. It may be noted that I am reviewing this work in a different chronological order than might be surmised from the dates of publication of the books mentioned. Abandoning strict order seems justified, for the kind of communications available today and the giving of brief reports at professional meetings means that the development of ideas to be found in published and accessible works is not straightforward.

22. H. S. Terrace, *Nim* (New York: Alfred A. Knopf, 1979).

23. The Hayeses spelled the name Viki. For consistency, I have changed Terrace's spelling throughout my description.

24. Terrace, p. 9.

25. Terrace, p. 10.

26. Terrace, p. 13.

27. Terrace, p. 18.

28. Terrace, p. 20.

29. Terrace, p. 31.

30. Terrace, pp. 32–33.

31. The Gardners capitalized words that they took to be the English equivalent of signs; Terrace italicized them. I have converted these to caps for consistency.

32. Terrace, p. 75.

33. Terrace, p. 81.

34. Terrace, p. 109.

35. Terrace p. 137.
36. It may be asked how these were recorded. The answer is that the teacher whispered a word when signed into a recorder. Presumably, the whisperings had no meaning for Nim.
37. Terrace, p. 184.
38. Terrace, pp. 194–195.
39. Terrace, pp. 196–197.
40. Terrace, p. 207.
41. Terrace, p. 207.
42. Terrace, p. 232.
43. Terrace, pp. 214–215.
44. Terrace, p. 220.
45. Terrace, p. 221.
46. Eugene Linden, who co-authored the description of Koko with Patterson, has written a book that describes the careers of Washoe, Koko, and Nim, among other signing-apes, as they were moved about from experimenter to owner and sometimes back again. The book provides information on what was done with these animals when their original investigators needed them no more. E. Linden, *Silent Partners* (New York; Times Books, 1986).

Chapter 12. Language and Meaning: Sarah and Lana, Sherman and Austin, Kanzi and Ali

1. D. Premack and A. J. Premack, *The Mind of an Ape* (New York: W. W. Norton, 1983), p. 4.
2. Premack and Premack, p. 5.
3. D. Premack, *Intelligence in Ape and Man* (New York: John Wiley, 1976), p. 1.
4. Premack and Premack, p. 39.
5. Premack, p. 29.
6. The reference is to W. Köhler, *The Mentality of Apes,* translated into English by Ella Winter (New York: Harcourt, Brace, 1927).
7. Premack and Premack, p. 57.
8. Premack and Premack, p. 51. The full report is G. Woodruff and D. Premack, "Intentional Communication in the Chimpanzee: The Development of Deception," *Cognition* 7 (1979): 333–362.
9. Premack and Premack, p. 11. Later, Premack wrote a concise and informative book on the issue of animal language: *"Gavagai! or the Future History of the Animal Language Controversy* (Cambridge, Mass.: MIT Press, 1986). In this book he takes up possible solutions to the Clever Hans problem in language training of apes (p. 12), the literature on training dolphins to speak, motor conditioning (p. 17), whether apes can form sentences and engage in communication (p. 29), and gives further information on the achievements of Peony and Elizabeth (pp. 63 and 152), and what he calls the "missing links" of ape language (p. 124).
10. D. M. Rumbaugh, *Language Learning by a Chimpanzee, the LANA Project* (New York: Academic, 1977).
11. Gill and Rumbaugh, in Rumbaugh, pp. 157–162.
12. Rumbaugh and Gill, in Rumbaugh, pp. 170, 173–174.
13. Rumbaugh and Gill, in Rumbaugh, pp. 190–191.
14. Savage and Rumbaugh, in Rumbaugh, p. 288.

15. E. S. Savage-Rumbaugh, *Ape Language, from Conditioned Response to Symbol* (New York: Columbia University Press, 1986).
16. Savage-Rumbaugh, p. 10.
17. Savage-Rumbaugh, p. 31.
18. Savage-Rumbaugh, p. 39.
19. Savage-Rumbaugh, pp. 63–64.
20. Savage-Rumbaugh, p. 75.
21. Nor, let it be said, is videotape the necessary evidence of truth, as we know from televised sports.
22. Savage-Rumbaugh, p. 101.
23. Savage-Rumbaugh, p. 113.
24. Savage-Rumbaugh, p. 115.
25. Savage-Rumbaugh, p. 117.
26. Savage-Rumbaugh, p. 139.
27. I compiled the data from Savage-Rumbaugh Table 8.1, p. 173.
28. Savage-Rumbaugh, p. 206. Lana, who returns to work, was able to learn the names of tools promptly, perhaps because of her extensive training with symbols. Evidence is presented, however, that Sherman and Austin used concepts such as "food," while Lana did not match their ability to conceptualize (p. 254).
29. Savage-Rumbaugh, p. 375.
30. R., Epstein, R. P., Lanza, and B. F. Skinner, "Symbolic Communication between Two Pigeons *(Columbia liva domestica),*" *Science* 207 (1980): 543–545.
31. Savage-Rumbaugh, p. 129.
32. From *Chronicle of Higher Education,* September 16, 1990, p. A5. Reported to be in *Language and Intelligence in Monkeys and Apes: Comparative Developmental Perspectives* (Cambridge, U.K.: Cambridge University Press, 1990).
33. E. S. Savage-Rumbaugh, J. Murphy, R. A. Sevcik, and D. M. Rumbaugh, "Language Comprehension in Ape and Child." *Monograph Series of the Society for Research in Child Development,* 1993. Number 233, 58, numbers 3–4.
34. T. Matsuzawa, *The Perceptual World of a Chimpanzee* (Kyoto, Japan: Primate Research Institute, March 1990).
35. Work with Al, Kanzi, and Koko continues. So far as I can tell, published work on the other apes mentioned has ended (as of June 1993).

References

Bailey, T. P. *Ethology: Standpoint, Method, Tentative Results.* Berkeley: University of California Press, 1899.

B.B.E. "Roger, A Record of the Performances of a Remarkable Dog." *Century Magazine,* 1907–1908, 59, 599–602.

Bradley, A. C. *Shakespearean Tragedy, Lectures on Hamlet, Othello, King Lear and MacBeth.* London: Macmillan, 1904. Also, London: Penguin, 1991.

Burroughs, E. R. *Tarzan of the Apes.* Cutcchogue, N.Y.: Buccaneer Books, 1977. (First edition, Frank A. Munsey Co., 1912.)

Buytendijk, F.J.J. *The Mind of the Dog.* New York: Houghton Mifflin, 1936.

Claparède, E. "Les Chevaux Savants d'Elberfeld." *Archives des Psychologie* 1912, 12, 263–304.

Claparède, E. "Encore Les Chevaux d'Elberfeld (Avec une Note de M. le Rd de Modzelewski." *Archives des Psychologie,* 1913, 13, 244–284.

Darwin, C. *On the Origin of Species by Means of Natural Selection, or the Preservation of Favoured Races in the Struggle for life.* London: John Murray, 1859.

Darwin, C. *The Descent of Man and Selection in Relation to Sex.* London: John Murray, 1871. (First U.S. edition, 2 vols. New York: Appleton, 1872.)

Darwin, C. *The Expression of the Emotions in Man and Animals.* London: John Murray, 1872. (First U.S. edition, New York: Appleton, 1872.) Also, Chicago: University of Chicago Press, 1965.

Eco, U. *Foucault's Pendulum.* Translated by W. Weaver. New York: Harcourt, Brace, Jovanovich, 1989.

Epstein, R., Lanza, R. P., and Skinner, B. F. "Symbolic Communication between Two Pigeons *(Columbia livia domestica).*" *Science,* 1980, 207, 543–545.

Farmer, P. J. *Tarzan Alive: A Definitive Biography of Lord Greystoke.* Garden City, N.Y.: Doubleday, 1972.

Fernald, L. D. *The Hans Legacy: A Story of Science.* New York: Erlbaum, 1984.

Feuerbach. (See Pies, H.)

Fossey, D. *Gorillas in the Mist.* Boston, Mass.: Houghton Mifflin, 1983.

Freud, S. "Analysis of a Phobia in a Five-year-old Boy." Translation of "Analyse der Phobie eines funfjahrigen Knaben" (1909). In *The Complete Psychological Works of Sigmund Freud,* vol. X, Two Case Histories. Translated and edited by James Strachey. London: Hogarth, 1955.

Freud, S. "Three Essays on the Theory of Sexuality." Translation of "Drei Abhandlugen zur Sexualtheorie." *The Standard Edition of the Complete Psycholog-ical Works*. Translated by J. Strachey. London: Hogarth, 1955, 12, 125.

Freud, S. "Two Case Histories." *The Standard Edition of the Complete Psychological Works*. Translated by J. Strachey. London: Hogarth, 1955, X: 1–149.

Furness, W. H. *The Home-life of Borneo Head-hunters*. Philadelphia: J. B. Lippincott, 1909.

Furness, W. H. "Observations on the Mentality of Chimpanzees and Orangutans." *Proceedings from the American Philosophical Society*, 1916, 55, 281–290.

Gardner, R. A., Gardner, B. T., and Van Cantfort, T. E. *Teaching Sign Language to Chimpanzees*. Albany: State University of New York Press, 1989.

Garner, R. L. *The Psychoscope*. Warrenton, Va.: The True Index Print, 1891.

Garner, R. L. *The Speech of Monkeys*. London: Heinemann, 1892; New York: Webster, 1892. (Translated by William Marshall as "Die Sprache der Affen," Dresden: Schultze, 1905.)

Garner, R. L. *Apes and Monkeys*. London: Osgood and McIlvaine, 1896.

Garner, R. L. *Apes and Monkeys: Their Life and Language*. Boston: Ginn & Co., 1896.

Garner, R. L. *Gorillas and Chimpanzees*. London: Osgood and McIlvaine, 1896. (Largely a reprint of *Apes and Monkeys: Their Life and Language*.)

Garner, R. L. *Nancy Bet. The Story of Sloomy Perkins and His Transaction in Real Estate*. Norfolk, Va.: Landmark Steam Printing House, 1899.

Gesell, A. *Wolf Child and Human Child*. New York: Harper, 1941.

Goodall, J. *The Chimpanzees of Gombe: Patterns of Behavior*. Cambridge, Mass.: Harvard University Press, 1986.

Gould, S. J. *Ontogeny and Phylogeny*. Cambridge, Mass.: Harvard University Press, 1977.

Grant, D. *The Life Work of Lord Avebury*. London: Watts, 1924.

Haggerty, M. E. "Imitation in Monkeys." *Journal of Comparative Psychology*, 1909, 19, 337–455.

Haggerty, M. E. *Children of the Depression*. Minneapolis: University of Minnesota Press, 1933.

Hahn, E. *On the Side of the Apes*. New York: Crowell, 1971.

Hahn, E. *Look Who's Talking*. New York: Crowell, 1978.

Hahn, E. *Eve and the Apes*. New York: Weidenfeld and Nicolson, 1988.

Hamilton, G. V. "A Study of Trial and Error Reactions in Mammals." *Journal of Animal Behaviour*, 1911, 1, 33–66.

Hamilton, G. V. "A Study of Sexual Tendencies in Monkeys and Baboons." *Journal of Animal Behaviour*, 1914, 4, 195–318.

Hamilton, G. V. *A Research in Marriage*. New York: Albert and Charles Boni, 1929.

Handke, P. Translated by Michael Roloff. *Kaspar and Other Plays*. New York: Hill and Wang, 1969.

Hanks, J., and Bringmann, W. G. "Whatever Happened to Little Hans?" *History of Psychology Newsletter*, 1989, 221, no. 34/4, 78–81.

Haraway, D. *Primate Visions, Gender, Race, and Nature in the World of Modern Science*. New York: Routledge, 1989.

Hayes, C. *The Ape in Our House*. New York: Harper, 1951.

Hayes, K. J., and Nissen, C. H. "Higher Mental Functions of a Home-raised Chimpanzee." In A. M. Schrier and F. Stollnitz, *Behavior of Nonhuman Primates*, vol. 4. New York: Academic, 1971.

Hobhouse, L. T. *Mind in Evolution*. London: Macmillan, 1926.

Howe, J. W., *The Massachusetts Philanthropist, Memoir of Dr. Samuel Gridley Howe.* Boston: Wright, 1876.

Hubel, D. H., and Weisel, T. N. "Receptive Fields, Binocular Interaction and Functional Architecture in the Cat's Visual Cortex." *Journal of Physiology,* 1962, 160, 106–154.

Huggins, M. L. *Kepler, a Biography.* Not cited further by Lubbock.

Humphrey, N. *Consciousness Regained.* Oxford: Oxford University Press, 1983.

Itard, J.-M.-G. *The Wild Boy of Aveyron.* Translated by G. Humphrey and M. Humphrey. New York: Century, 1932.

James, W. *Principles of Psychology.* New York: Henry Holt, 1890.

Johnson, H. M. "The Talking Dog." *Science,* 1912, 35, 749–751.

Joncich, G. *The Sane Positivist: A Biography of Edward L. Thorndike.* Middletown, Conn.: Wesleyan University Press, 1968.

Kant, I. *Kritik der reinen Vernunft [Critique of Pure Reason.].* Riga: Johann Friedrich Hartknoch, 1781.

Kellogg, W. N. "Communication and Language in the Home-raised Chimpanzee." *Science,* 1968, 162, 423–427.

Kellogg, W. N., and Kellogg, L. A. *The Ape and the Child: A Study of Environmental Influence upon Early Behavior.* New York: McGraw-Hill, 1933.

Köhler, W. *The Mentality of Apes.* English translation by Ella Winter of *Intelligenzprüfungen an Menschenaffen.* Berlin: Springer, 1917.

Krall, K. *Denkende Tiere, Beitrage zur Tierseelenkunde auf Grund Eigener Versuche, der Kluge Hans und Meine Pferde Muhamed und Zarif.* Leipzig: Freidrich Engelmann, 1912.

Krall, K. "Berto, das blinde rechnende Pferd." *Gesischte für Tierpsychologie,* 1913, 1, 10. (Unverified: taken from E. Claparède, "Sur l'instruction de Berto"; see Claparède, 1912, 1913.)

Kyes, R. C., and Candland, D. K. "Baboon *(Papio hamadryas)* Visual Preferences for Regions of the Face." *Journal of Comparative Psychology,* 1987, 101, 345–348.

Lane, H. *The Wild Boy of Aveyron.* Cambridge, Mass.: Harvard University Press, 1976.

Lenoir, T. *The Strategy of Life, Teleology and Mechanics in Nineteenth Century German Biology.* Chicago: University of Chicago Press, 1989. (Reprint of 1982 edition.)

Linden, E. *Silent Partners: The Legacy of the Ape Language Experiments.* New York: Times Books, 1986.

Lofting, H. *The Voyages of Doctor Dolittle.* Philadelphia: J. B. Lippincott, 1950.

Lorenz, K. "Physiological Mechanisms of Animal Behaviour." In *Symposia of the Society of Experimental Biology.* New York: Academic, 1950.

Lorenz, K. "Der Kumpan in der Umwelt des Vogels." *Journal of Ornithology,* 1935, 83, 137–213. (Translation is "Companionship in Bird Life, Fellow Members of the Species as Releasers of Social Behavior.") In C. H. Schiller, *Instinctive Behavior: The Development of a Modern Concept.* New York: International Universities Press, 1957.

Lovejoy, A. O. *The Great Chain of Being, a Study of the History of an Idea.* Cambridge, Mass.: Harvard University Press, 1936.

Lowenfeld, H. *Handbuch des Hypnotismus.* Wiesbaden, 1901. (Cited by Rosenthal, p. 19, in translation of Pfungst, p. 25.)

Lubbock, J. (Lord Avebury). *On the Senses* [sic], *Instincts, and Intelligence of Animals with Special Reference to the Insects.* New York: Appleton, 1908.

Lubbock, J. "Teaching Animals to Converse." *Nature,* 1884, 2, 547–548.

Maclean, C. *The Wolf Children.* New York: Hill and Wang, 1978.

"Manors & Morals: Detective Story." *Time,* 1952, 60: 21.

Matsuzawa, T. *The Perceptual World of a Chimpanzee.* Kyoto: Japan. Kyoto University Primate Research Institute, March 1990.

Mills, T. W. *The Nature and Development of Animal Intelligence.* New York: Macmillan, 1898.

Mills, T. W. "The Nature of Animal Intelligence and the Methods of Investigating It." *Psychological Review,* 1899, 6, 262–274.

Morgan, C. L. *Animal Life and Intelligence.* Boston: Ginn & Co., 1891.

Muybridge, E. *Animals in Motion.* New York: Dover, 1957. Original, 1887.

Ogburn, W. F., and Bose, M. K. "On the Trail of the Wolf Children." *Genetic Psychology Monographs,* 1959, 60, 117–193.

Ormond, R. *Catalog: Sir Edwin Landseer.* Philadelphia and London: Philadelphia Museum of Art and the Tate Gallery, 1981.

Patterson, F., and Linden, E. *The Education of Koko.* New York: Holt, Rinehart and Winston, 1981.

Percy, W. *The Thanatos Syndrome.* New York: Farrar, Straus, and Giroux, 1987.

Pies, H. *Kaspar Hauser, Augenzeugenberichte und Selbstzeugnisse, herausgegben, eingeleitet und mit fussnoten versehenn.* 2 vols. Stuttgart, G.m.b.h., 1928. Contains reprint of P. J. A. Feuerbach, "Kaspar Hauser, Beispiel eines Verbrechens am Seelenleben des Menschen" Ansbach, Germany, 1832. (Translated as "Kaspar Hauser, an Account of an Individual Kept, in a Dungeon Separated from all Communication with the World, from Early Childhood to about the Age of Seventeen.") Translated by Lieber. London: Simpkin and Marshall, 1833.

Pfungst, O. *Clever Hans, The Horse of Mr. von Osten.* New York: Henry Holt, 1911.

Pfungst, O. *Das Pferd des Herr von Osten (der Kluge Hans), eine beitrag zur experimentellen tier und menschen-psychologie.* Leipzig: Barth, 1907.

Plato. *The Timaeus.* In F. MacD. Cornford, *Plato's Theory of Knowledge.* London: Routledge and Kegan Paul, 1935.

Premack, D. *Intelligence in Ape and Man.* New York: John Wiley, 1976.

Premack, D. *Gavagai!, or the Future History of the Animal Language Controversy.* Cambridge, Mass.: MIT Press, 1986.

Premack, D., and Premack, A. J. *The Mind of an Ape.* New York: W. W. Norton, 1983.

Rahn, C. L. Translation of Pfungst, O., *"Das Pferd des Herr Von Osten (der Kluge Hans), eine beitrag zur experimentellen tier und menschen-psychologie.* (Translated as "Clever Hans, the horse of Mr. von Osten.") New York: Henry Holt, 1911. (Reprinted in Rosenthal, 1965, but without credit to Rahn, the translator.)

Rauber, A. A. *Homo Sapiens Ferus; oder, Die Zusthande der Verwilderten und ihre Bedeutung für Wissenschaft, Politik und Schule.* Leipzig: Denicke,1855.

Reynolds, V. *Budongo, an African Forest and Its Chimpanzees.* Garden City, N.Y.: Natural History Press, 1965.

Reynolds, V. *The Apes.* New York: E. P. Dutton, 1967.

Rhine, J. B., and Rhine, L. 1929a. "An Investigation of a 'Mindreading' Horse." *Journal of Abnormal and Social Psychology,* 1929, 23, 449–466.

Rhine, J. B., and Rhine, L. 1929b. "Second Report on Lady, the "Mindreading" Horse." *Journal of Abnormal and Social Psychology,* 1929, 24, 287–292.

Richards, L. E. *Samuel Gridley Howe*. New York: Appleton, 1935.

Rizzo, F. "Memoirs of an Invisible Man, I, II, III, IV." *Opera News*, 1972, 36: 24–28, 26–29.

Romanes, G. J. *Animal Intelligence*. London: Routledge and Kegan Paul, 1882. (First U.S. edition, New York: Appleton, 1883.)

Romanes, G. J. *Mental Evolution in Man*. London: Routledge and Kegan Paul, 1888.

Rosenthal, R., ed. *Clever Hans (the Horse of Mr. Von Osten) by Oskar Pfungst*. New York: Henry Holt, 1965.

Rumbaugh, D. E., ed. *Language Learning by a Chimpanzee, the LANA Project*. New York: Academic, 1977.

Savage-Rumbaugh, E. S. *Ape Language, from Conditioned Response to Symbol*. New York: Columbia University Press, 1986.

Savage-Rumbaugh, E. S., Murphy, J., Sevcik, R. A., and Rumbaugh, D. M. "Language Comprehension in Ape and Child." *Monograph Series of the Society for Research in Child Development*, 1993. Number 233, 58, numbers 3–4.

Schiller, C. H., trans. and ed.. *Instinctive Behavior: The Development of a Modern Concept*. New York: International Universities Press, 1957.

Schwartz, H. *Samuel Gridley Howe*. Cambridge, Mass.: Harvard University Press, 1956.

Sebeok, T., and Sebeok, A. U. *Speaking of Apes*. New York: Plenum, 1980.

Shapiro, K. Letter in *Monitor*, a publication of the American Psychological Association, Washington, D.C.

Shattuck, R. *The Forbidden Experiment, the Story of the Wild Boy of Aveyron*. New York: Washington Square Press, 1981.

Singh, J.A.L., and Zingg, R. M. *Wolf-Children and Feral Man*. New York: Harper and Brothers Publishers, 1939.

"Talking Horse." *Life*, 1952, 33: 20–21.

Temerlin, M. K. *Lucy: Growing up Human*. Palo Alto, Calif.: Science and Behavior, 1975.

Terrace, H. S. *Nim*. New York: Alfred A. Knopf, 1979.

Thorndike, E. L. "Animal Intelligence: An Experimental Study of the Associative Processes in Animals." *Psychological Review, Monograph Supplement*, no. 8, 1898, 2, 1–109.

Thorndike, E. L. "The Intelligence of Monkeys." *Popular Science Monthly*, 1901, 59, 273–279.

Thorndike, E. L. *Animal Intelligence, Experimental Studies*. New York: Hafner, 1965. (Reprint of *Animal Intelligence*, New York: Macmillan, 1911.)

Tuttle, R. H. *Apes of the World*. Park Ridge, N.J.: Noyes, 1986.

"Two Men on a Horse." *Newsweek*, 1948, 32: 29.

Von Uexküll, J. *Im Kampf um die Tierseele*. Wiesbaden: Leitfader, 1934.

Von Uexküll, J. *Streifuze durch die Umwelten von Tieren and Menschen*. Berlin: Springer, 1934. (Translated by C. L. Schiller as "A Stroll Through the Worlds of Animals and Men, a Picture Book of Invisible Worlds.")

Watson, J. B. *Psychology, From the Standpoint of a Behaviorist*. Philadelphia: J. B. Lippincott, 1919.

Watson, J. B. *Behaviorism*. New York: W. W. Norton, 1925.

Woodruff, G., and Premack, D. "Intentional Communication in the Chimpanzee: The Development of Deception." *Cognition*, 1979, 7, 333–362.

Witmer, L. "The Restoration of Children of the Slums." *Psychological Clinic*, 1909, 3, 266–280.

Witmer, L. "A Monkey with a Mind." *Psychological Clinic,* 1909, 3 (December 15, 1909), 179–205.

Yerkes, R. M. "The Behavior of 'Roger,' Being Comment on the Foregoing Article Based on Personal Investigation of the Dog." *Century Magazine,* 59, 1907–1908, 602–606.

Yerkes, R. M. *Almost Human.* New York: Century, 1925.

Yerkes, R. M. *Chimpanzees, a Laboratory Colony.* New Haven, Conn.: Yale University Press, 1943.

Yerkes, R. M., and Yerkes, A. *The Great Apes, a Study of Anthropoid Life.* New Haven, Conn.: Yale University Press, 1929.

Illustration Credits

Chapter 1. Figure 1.1, Blumenbach, *Beyträge zur Naturgeschichte* (1811). Reprinted in Singh, J. A. L., and Zingg, R. M. *Wolf-Children and Feral Man*. New York: Harper and Brothers Publishers, 1939, p. 183. Courtesy of the Centennial Museum; 1.2, Itard, J.-M.-G. *The Wild Boy of Aveyron*. Translated by G. Humphrey and M. Humphrey. New York: Century, 1932.

Chapter 2. Figure 2.1, Feuerbach, as reproduced in Singh and Zingg, 1939, p. 278. Courtesy of the Centennial Museum; 2.2, 2.3, 2.4, 2.5, 2.6, 2.7, from Singh, J. A. L., and Zingg, R. M. *Wolf-Children and Feral Man*. New York: Harper and Brothers Publishers, 1939, pp. 13, 26, 28, 30, 34, 79. Courtesy of the Centennial Museum.

Chapter 3. Figure 3.1, Ormond, R. *Catalog: Sir Edwin Landseer*. Philadelphia and London: Philadelphia Museum of Art and the Tate Gallery, 1981, p. 111. Courtesy of the British Museum.

Chapter 4. Figures 4.1, 4.2, Freud, S. "Analysis of a Phobia in a Five-year-old Boy." Translation of "Analyse der Phobie eines funfjahrigen Knaben" (1909). In *The Standard Edition of the Complete Psychological Works of Sigmund Freud*, vol. X, Two Case Histories. Translated and edited by James Strachey. London: Hogarth, 1955, pp. 13, 44. Reprinted by permission of Sigmund Freud Copyrights, The Institute of Psycho-Analysis, The Hogarth Press, and HarperCollins.

Chapter 5. Figures 5.1, 5.3, Pfungst, O. *Clever Hans, The Horse of Mr. von Osten*. New York: Henry Holt, 1911, frontispiece, p. 57; 5.2, Reprinted by permission of the *Quincy Patriot Ledger*. Photograph taken by Laban Whittaker.

Chapter 6. Figures 6.1, 6.2, 6.3, 6.4, 6.5, 6.6 from Krall, K. *Denkende Tiere, Beiträge zur Tierseelenkunde auf Grund Eigener Versuche, der Kluge Hans und Meine Pferde Muhamed und Zarif*. Leipzig: Friedrich Engelmann, 1912, Illustrations 13, 28, 61, 72, 90, 94.

Chapter 7. Figure 7.1 from *Century Magazine*, 1907–8, 59, p. 600; 7.2, 7.3, Von Uexküll, J. *Streifzüge durch die Umwelten von Tieren und Menschen*. Berlin: Springer, 1934. Translated and edited by Schiller, C. H. *Instinctive Behavior: The Development of a Modern Concept*. New York: International Universities Press, 1957, p. 28. Reprinted by permission of International Universities Press, Inc.

Chapter 8. Figures 8.1, 8.2, 8.3 from Romanes, G. J. *Mental Evolution in Man*. London: Routledge and Kegan Paul, 1888, frontispiece; 8.4, 8.5, 8.6 from Witmer, L. "A Monkey with a Mind." *Psychological Clinic*, 1909, 3 (December 15, 1909), pp. 180, 199, 193; 8.7, 8.8 from Garner, R. L. *Apes and Monkeys: Their Life and Language*. Boston: Ginn Co., 1896, pp. 24, 133.

Text Credits

We gratefully acknowledge the following publishers and persons who generously granted us the permission to quote from the following works:

Hayes, C. *The Ape in Our House*. New York: Harper & Brothers, 1951, pp. 95, 154–155, 240–241. Copyright © 1951 by Catherine Hayes. Copyright © renewed 1979 by Catherine Hayes Nissen. Reprinted by permission of McIntosh and Otis, Inc.

Patterson, F., and Linden, E. *The Education of Koko*. New York: Holt, Rinehart and Winston, 1981, pp. 27–28, 63, 72–73, 77–80, 83, 92, 99–101, 191, 140, 209–210. Reprinted by permission of Russell & Volkening as agents for the author. Copyright © 1978 by Francine Patterson and Eugene Linden.

Premack, D., and Premack, A. J. *The Mind of an Ape*. New York: W. W. Norton, 1983, pp. 4, 5, 11, 39, 51, 57. Reproduced from *The Mind of an Ape* by David Premack and Ann James Premack, by permission of W. W. Norton & Company, Inc. Copyright © 1983 by Ann J. Premack and David Premack.

Rumbaugh, D. M., ed. *Language Learning by a Chimpanzee, the LANA Project*. New York: Academic, 1977, pp. 157–162, 170, 173–174, 190–191, 288. Reprinted by permission of D. M. Rumbaugh.

Savage-Rumbaugh, E. S. *Ape Language, from Conditioned Response to Symbol*. New York: Columbia University Press, 1986, pp. 10, 31, 39, 63–64, 75, 101, 113, 115, 117, 129, 139, 173, 206, 254, 375. Reprinted by permission of Sue Savage-Rumbaugh.

Singh, J. A. L., Zingg, R. M. *Wolf-Children and Feral Man*. New York: Harper & Brothers, 1939, pp. xv, xxvi–xxvii, 3–9, 11–17, 39–40, 52–54, 97, 103–104, 113, 182–187, 193, 284–286, 288–289, 291–292, 296–297, 315, 317–318, 341–343, 346–347, 353–354, 359–365. Reprinted by permission of the Centennial Museum.

Terrace, H. S. *Nim*. New York: Alfred A. Knopf, 1979, pp. 9, 10, 13, 18, 20, 31, 32–33, 75, 81, 109, 137, 184, 194–195, 196–197, 207, 214–215, 220–221, 232. Reprinted by permission of Alfred A. Knopf, Inc. and Methuen London.

Index